ELECTRON PROBE MICROANALYSIS IN BIOLOGY

Edited by

David A. Erasmus, D.Sc.

Reader in Parasitology, Department of Zoology
University College, Cardiff

London
CHAPMAN AND HALL
A Halsted Press Book
John Wiley and Sons, New York

First published 1978
by Chapman and Hall Ltd
11 New Fetter Lane, London EC4P 4EE

Typeset and printed in Britain by
The Cambridge University Press

SBN 412 15010 7

Distributed in the U.S.A. by Halsted Press
a Division of John Wiley & Sons, Inc., New York

Library of Congress Cataloging in Publication Data

Main entry under title:
Electron probe microanalysis in biology.
'A Halsted Press book.'
Includes bibliographies.
1. Electron probe microanalysis. 2. Biology—
Technique. I. Erasmus, David A.
QH324.9.E38E43 578'.4 77-28247
ISBN 0-470-26303-2

Contents

Contributors

T. C. Appleton	*Physiological Laboratory, Cambridge CB2 3EG, U.K.*
I. D. Bowen	*Department of Zoology, University College, P.O. Box 78 Cardiff CF1 1XL, U.K.*
J. A. Brown	*Department of Zoology, The University, Sheffield S10 2TN, U.K.*
J. A. Chandler	*Tenovus Institute, Welsh National School of Medicine, The Heath, Cardiff CF4 4XX, U.K.*
T. W. Davies	*Department of Zoology, University College, P.O. Box 78 Cardiff, CF1 1XL, U.K.*
D. A. Erasmus	*Department of Zoology, University College, P.O. Box 78 Cardiff, CF1 1XL, U.K.*
H. O. Garland	*Department of Zoology, The University, Sheffield S10 2TN, U.K.*
I. W. Henderson	*Department of Zoology, The University, Sheffield S10 2TN, U.K.*
A. J. Morgan	*Department of Zoology, University College, P.O. Box 78 Cardiff CF1 1XL, U.K.*
J. C. Russ	*Edax International Inc., P.O. Box 135, Prairie View, Illinois, 60069, U.S.A.*
T. A. Ryder	*Department of Zoology, University College, P.O. Box 78 Cardiff CF1 1XL, U.K.*

I

D. A. ERASMUS

Introduction

The incorporation of X-ray detectors into electron optical columns has provided the basis of a technique – X-ray microanalysis, which has revolutionized the way we look at ultrastructure and the function of subcellular components. This technique has enabled microscopists to link directly chemical analysis with morphology at a subcellular level. There is no other way by which this immediate association can be made and X-ray microanalysis provides the essential links between morphology and biochemistry which have been missing for many years and therefore limited, to some extent, the integration of the two disciplines. This new approach provides a considerable challenge to the electron microscopist in biology as it necessitates an awareness of the physics of the analytical procedure, the chemistry of the elements investigated, as well as the ability to interpret the data in a biological context. The biologist, for most of the time, has to think of his specimen as a collection of atoms rather than an optical representation of morphology. Every stage of the technique from recovery of the sample, processing, through to final analysis has really to be thought out in terms of atomic features. Established standards of specimen preparation and structural interpretation, which are familiar to the investigator of ultrastructure, have to be replaced by completely different and generally more demanding criteria.

All types of electron optical systems can be provided with analysis systems so that a new category of TEM, SEM and STEM instruments has appeared. The field is therefore enormous and in order to follow certain aspects in depth, the contents of this book have been restricted essentially to the analysis of specimens in the transmission electron microscope.

As well as providing a facility which enables chemical composition to be linked to morphology, X-ray microanalysis has released certain established techniques from practical constraints. In many histochemical techniques performed at the ultrastructural level, the end product must be one which has considerable opacity to the electron beam. It is now possible to rethink the entire procedure in that the end product no longer needs to be electron opaque. The presence of a characteristic, easily identifiable element is all that is desired. This change in approach can also be applied to several immunochemical tests. An additional novel contribution is that the analyst need not confine his attention to the study of sectioned material. It is quite easy to analyse fluid samples obtained directly from a biological environment or arising from the solubilization of tissue samples. This facility widens even further the horizon of biological investigations.

It is relatively simple to convert most electron optical systems into analysis instruments. Spectrometers may be either wavelength dispersive or energy dispersive and each has its advantages and disadvantages. My personal preference is for energy dispersive systems for biological analysis. Although ED systems do not have the spatial resolution and high sensitivity of wavelength spectrometers they do have particular advantages for the biologist. Energy dispersive systems provide a simultaneous display of all elements present in the sample and although the investigator might have intended to study selected elements, a complete spectrum frequently produces valuable information which was not anticipated. Furthermore, the simultaneous nature of the display means that repetitive, sequential analysis of the same area to record a total elemental spectrum is not necessary. Rapid analysis time does reduce specimen damage to a considerable extent.

In spite of these advantages, the biologist is working with the X-ray emission from a restricted irradiated volume of tissue which contains low concentrations of elements. The elements of most frequent interest to the biologist are sodium, magnesium, phosphorus, sulphur, chlorine, and potassium and because some of these emit X-rays of low energy, efficiency of collection and detection becomes vitally important. The ratios of Peak: Background are the basis of all quantitation and it is essential that the user has an analysing system which provides optimum ratios. The design of microscope specimen chambers, specimen holders, and detector position needs to be critically examined before serious X-ray analysis can be undertaken. An additional facility which is essential for any comparative work, either qualitative or quantitative in nature, is the ability to monitor all operating conditions quite precisely, especially beam current. Valid comparisons between different parts of the same specimen and certainly between different specimens can only be achieved if operating conditions are defined so that each analysis is conducted under identical conditions.

Sooner or later, most investigators will require quantitation, relative or absolute, in the expression of their results. Fortunately, the analysis of biological specimens in the form of thin sections is relatively simple in that complex correction factors are unnecessary. However, problems do exist in the choice of standards with a 'biologically-equivalent' matrix, and also in that matrix mass, as assessed by Bremsstrahlung, may not be precisely determined because of the heterogenous origin of the Bremsstrahlung.

Recent observations suggest that the process of quantitation of freeze-dried sections should also include consideration of the different degrees of hydration which the structural components of a biological system exhibit.

Undoubtedly, specimen preparation remains the most difficult obstacle the biologist has to overcome. Any process which follows the path of conventional techniques will be full of difficulties and artefacts. In terms of elemental retention, both quantitative and spatial, dry cryosectioning seems, at present, the superior method, although in some instances freeze-substitution has advantages. The relative merits of maintaining cryosections in either a frozen-hydrated state or subjected to the additional treatment of freeze-drying have not been clearly resolved. The major disadvantage of the frozen hydrated specimen is the poor quality of the structural detail which can be observed and this inability to identify precisely the irradiated volume of tissue invalidates one of the major advantages of the analytical technique. The inability to perceive clearly morphological detail in even freeze-dried cryosections is a considerable disadvantage. In this context, the advent of STEM could improve matters enormously. The facility to obtain an image under conditions of minimum irradiation damage prior to analysis, as well as the ability to process electronically the image signal so that specimen contrast can be improved are features which could revolutionize the examination of cryosections.

Even when ideal specimen preparation techniques have been established and accurate quantitation achieved, the biologist is left with the perplexing problem as to the biological significance of his findings. Reference to background data in the literature is often of little assistance as the information is generally derived from large samples (compared to the irradiated microvolume of a thin section), using techniques which are completely different in principle from X-ray analysis, e.g. spectrophotometry. Furthermore, the X-ray analyst is able to detect absolute and relative changes often below the sensitivity of more conventional techniques. A most valuable 'bridging' technique in this context is the ability to ash and solubilize small quantities of tissue (0.04 mg) and spray the solution on to grids for X-ray analysis in the TEM. In this way, small compartments of a large organ can be sampled, and composition changes, experimentally or otherwise initiated, detected. Because the technique of preparation is similar to that employed in spectrophotometry and because a greater mass of tissue is analysed, a reasonably fair comparison can be made with basic physiological analysis. On the other hand, because the procedure requires the analysis of droplets on a grid in a TEM, it lends itself to fair comparison with thin section analysis.

In the study of a biological problem by X-ray microanalysis it is important to realize that there is no universal preparative technique. The investigator must clearly define his problem and then adopt the most appropriate approach for that particular problem. As in other fields, a multidisciplined approach is the one most likely to succeed. The analyst should therefore consider and test several methods of specimen preparation as well as resorting to bulk analysis where appropriate. Caution should be exercised in the comparison of data derived from the analysis of specimens in SEM with those derived from TEM analysis, whether in scanning or transmission modes. The nature of the specimen, the operating parameters and the corrections necessary in quantitation differ considerably between the two approaches. The fact that an energy spectrum can be obtained within a few hundred seconds of analysis time lulls the operator into a false sense of optimism. Analysis is a slow, tedious procedure if all operating parameters are to be clearly defined and statistically significant data obtained.

This text attempts to deal with selected aspects of the subject. Initial chapters deal with basic analytical theory as well as a review of progress in the field. Subsequent chapters deal in depth with specimen preparation, cryosectioning, fluid analysis, and histochemical techniques. It is hoped that the information included will be of interest to undergraduates in their final year and to researchers with a basic orientation towards electron microscopy, but wishing to broaden their approach. For those who wish to relate composition directly to ultrastructure, X-ray microanalysis in electron optical systems is an inevitable, exciting venture which will disclose horizons previously unrecognized, and it is hoped that they will obtain some guidance from the comments made in the ensuing chapters.

I am greatly indebted to my colleagues, Mr T. W. Davies and Dr A. J. Morgan, for innumerable stimulating and perspicacious discussions which have helped crystallize the approach adopted in this book.

2

J. C. RUSS

Electron probe X-ray microanalysis–principles

2.1 Introduction

Ever since the introduction of the electron microscope four decades ago, efforts have been made continually to use it to obtain a better understanding of the microstructure and the relationship between structure and function in biological specimens. For the present purpose, this category can be expanded to cover a multitude of fields related by an interest in the study of organic samples. The initial promise of the electron microscope was the ability to extend resolution beyond that of the light microscope to show finer details of morphology. The limitations of the use of the method have, from the earliest days, resided in the specimen preparation area, and we will see that in large measure this is still the case.

The electron microscope itself and its morphological applications, as well as the electron microprobe and its analytical capabilities, and the more recent developments in combining and extending these methods, have met with very rapid acceptance and use in the fields of materials science – including such diverse disciplines as geology, metallurgy, solid state electronics, and some chemistry. The extension of the techniques to the study of biological specimens has been slow and largely dependent on the work of a handful of physicists and biologists who recognized the potential of the method and made the rather considerable effort to talk to each other and to understand each others' fields.

The delineation and classification of internal structure in biological samples has, as its basis, a desire to understand the functions of tissue at the cellular and subcellular level. The use of chemical treatment (fixation and staining) to reveal or mark certain structures was inherited from the earlier work in light microscopy, and has been used and somewhat refined in an effort to identify the chemistry at work in the tissue. In fact, a great deal of our present knowledge of the chemistry of biological systems has been derived from electron microscopy – either by virtue of the use of chemical stains and markers, or the 'grind and find' technique of tissue fractionation, separation, identification, and bulk analysis of the isolate. Both methods leave much to be desired in terms of specificity and relation to the living condition.

The limitations of histochemistry include numerous (some perhaps unrecognized) artifacts, and false positive reactions. The lack of accuracy is compounded by irreproducibility due to strong dependence on such variables as tissue processing, infiltration techniques, temperature, fading or oxidation of stains, and the human variable of the technician performing these many steps.

Stated simply, we would like to be able to describe the complete chemistry and reactions taking place, on a subcellular level, at any instant of time. The method of X-ray microanalysis does not provide that, but coupled with proper specimen preparation to preserve the activity taking place, it can analyse and localize the elemental distribution in the specimen, and with the development of new staining methods to be described (see Chapter 6) it may be applicable to the determination of reaction sites as well. It will be very important for the reader to remember that this method gives *elemental* analysis only, and the inference of the chemical form, or activity of the elements or the compounds that may be present, must be based on the user's judgment from other independent knowledge.

2.2 Electron excitation of X-rays

The process of electron-excited excitation of X-rays is quite straightforward in basic principle. The high velocity incident electrons strike the bound, inner shell electrons of the atoms in the material, knocking them out of their orbits or energy levels. This is called ionization, and creates a vacancy in the energy levels of the atom which is filled almost immediately by an electron that was in a higher energy level. As this electron drops into the inner shell vacancy, the excess energy is emitted in the form of an X-ray photon. The difference between the two energy levels is the X-ray energy, which is thus characteristic of the element. If the vacancy is created in the innermost shell (the K shell) it is called a K X-ray, and since several different outer shell electrons can fill the vacancy we can have $K\alpha_1$, $K\beta_3$, or other lines, the subscript identifying, in a poorly organized manner, the specific transition involved. A vacancy in the second shell produces an L X-ray, and in the third, an M X-ray. Beyond that the energies are too low to be useful for this type of analysis.

The incident electron must have enough initial energy to knock the bound electron out of its energy level; this is called the absorption edge energy E_c. A single electron may give up all or some of its energy in a single collision; usually it will undergo several such events as it travels through the sample, until finally it is reduced to thermal velocity. The other types of interactions the electron may have include elastic scattering in which no energy is lost but the direction of travel is changed, and Bremsstrahlung production in which the electron is deflected and slowed in the electrostatic field of an atom's nucleus. In this latter case, the lost kinetic energy is radiated as an X-ray photon, but the energy is not characteristic of the elements present but of the geometry of the encounter between electron and nucleus. The total Bremsstrahlung production is proportional to the matrix atomic number and to the square of the accelerating voltage used, and appears as the background in the spectrum, underlying the peaks corresponding to K, L, or M X-rays from the elements present.

Since Castaing (1951) introduced the electron probe microanalyser, an understanding has been built up of the physics of the interactions of electrons with matter and the generation of X-rays for analysis. The understanding is inexact and, in the interest of obtaining equations that can be handled mathematically to give useful results, a number of approximations have been made. Since most of this work has been used for inorganic applications, the models have been refined with that in mind and contain errors and bias when extended to organic matrices. This source of error may be tolerable, and in any event, can be lessened as more attention is devoted to this field and more measurements are made.

As most of the historical applications of these methods, and indeed most of the current ones as well, are in the non-biological fields, and deal with bulk samples, we must begin our efforts to understand the methods with this in mind. The advantages of using thin sections rather than bulk samples are very significant for many biological applications, and will be discussed shortly. The difference between using a TEM or STEM for this work rather than the conventional SEM or microprobe will also be dealt with. But first, it is useful to understand the process of X-ray generation and emission.

Of the various models used to conceptualize the process of electron excitation of X-rays, the most widely known is the so-called ZAF approach, which considers three processes each more or less independently of the others: (1) the generation of X-rays within the sample varies with the matrix composition and the incident electron energy, and is called the 'atomic number effect' because of its dependence on the average atomic number of the matrix; (2) the absorption of the generated X-rays is controlled by their depth distribution and the absorption coefficients of each of the elements present in the matrix; (3) secondary fluorescence of additional X-rays from some of the elements present can be produced by the absorption of the primary generated X-rays. Each of these effects depends to some extent on the others, of course, and in addition the solution of the equations for each requires the concentration of each element in the matrix, which is unknown. The concentration is simply proportional to the X-ray intensity for the element divided by the correction factors CZ, CA, and CF for the atomic number, absorption, and fluorescence effects. This leads to an iterative calculation and, because of the complexity of the mathematics, this is usually carried out on a computer.

2.2.1 THE ATOMIC NUMBER EFFECT

The generation, or atomic number effect (CZ), deals with the penetration and scattering of the electrons as they ionize in the sample. The details of the scattering and the zig-zag course of each electron can only be dealt with in detail by using probabilistic calculations. This can be done using a Monte-Carlo type of calculation in a computer to sum up the trajectories and the X-rays produced by many electrons. The atomic number effect summarizes these detailed results into two factors, R and S($CZ = R/S$).

R is the fraction of the incident electron energy that remains in the sample. To a first approximation for a bulk sample (one thick enough so that no electrons reach the bottom), it is $(1-\eta)$ where η is the fraction of the incident electrons that scatter through a large angle and leave the sample (backscatter), but in practice the R value that is used is also adjusted to account for the fact that before they leave the sample, these backscattered electrons will lose some of their initial kinetic energy and thus produce some ionizations along their path. The value of R in general decreases with increasing atomic number, because the heavier atoms cause more electron backscattering, and also it decreases with increasing electron voltage because more energetic electrons are more likely as they scatter to leave the sample eventually.

The term S describes the rate of energy loss of the electrons, and is called the electron stopping power. The principal cause of electron energy loss is ionization of atoms and in fact other effects are generally ignored, although they can become significant for very low atomic number matrices such as biological tissue. In general, the stopping power is greatest for low atomic number matrices (although the variation with Z is not smooth), and for low electron energies. This means that S and R vary together with atomic number, and this is fortunate since it means that the ratio R/S, which is the 'atomic number correction', is not so sensitive to accelerating voltage or composition, and that the inaccuracies and errors in the calculation of the R and S terms tend to cancel.

The calculation of R generally requires only the electron voltage V, which enters the equation as the 'overvoltage ratio' V/E_c where E_c is the absorption edge energy for the elements (the energy required to produce ionization), and the average atomic number of the matrix. The latter value is obtained by summing all of the individual atomic numbers for the elements present times their respective concentrations, or $\bar{Z} = \sum_j C_j Z_j$. Then the value of R as a function of Z and U can be calculated. The equations, like many we shall encounter, are not theoretically based, but are a mathematical fit to experimentally determined data points. An excellent summary of various mathematical models used for quantitative analysis has been published by Martin & Poole (1971).

Since the number of electrons that backscatter and their energy distribution (and hence the amount of energy they deposit in the sample) vary with specimen tilt, a term should be included in the equation to handle this case. As yet, no satisfactory data have been accumulated to make this feasible. It should also be noted that the majority of the data used to fit the function we use to calculate R were measured on elements with atomic numbers from about 13 up. The extrapolation to Z values representative of biological tissue is therefore of doubtful accuracy.

The stopping power S presents a greater difficulty, since it is not proper to use a mean or average value of the matrix atomic number Z. As mentioned before, S does not vary smoothly but must be calculated for each element and then be averaged. Also, since S is a function of electron energy we should perform an integration of the values of S as the electron loses energy in the sample. This type of calculation becomes prohibitively complex and so simpler equations are frequently used to obtain values with reasonable accuracy. The most common models assume, as a simplifying description, a constant 'effective' electron energy half-way between the initial energy and the absorption edge energy (below which it can no longer excite the element). This assumption becomes poor when the difference between the accelerating voltage and elemental absorption edge becomes large, which may be the case for light elements and high voltages.

Again, it is important to realize that the approximations have been made to give acceptable accuracy for matrices much heavier than biological tissue. The rate of energy loss due to ionization (knocking out of electrons) is proportional to the number of electrons encountered [the product of (atomic number Z/atomic weight A), density ρ, and Avagadro's number N_o], and depends on the mean ionization potential. But for low Z elements it is possible to get a significant amount of energy loss due to other mechanisms. One approach to take this into account is to introduce a fudge factor into the value for the mean ionization potential so that it deviates markedly at low Z. This empirical adjustment has been tested, however, only with inorganic materials and may not be accurate or appropriate for tissue.

In conclusion, it is fair to say that of all the correction factors, the CZ or atomic number factor is on the poorest theoretical ground and hence can be extended to biological tissue the most poorly. On the other hand, the fact that the two major terms R and S vary together serves to keep their ratio near unity, and this makes errors tolerable. The magnitude of the CZ factor will not become excessive provided that we do not work with either very high or very low ratios of accelerating voltage to absorption edge energy. A useful rule of thumb is to keep CZ between 0.5 and 1.8, and preferably in the 0.8–1.25 range. This will frequently argue against the use of accelerating voltages below about 2–2.5 times the absorption edge energy (which also produces poor excitation of the lines), or above about 7–10 times the absorption edge energy.

2.2.2 THE ABSORPTION CORRECTION

Once generated within the sample, the X-rays radiate in all directions; it is only those that travel through the sample to reach the surface and then enter the X-ray detector that are useful for analysis. Whenever X-rays pass through matter they are absorbed exponentially with distance: $I/I_o = e^{-\mu\rho t}$ where I_o is the original intensity or number of photons, t is the distance, ρ is the density, and μ is the mass absorption coefficient. The mass absorption coefficient for each element is a function of energy, since it describes the probability that a given atom will absorb a photon with a particular energy. This probability is high when the photon has just the right amount of energy to knock out a bound electron to create an ionized atom, and decreases exponentially for higher energies. The values of μ are tabulated for all elements for the principal X-ray emission energies of the elements in tables such as those of Heinrich, although the accuracy of the present values for light elements is poor.

When the material through which the X-rays are passing is a mixture of several elements, the value of μ for the matrix is just the sum of the values, of μ for each individual element for that energy X-ray added in proportion to their concentrations C_j. We can calculate the matrix absorption of the X-rays being generated, to account for their reduction in intensity. However, since the matrix absorption coefficient depends on the concentration of each of the elements, which of course, we do not know, the solution of these complex relationships will require assuming a value for each elemental concentration, calculating an answer, seeing if it agrees with our assumption, and then trying again; this is known as an iterative solution.

The X-rays generated by the electron beam are not all at the same depth in the sample, but rather have a distribution with depth. This distribution is generally expressed as $\phi(\rho\xi)$ where ξ is the depth and ρ is the density, so that the same function ϕ can be used for samples of different density. The distance the X-rays must travel is greater than ξ, since the X-ray detector is not directly above the specimen surface but is at an angle. The X-ray takeoff angle (ϕ) is the angle between the path to the detector and the sample surface, and the electron incidence angle (ψ) is the angle the sample surface is inclined. From simple trigonometry we can then find that the distance the X-rays travel (and are absorbed) in the sample is related to the depth ξ as $\xi . \sin \psi / \sin \phi$.

If we then integrate all of the X-rays for an element from various depths, as given by the $\phi(\rho\xi)$ curve, and take into account the absorption of each in terms of the matrix absorption coefficient, we can get the measured intensity corrected for absorption. The integral can be written as function $F(\chi)$ where $\chi = \mu \sin \psi / \sin \phi$. In other words, we have replaced the function $\phi(\rho\xi)$, which has a clear physical significance, with another function $F(\chi)$ whose physical meaning is less obvious, but which includes the absorption and geometry terms and is in fact related to ϕ by a mathematical convention called the Laplace Transform, which is encountered in many other fields of mathematics and physics.

By making some assumptions for the shape of ϕ, Philibert (1963) was able to solve this expression for F in a simple form which can be calculated without much difficulty. This value of $F(\chi)$ for each element in the sample, evaluated for the matrix absorption, is just the reciprocal of the absorption correction CA. The absorption of X-rays in the matrix depends very strongly on its composition as well as the accelerating voltage used. If the voltage is very high, the depth of X-ray production $\phi(\rho\xi)$ will be very deep, and so the absorption distance will be greater. Hence, although more X-rays may be produced, the size of the absorption correction will be large, and the errors introduced by the approximation used in deriving the expression for $F(\chi)$, and the errors in the measured and tabulated values of mass absorption coefficients μ will cause the overall accuracy to suffer. A value of CA less than about 0.5 is a clear indication that the voltage is too high for that element and that the results are likely to be in error.

2.2.3 SECONDARY FLUORESCENCE

The absorption of the X-ray photons from an element in the sample results in the excitation of another element. The process of absorption is that photon energy is used to ionize an atom by ejecting an electron from its energy level. Then this vacancy is filled by another electron and an X-ray characteristic of that atom is emitted. This means that the X-ray intensities measured from some elements will be increased because of this secondary excitation. Since the photon must be greater than the electron energy level in order to produce an ionization, only higher energy lines can excite lower energy ones, not vice versa.

The extent to which a given element's X-ray intensity is increased by this process of secondary fluorescence is a function of the other elements in the sample, and in general, is great when there are elements with emission lines just above its absorption edge, since these have the greatest probability of being absorbed. The increase in intensity γ due to secondary fluorescence results in a value of $CF = 1 + \gamma$. It includes terms for relative excitation probabilities, fluorescent yields, and other parameters which are available in published tables. It also takes into account the depth distribution of the primary characteristic X-rays, but it ignores the possibility of secondary fluorescence due to the Bremsstrahlung X-rays generated in the sample. This can, of course, take place but it contributes a negligible amount of additional secondary fluorescence except in the case of a very high atomic number matrix, which generates the greatest amount of Bremsstrahlung. Hence, the assumption is certainly acceptable for primarily organic matrices, and in fact for such samples, the low concentrations of the elements means that the number of photons available to cause secondary fluorescence is small, and so this effect can often be ignored completely in analysing biological samples.

2.2.4 THE CONSTANT OF PROPORTIONALITY

The three correction terms so far described compensate the effects of specimen composition on the measured intensity for each element. However, the original relationship expressed the concentration C as proportional to the intensity I divided by the correction factors CZ, CA, and CF; to obtain quantitative results we need the constant of proportionality.

In principle, we could use any sample of known composition as a reference; by measuring the intensity and calculating the CZ, CA and CF factors we could obtain the proportionality constant and hence proceed to obtain percentage composition values for the unknown. In practice, however, there are many elements for which we cannot find materials that are stable under the electron beam and homogeneous on a submicron scale. This latter requirement is necessary because we are calculating the effects of the elements on each other's X-radiation and so they must be intimately mixed together. Some minerals or metal alloys can be used as standards for several elements at once, but the most usual method is to use pure elements as standards. This has the further advantage that the CZ and CA factors can be readily calculated for the pure elements, and there is no interelement fluorescence (only one element being present) so that CF can be omitted.

The disadvantages of using pure elements are that a separate measurement of a standard is required for each element in the unknown, and furthermore that for some elements of interest the pure element is not available as a stable solid. In fact, for quite a few elements there are no really satisfactory oxide or binary standards at all.

The requirement for standards for each element can be relaxed or eliminated by expressing in mathematical form the relationship between the intensities of different pure elements, under given conditions of excitation and geometry. The intensity ratio between any two elements, for any two emission lines, is the ratio of the relative excitation of the elements, which depends on several factors: the probability R that the electron stays in the sample rather than backscattering, the overvoltage $U = V/E_c$ (this determines the likelihood of the electron ionizing the atom); the fluorescent yield W (the probability that an ionized atom emits an X-ray, rather than reabsorbing the photon immediately to liberate an Auger electron); the relative intensity L of the lines being measured; the absorption of each line within the sample $F(\chi)$; and finally the probability T that the X-ray is detected and counted by the X-ray spectrometer. These terms are all readily calculated or obtained from tables. Since U depends on the accelerating voltage and $F(\chi)$ depends on both the voltage (which determines the depth distribution of X-ray production) and the geometry of specimen tilt (which affects the absorption of the X-rays) the ratio must be determined for the actual analysing conditions being used.

In principle then we could use as a reference standard, any pure element that was handy, perhaps iron or copper, and calculate all of the proportionality constants from first principles, thus obtaining quantitative percentage results for our unknown. In practice it usually produces better accuracy to choose a standard element as near as practical in the periodic table to the elements in the unknown, since this makes the ratio nearer to one and thus reduces the magnitude of errors introduced by the use of inexact constants. It may also be wise in some cases to use several standards, for different groups of elements (e.g. heavy and light).

2.3 Instruments and specimens

2.3.1 THE SEM AND THE TEM

Transmission electron images from thin specimens can be obtained with either the scanning electron microscope or the transmission electron microscope. The image in each case is qualitatively the same, but there are several important differences. The image resolution in the SEM (or more properly the STEM when used in this mode of operation) is determined largely by the diameter of the incident electron beam. With most commercial SEMs this means the image resolution will be in the 5–20 nm range, whereas most common TEMs can readily achieve well under 1 nm and often under 0.5 nm. At is important to remember, however, that not all biological samples are capable of retaining information at this fine a scale anyway. Also, if the use of the transmission image is just to locate the features to be analysed then 10 nm may be more than adequate.

Another consequence of the fact that image resolution is determined by the STEM beam diameter is that the image resolution stays relatively constant with specimen thickness. In the TEM the electrons passing through a thick specimen have lost energy by being scattered and thus cannot be sharply focused by the lenses, which have unavoidable chromatic aberrations. Since no lenses are used to form the image in the STEM, the only degradation in resolution with thickness is due to low probability multiple scattering events and gradual loss of contrast and signal strength. It is quite feasible to obtain transmission scanning electron images from sections several micrometres thick, although we will see that these specimens are not ideal for microanalysis.

The STEM image also has an advantage in contrast over the normal TEM image, which results from the point-by-point method of imaging and the ability to amplify the signal electronically. This is of particular importance because the reduction or elimination of staining chemicals, which provide image contrast but deposit elements that make the analysis more difficult, is desirable when microanalysis is to be performed.

The desirability of using the scanning transmission electron image has led many manufacturers of both TEMs and SEMs to provide this capability, and in fact the availability of transmission attachments for the SEM and scanning attachments for the TEM suggests a convergence of the two instruments. In principle, it makes no difference which instrument might be used for microanalysis of thin sections; but in practice each one offers some advantages at the present time. The SEM has more space in the specimen chamber which makes it easier to attach X-ray detection equipment and locate it close to the specimen for optimum efficiency. It is also possible to construct special stages to hold thin specimens, constructed from low atomic materials such as beryllium or graphite to minimize background, and possibly designed to hold the specimens at low temperatures. Low temperature holders may be needed when rapid freezing is used as a specimen preparation technique as will be discussed in Chapters 3, 4 and 5.

The TEM, on the other hand, has in general a higher accelerating voltage available, and thus a brighter electron gun. Both high electron voltages and high beam current densities are desirable for thin section microanalysis. High voltages by themselves do not improve, and may degrade, the analytical results (primarily because of increased background due to scattered electrons), but are often necessary to penetrate the somewhat thicker sections used for analysis. The TEM also has a more flexible electron optical system that allows excellent control of the electron beam and permits the operator to magnify the beam again to observe its exact size, shape, and location. Both instruments can be well used for thin section microanalysis, but it seems likely that attention to the design of special specimen stages and detectors for the TEM will produce the most desirable combined instrument.

2.3.2 SPATIAL RESOLUTION

Bulk specimens

The resolution of the normal SEM image formed by secondary electrons from the surface of a bulk specimen can be better than 10 nm, because these electrons have a very low energy and only escape to be detected if they originate in a thin layer near the surface. This close to the surface, the incident electrons have not yet had much chance to scatter transversely, and so the secondary electrons can form an image with a spatial resolution nearly as good as the incident beam diameter.

The X-ray photons come from essentially the entire electron capture volume, until the electrons lose so much energy by scattering that their energy drops below E_c and they can no longer excite the element of interest. Since this volume is much larger than the incident beam, it is obvious that the X-ray resolution is poor compared with the secondary electron image resolution.

The consequence of this is that it is not possible to simply select a feature in the secondary electron image, position the beam stationary on that point, and analyse it. The user must always keep in mind the larger volume contributing X-rays. If the feature of interest is large enough in the surface image it may still not extend as far into the sample as the electrons penetrate, and so it will not be selectively analysed. To visualize the magnitude of the size of the excited volume in biological tissue, for an accelerating voltage of 15 kV, for instance, the excited volume for calcium analysis is about 7 μm deep and more than 3.5 μm in its widest point. Furthermore, the greatest fraction of the X-rays comes from a depth of more than 3 μm and so may bear no relationship whatever to the surface features. Higher accelerating voltages will create an even larger and deeper volume of excitation. Many structures of biological interest are a great deal smaller than 1 μm, and furthermore, the surface image of biological specimens is in any event hard to use to locate or recognize internal structures. The transmission electron microscope produces an image that shows internal structure very well, and the same type of image can be formed with a scanning transmission electron microscope, although there are some differences in image resolution, contrast, and specimen thickness requirements. In either microscope the use of the thin section as a sample also provides the capability to obtain more selective microanalysis on a finer scale.

Thin sections

In describing the size of the X-ray excited volume in bulk samples, it was necessary to follow the electrons as they scattered in the material. The total capture volume is roughly tear-drop-shaped and many times larger than the incident electron beam. The situation is quite different for a thin section. The electrons leave the specimen with most of their initial energy and most of them have not scattered far in a transverse direction. The excited volume is roughly cone-shaped, with a gradually increasing angle as specimen thickness increases. The angle is fairly consistent for specimens embedded in plastic or thin sections of tissue containing the original water in the form of ice crystals. For specimens prepared by cryo-sectioning and freeze-drying the removal of the bulk of the specimen mass reduces the electron scattering proportionately and the angle of spread of the cone is less. The angle decreases as electron energy increases, but the magnitude of this change is not as great as the variation due to specimen density, etc. over the range from 40 to 100 kV.

For samples embedded in plastic the spatial resolution can be roughly approximated as about half of the thickness of the section. This means that selective analysis of features smaller than 100 nm is readily possible, as compared with several μm for bulk material. The use of very thin sections would give the best spatial resolution, but this is rarely practical because the small amount of material present will give an extremely small signal and require a long analysis time, during which the electron beam may damage or contaminate the specimen. On the other hand, specimen thicknesses greater than about 0.5 μm tend to degrade spatial resolution severely, because multiple electron scattering increases the angle of the cone-shaped excited volume. Also, the principles of the method to be described for quantitative analysis require that sections less than about 0.5 μm be used.

A good thickness range for plastic-embedded or frozen (not dried) sections is about 150–250 nm, and for frozen and dried sections, about 200–400 nm. Thinner or thicker sections can be employed for specific situations, of course.

2.3.3 DETECTION LIMITS

With any analytical method a question of primary importance is the practical limits of detection. Electron excited X-ray analysis is generally capable of detecting elemental concentrations in bulk material down to the 0.1 to 0.01 % concentration level. The limitation is the ability to distinguish statistically a small peak from the Bremsstrahlung background. This level of detection applies to most of the elements in the periodic table from about sodium ($Z = 11$) up, falling off gradually for the heavier elements because the available accelerating voltages, especially in the SEM, are not usually high enough to excite them efficiently. The detection limit becomes rapidly poorer for lighter elements such as carbon, nitrogen, and oxygen which emit fewer X-rays and whose X-rays are lower in energy and hence are more quickly absorbed.

Since in general the microanalysis of biological tissue in bulk presents a favourably low matrix absorption, most elements from sodium up can be readily detected down to less than a thousand parts per million, and with favourable conditions down to a few hundred parts per million. But in a bulk sample the total analysed volume can easily be greater than 10 μm^3 – and can sometimes approach 50 μm^3 – so that the detection of a small amount of the element localized on a structure within that volume is difficult and often impossible.

For the case of a thin section of tissue, the detection limit approaching a few hundred parts per million again serves as a starting point but now the total volume and mass of the analysed volume can be 10^{-3}–10^{-4} μm^3. For a nominal density of about 1 g/ml, this would suggest a minimum detectable mass for an element of as little as 10^{-20} g – only a few hundred atoms. Unfortunately, while this is mathematically possible, the X-ray intensity from such a few atoms would be so low that it would take many hours to detect enough to have any statistical significance. In the case of thin sections the limit of detection is determined not by concentration but by signal strength, which is proportional to elemental mass. Detection limits of 10^{-17}–10^{-18} g are realizable in an analysis time of a few hundred seconds, which is tolerable to the specimen. The advent of brighter electron guns with lanthanum hexaboride or field emission sources should improve these limits in the future. The ultimate limit to the increase in signal strength with increasing the electron beam current is the ability of the specimen to withstand heating effects when high current densities are used.

2.4 X-ray detection

2.4.1 ENERGY-DISPERSIVE ANALYSIS

There are two methods that can be used to measure and count the X-rays, generally known as energy-dispersive and wavelength-dispersive based on the method used to distinguish the X-rays from different elements. In the energy-dispersive X-ray analyser, the X-rays enter a detector made of very pure silicon with as nearly a perfect a crystal structure as possible and with any remaining imperfections such as dislocations or vacancies compensated for by diffusing in small, mobile lithium atoms.

Within the detector, which is a semiconductor, each X-ray is absorbed, and its energy is used to raise electrons to the conduction band. The number of these electrons is proportional to the X-ray energy, since 3.8 electronvolts are needed to produce each conduction electron. The electrons are then collected by an applied high voltage (typically 500–1000 V) and the detector is ready within a fraction of a microsecond for the next X-ray to enter.

The electron charge from each X-ray is accumulated in a field-effect transistor closely connected to the detector, and produces a small voltage pulse. Both the detector and the transistor are cooled to liquid nitrogen temperature to reduce the stray voltages caused by thermal effects, which would make it difficult to measure these extremely small pulses. The pulses are amplified, each pulse having a magnitude linearly proportional to the energy of the X-ray that produced it. The height of each pulse is measured by allowing it to charge up a small capacitor, and measuring the time required for it to discharge on a very high frequency clock. This time is again proportional to the X-ray energy. The time is used as an address in a memory to store a count for that X-ray, so that a complete spectrum is built up with X-ray energy as the horizontal scale and the number of photons counted at each energy on the vertical axis. The location of peaks in the spectrum thus identifies their energy and hence the element from which the X-rays came, and the size of the peak gives the number of X-rays from that element which can be used to obtain quantitative information.

The spectrum is normally displayed on a cathode ray tube or television screen, along with energy markers and other information to aid the operator's interpretation. The system may also include a minicomputer programmed to operate on the spectrum to determine the total X-ray counts for various elements and calculate compositional results from that data.

Since the X-ray photon energy is represented by the number of electrons collected from the detector, all that should be needed is a way to count them. The very small pulses that are produced must be highly amplified before they can be measured, however; the amplification process creates several problems. The first is calibration: since it is impractical to build an amplification and measurement system that is exact, it is necessary to adjust the electronics of each unit, using X-rays of known energy, to make the peaks fall at a known point on the horizontal (energy) scale of the display. This process requires using a sample with two or more known elements that can be excited with the electron beam, and the adjustment of amplifier gain and zero. These adjustments must be checked from time to time, since component ageing or slight changes in ground connections may cause peaks to shift. These shifts may be tolerable to an operator used to identifying elemental peaks visually, and knowing roughly what elements to expect. However, when computer processing is used, exact calibration is extremely important.

The amplifier is also the source of most of the limitations in count rate that energy-dispersive systems suffer. The very high gain required and the long amplifier time constants (many microseconds, as contrasted to 0.25–0.5 microseconds for amplifiers used in wavelength-dispersive systems) limit the number of pulses that can be processed. The long time constants are used to 'average out' electronic noise that would otherwise distort the pulses and vary their height. There is still some random variation in pulse height that occurs, due to two causes. In the silicon detector itself, the production of electrons is an inherently statistical process. Hence many photons of the same energy will produce slightly different numbers of electrons. The average number will be constant, but the fluctuations about that average give rise to part (about two-thirds) of the width of the peak observed in the final spectrum. The rest of the peak width is due to fluctuations in the pulses introduced by electronic noise in the amplification process. The total width of each peak depends on energy and should be constant for a given system.

The long amplifier time constants used mean that at high count rates the pulse from one X-ray event may partially overlap the pulse from the next one. When this 'pile-up' occurs, the height of one or both pulses may be affected.

In this case, special circuitry is used to detect the pile-up and so the pulses are not measured. This prevents putting counts into the spectrum that would be erroneous, but it means that information is lost. In fact, as the count rate of X-rays entering the detector increases, the probability of pile-up events increases until there are essentially no measurable pulses, and the system is paralysed. This makes it important to monitor the input count rate of the system to be sure that it does not exceed the amplifier's capability. Most systems allow for this, either by displaying the input count rate directly (the rate at which pulses enter the amplifier), or by a 'dead-time' meter that indicates what percentage of the time the electronics are busy (and hence 'dead' to an incoming X-ray event).

Energy-dispersive X-ray detectors are cooled with liquid nitrogen. This serves two purposes – reducing thermal noise in the electronics, and stabilizing the lithium diffused into the silicon crystal (which would gradually diffuse out at room temperature). Since the loss of lithium is very slow, occasional detector warm-ups should not affect the performance of the system. Unfortunately, practical design considerations may give problems from accidental warm-ups. The first of these is damage to the field-effect transistor used as the first stage of the amplifier, if the full bias voltage is applied across it as the detector resistance drops with temperature increase. In most modern systems this is prevented either by circuit design or liquid nitrogen level sensors that turn off the voltage. Longer term damage can occur because the vacuum system in the cryostat is sealed, and after several years of operation the 'getter' used to keep the vacuum clean has accumulated material which, on warm-up, may redeposit on the detector and conduct current. This would increase the electrical noise, and degrade system resolution. Another source of long term damage is to the LN vessel itself, due to corrosion from exposure to liquid water condensed from moist, cold air. For these practical reasons, it is wise to keep the detecting unit cooled with liquid nitrogen at all times, whether it is in use or not.

2.4.2 WAVELENGTH-DISPERSIVE ANALYSIS

The older method for discriminating between X-rays of different energies makes use of the fact that X-rays, like all photons, can be thought of as having a wave nature as well as a particle behaviour. The wavelength of an X-ray photon is inversely proportional to its energy ($\lambda = 1.2398/E$, where the wavelength λ is in nanometres and the energy E is in kiloelectronvolts). When photons of a particular wavelength strike a crystal with a set of atomic planes regularly spaced at a distance d (a characteristic of the crystal and the angle at which it is cut), diffraction will occur if the angle of the crystal is exactly right. This phenomenon occurs because the incoming photons re-radiate in all directions from the atoms in the crystal, and their wave nature causes constructive interference to occur in a particular direction. The critical relationship between the angle of incidence of the photons (ϕ) and their wavelength (λ), and the spacing of atomic planes in the crystal (d), is the well known Bragg Law ($n\lambda = 2d \sin \phi$ where n is any integer).

By arranging a mechanism to hold a suitable crystal and position it at various angles, and a detector such as a gas-flow counter that gives an electrical pulse for each X-ray that strikes it, it is possible to select X-rays from a particular element and count them. In practice, several such crystals are needed with different atomic spacings to cover the range of elements in the periodic table. Since the selection takes place on the basis of wavelength, the method is called wavelength-dispersive X-ray analysis.

Diffracting crystals are chosen for their 'd' spacing, which defines the wavelength range for which they can be used and the intensity of their diffraction. The wavelength range of interest is typically from 0.1 nm (about the wavelength of the Kα line of atomic number 35, or the Lα line of atomic number 88) up to about 1.0 nm (about the wavelength of the Kα line of atomic number 12, or the Lα line of atomic number 33). As can be seen, this will allow the analysis of most of the periodic table by either the principal K or L line. Longer wavelengths in the 1–10 nm range must be diffracted to analyse lighter elements down to atomic number 4.

Bragg angle settings from about 10 to 70 degrees can be practically achieved with most spectrometer designs, which means that crystals with 'd' spacings from 0.2–6 nm are required. The most commonly used are:

LiF (lithium fluoride)	0.201 nm
quartz	0.33 nm
PET (pentaerythritol)	0.44 nm
ADP (ammonium dihydrogen phosphate)	0.53 nm
KAP (potassium acid phthalate)	1.33 nm
RAP (rubidium acid phthalate)	1.3 nm

for the 0.1–2 nm range, and pseudo-crystals from soap films such as lead stearate (3.6 nm) for the longer wavelength range. A selection of about four crystals, which can be externally selected, will generally give reasonable coverage of the range in a single spectrometer. Multiple spectrometers can also be used when it is desired to monitor more than one element at a time (e.g. because the specimen may be damaged by long exposure to the beam), or to reduce the time needed to scan through the wavelength range to identify an unknown.

In addition to choosing the proper crystal and angle, the alignment of the X-ray source (the point at which the electrons strike the sample), the diffracting crystal, and detector slit is quite critical to obtain proper results. The intensity of the measured peak, and the peak-to-background ratio depend on spectrometer alignment. This means the specimen position must be adjusted to the proper alignment point, which is difficult for some samples and microscopes.

The WD spectrometer's high resolution and resulting high peak-to-background ratios give good detection limits (often approaching 100 ppm), and few peak interferences between elements, at the penalty of detecting a very small fraction of the generated X-rays and hence requiring large beam currents.

2.4.3 COMPARISON OF ENERGY AND WAVELENGTH METHODS

Energy-dispersive X-ray analysis has the advantage that it processes simultaneously X-rays of all energies, and accumulates a spectrum showing the peaks from all of the elements present. The presence of elements not anticipated is made evident, and the rapid comparison of several elements is thus facilitated. When wavelength-dispersive methods are used it is necessary to scan each of several crystals through a range of angles to obtain an analysis of all elements, and this can take from 15 minutes to 1 hour, depending on the sensitivity required and the number of spectrometers attached to the microscope or microprobe. During this time many biological materials may be damaged by the electron beam.

On the other hand, because the electronics must simultaneously measure and count the pulses corresponding to the X-rays of all energies from the sample rather than only counting those for a single element, the ability of the wavelength-dispersive method to handle very high X-ray fluxes is superior to the energy-dispersive method. Such high X-ray fluxes are produced only with bulk samples of high average atomic number and high electron beam currents as encountered in the electron microprobe, and in fact for the analysis of biological thin sections the problem is usually that the small amount of mass present and the low beam currents that can be compressed into fine beams and withstood by the specimen produce very low X-ray fluxes. In this case, the energy-dispersive method has the advantage that the detector can be placed quite close to the sample to intercept and count a large fraction of the X-rays, which are radiated equally in all directions. The geometrical restrictions of X-ray diffraction and the size of the mechanism required to position the crystal and detector makes the solid angle of X-rays collected very small in the wavelength-dispersive method, and in addition 80–95% of the X-rays are lost in the process of diffraction.

The greatest advantage of the wavelength-dispersive method is that it produces very sharp peaks in the spectrum, so that there are few instances of interferences between peaks from different elements. This is called the spectral resolution, and should not be confused with spatial resolution of the analysis, which is unrelated. In the energy-dispersive spectrum the peaks are nearly ten times broader than for a wavelength-dispersive spectrum. There is still ample ability to separate the principal peaks from different elements, but there are cases in which a peak from one element may overlap partially or completely another peak from a different element. Fortunately, these interferences are not common for the elements of principal biological interest, and can be sorted out satisfactorily either by referring to the principal peaks of the elements or by using a suitably programmed minicomputer to strip the overlapping peaks.

Both types of X-ray spectrometers can be extended to the analysis of light elements (e.g. carbon, nitrogen, or oxygen). The wavelength-dispersive method requires the use of special crystals with large atomic spacings and an X-ray detector with a special thin window to admit the X-rays. The energy-dispersive detector normally is protected by a beryllium window to separate it from the microscope vacuum and prevent it from acting as a cold trap (since it is liquid nitrogen-cooled). This window must be removed to detect low energy X-rays from light elements, and other devices such as cold traps and electron deflectors must be used to protect the detector. Fortunately, the analysis of elements lighter than about atomic number 11 (sodium) is not usually required for biological samples (where their presence does not yield much useful chemical information) and so these special detector modifications can be omitted for this application.

One other point of comparison between the energy- and wavelength-dispersive methods is that since the energy-dispersive spectrometer does not have moving parts that require alignment, and since it presents a complete set of information simultaneously to the user, it is generally much easier to use, needs much less maintenance, and requires less specific operator skill or training to use.

For all these reasons the energy-dispersive method, although available for a much shorter period of time than the wavelength-dispersive method, is much more widely used (outnumbering it by nearly 3:1) in all fields of application (Russ, 1970). This is especially true in the analysis of biological thin sections where its advantages are particularly significant.

2.5 Quantitative results

The use of thin sections yields immediately a great simplification for quantitative analysis. The electrons lose only a small fraction of their energy in passing through the sample and so the stopping power S does not change. Also, electron backscattering is negligible and can be ignored. Hence the only X-ray generation effect that needs consideration is the effect of accelerating voltage on the probability of ionizing an atom of a given element. This probability is called the ionization cross-section and depends on the accelerating voltage and the element's absorption edge energy. The probability rises with voltage to a maximum for an accelerating voltage between 2.5 and 3 times the absorption edge energy and then gradually decreases beyond that. However, since the electron gun brightness continues to increase with voltage, the total excitation of a given element generally shows a continuing upward trend with voltage, and the greatest generation of X-rays is usually obtained at or near the highest voltage availability (e.g. 80 or 100 kV in a normal TEM). Since background also rises with voltage, the best detection limits may be achieved at somewhat lower voltages, however.

In a thin section the absorption and secondary fluorescence of X-rays turns out to be negligible. This can be predicted from a consideration of absorption, which as described before is just $e^{-\mu\rho t}$ where μ is the absorption coefficient, ρ is density, and t is the distance (in this case we can consider it to be a measure of the section thickness). For all except the very light elements, the absorption of X-rays in a distance of even 0.5 μm in organic material is negligible. The X-rays of elements lighter than sodium are absorbed more but these elements present very many other problems for analysis as well and are generally not used either for qualitative or quantitative purposes. The presence in tissue of oxygen, carbon, and hydrogen, etc., is not of direct interest in most cases since these elements form the bulk of the sample. It is the presence of heavier elements that can be used to understand the chemical activity in the tissue.

The fact that absorption is negligible in thin sections also means, of course, that secondary fluorescence can be ignored since it is a consequence of the absorption of primary X-rays. The confirmation of the possibility to ignore these effects has been obtained experimentally by measuring the X-ray intensity from sections of varying thickness for an element present in a constant amount. The data (Jacobs & Baborovska, 1972) follow a straight-line linear relationship, which means that for specimens that we will call 'thin' the characteristic elemental intensity is a linear function of the mass of the element present. As a side note, the thickness range of specimens for which this is true includes nearly all specimens thin enough to form a viewable image in the conventional TEM. Because the use of an STEM image (in either TEM or SEM) can provide a satisfactory transmission image for much thicker sections, an adequate criterion in this mode must be the actual section thickness as discussed above.

2.5.1 THE P/B MODEL

The fact that the elemental X-ray intensity gives us a measure of the mass of the element present is useful in some circumstances, as for example the analysis of small particles or granules. By making suitable standards, calibration curves can be prepared and used to obtain quantitative results.

However, in most cases, the result of interest is not the mass of the element but its concentration. Knowing the mass of the element, to obtain the concentration we need to measure the total mass of the excited volume.

A convenient method for this purpose is to use the Bremsstrahlung X-ray intensity from the specimen. As described before, the generation of Bremsstrahlung radiation is due to the deceleration of the electrons in the field of the nucleus of the atoms, and the amount of deceleration and hence the total Bremsstrahlung production is proportional to the amount of mass through which the electrons are passing.

Hence the total Bremsstrahlung signal gives a measure of the total excited mass, independent of the exact shape of the excited volume or the specimen thickness. The ratio of the characteristic elemental intensity to the total Bremsstrahlung intensity [or Peak-to-Background (P/B) ratio] is thus proportional to the ratio of elemental mass to total mass, or in other words, to the elemental concentration (Hall & Werba, 1968).

When the 'background' is used as a means of assessing section thickness, it is important to separate the intensity of the Bremsstrahlung radiation from the excited volume of the section from other sources of background radiation. These include continuum generated in the section or in the supporting grid and holder by scattered or stray electrons, and pulses produced in the detector itself by scattered electrons that strike it. Usually, the best way to determine this 'extra' background is by making blank measurements with the beam on a control area or in a small hole in the sample.

The constant of proportionality can be determined by constructing a straight-line calibration curve using standards (for example, made by dissolving known elemental concentrations in gelatin) or by calculation from standards consisting of mineral chips or even bulk materials.

2.5.2 THE RATIO MODEL

The major shortcoming of the P/B model is that few thin section specimens contain elemental concentrations representative of the living condition. In most preparation methods, even if the heavy elements of interest can be assumed to stay in place (this, of course, is necessary if meaningful results are to be obtained), there is removal and/or addition of mass in the form of water, plastic, or other chemicals. And even with the method of rapid freezing and maintaining the sample at or near liquid nitrogen temperature with the ice present, there may be a loss of organic mass under the action of the electron beam. For these reasons, it is usually more meaningful to measure the relative concentration of the elements in the material, rather than their absolute concentration. Then if the concentration of one element is known from bulk analysis or by adding it as an internal reference standard (the chlorine in most embedding plastics can sometimes be used for this purpose), the relative elemental concentrations can be converted to the original concentrations of the elements in the tissue.

Relative elemental atomic concentrations can be directly calculated from their relative intensities, by dividing the intensities by a proportionality constant P. No standards are needed, and there is no need to include the Bremsstrahlung in the calculation because the analysed volume is the same for each element and thus cancels (Russ, 1973).

The value of P is simply the product of four probabilities: (1) the probability that an electron excites (ionizes) an atom of the element, which as described depends on the electron voltage and absorption edge energy of the element; (2) the probability that the atom emits an X-ray, which is the fluorescent yield described before; (3) the probability that the X-ray is in the particular characteristic line being counted; and (4) the probability that the X-ray enters the detector and is counted. Tables of values can be prepared for selected accelerating voltages, and used as needed to obtain quantitative results. (See Table 2.1 for a summary of quantitation models.)

2.6 Practical usage

2.6.1 SPECIMEN PREPARATION

The use of the electron microscope in combination with X-ray microanalysis of biological thin sections can be divided into three more or less distinct areas of application: (1) the analysis and localization of naturally occurring elements, such as electrolytes and enzyme co-factors; (2) the detection of elements associated with a pathological condition, either normal elements in unusual concentrations or elements such as lead and mercury not normally present in tissue; and (3) the localization of elements intentionally added as markers or stains to identify the site of some specific chemical activity. The methods of specimen preparation for each of these purposes are somewhat different, and require the development of new techniques not widely used at present for routine thin section preparation.

The most severe application in terms of the specimen preparation is the analysis of naturally mobile ions such as sodium and potassium. In order to preserve the distribution of these elements, it appears that the only suitable method must begin with very rapid freezing of the tissue. Normal chemical fixation methods alter membranes and allow complete redistribution of the elements to take place. Even normal freezing rates produce membrane disruption, and it is only such drastic cooling rates as freezing small tissue fragments in melting nitrogen that have so far given adequate results.

After being frozen, the tissue can be cryosectioned and kept stable with a suitable cold stage during examination and analysis, or the water can be removed by sublimation *in vacuo*. The latter method will cause some

Table 2.1 A summary of quantitation models for the X-ray microanalysis of thin sections (after Hendriks, 1975)

	Direct mass method					
	absolute	*relative*	*Mass fraction method*	*Concentration ratio method*		*Direct concentration ratio method*
Principle	The absolute mass of some element in a particle can be determined with the aid of thin standards of that element	The mass ratio of two elements in a thin specimen can be determined with the aid of thin standards of those elements	The mass fraction of some element in a biological specimen can be determined with the aid of a standard containing that element	The weight concentration ratio of two elements in a thin section can be determined with the aid of standards of those two elements		The weight concentration of two elements in a thin section can be determined without the use of standards
Application	Discrete particle analysis	Discrete particle +thin foil analysis	Biological thin sections	Homogeneous	Thin section	Homogeneous thin sections
Standards	Discrete particle	Discrete particles or thin foils	Homogeneous thin sections	Thin sections	Bulk samples	*No* standards
Formula	$M_i^{sp} = \frac{N_{pi}^{sp}}{N_{pi}^{st}} . M_i^{st}$	$\frac{M_i}{M_j} = C_i \frac{N_{pi}}{N_{pj}}$	$C_i = A_i \frac{(N_{pi}/N_B)^{sp}}{(N_{pi}/N_B)^{st}} \left\{ \frac{N_i}{\sum_{j=1}^{n} N_j Z_j} \right\} (Z^2/A)^{sp}$	$\frac{C_i}{C_j} = \frac{N_{pi}}{N_{pj}} . \frac{f_j}{f_i}$	$\frac{C_i}{C_j} = \frac{N_{pi}}{N_{pj}} . \frac{F_j}{F_i}$	$\frac{C_i}{C_j} = \frac{N_{pi}}{N_{pj}} . \frac{f_j}{f_i}$
Reference			Hall[a, b]	Philibert[c, d, e]		Russ/Duncumb[f, g]
Accuracy	Mainly depending on specimen and standard composition		10% For elements up to $Z = 20$			10–20%
Advantage	Simple technique		Rather simple technique	Rather simple technique	Bulk samples used as standards are easy to get	No standards have to be used
Disadvantage	Standards not always fully comparable with specimen		The overall composition of the specimen has to be known	Thickness of standards has to be known	Rather complex technique	This model is rather new and therefore not fully tested

Computer programme	Not necessary	Not necessary	Not necessary	Recommended	Available
Remarks	Particles must be completely covered by electron spot. Specimen and standards have to be measured under the same conditions. For thin foil analysis the specimen as a whole can be used as standard if the overall composition is known	Specimen and standards have to be measured under the same conditions. Care must be taken in quantitative analysis because biological specimens are easily destroyed by electron bombardment causing serious errors	f is the relative intensity (= int per mass thickness) per element. This value, once determined can be used to automate the analysing system	f_i includes the standard X-ray intensity with some mathematical correction factors	f_i is the relative intensity per element calculated by computer

References:

a T. Hall, *J. Microsc.* **99** (1973), 177

b T. Hall, *Micron* **3** (1972), 93

c J. Philibert, *J. Appl. Phys.*, okt. (1970), L70

d J. Philibert, *Brit. J. Appl. Phys.* **2**, 1 (1968), 685

e J. Philibert, *5th Int'l Congress on X-ray optics and microanalysis* (1968), p. 114

f J. Russ, *EDAX Editor* **2**, 1 (1973), 8

g J. Russ, *EDAX Editor* **2**, 3 (1973) 1

Nomenclature:

A = atomic weight

C = weight fraction (concentration)

f = relative intensity

F = N, Q (R/S)

i, j = subscripts denoting element i resp. j.

M = absolute mass

n = number of elements in sample

N_p = integral X-ray peak intensity

N_B = integral white background intensity

Q = ionization cross section

R = electron backscatter factor

S = stopping factor

sp = subscript denoting specimen (sample with unknown composition)

st = subscript denoting standard (sample with known composition)

Z = atomic number

redistribution of the ions in the ice crystals, but only to the nearest surface where they can still be analysed. The drying method also improves spatial resolution by removing mass that would increase electron scatter. Another possible method after freezing is freeze-substitution, in which tissue water is slowly replaced by an organic solvent at low temperature, following which the specimens can be warmed and analysed.

Some elements in tissue are naturally bound to protein molecules, and since they are thus less mobile and more stable, more conventional methods of fixation and embedding can be used. This is also true of most foreign elements, especially heavy atoms that are involved in many pathological conditions. When using chemical fixation, however, it is important to avoid staining the sections with elements whose X-ray peaks can mask or interfere with the elements of interest. The best policy is usually to eliminate the stain altogether and to rely on the higher contrast of the STEM image to recognize and locate features to be analysed.

One of the most exciting possibilities for microanalysis lies in the development of an entirely new class of staining chemicals that will mark particular chemical or reaction sites with great specificity. The present chemical stains deposit electron-dense precipitates that are not very selective. The new stains need not be electron dense, but instead can incorporate any element not normally present in the tissue that can be precipitated in an analysable amount. This promises to become a fertile field for the imaginative biochemist.

2.6.2 OPTIMUM OPERATING PARAMETERS

In addition to the critical steps of specimen preparation to preserve not only the specimen morphology but also the elemental distribution in the tissue, there are several operating parameters that the user should keep in mind to obtain the best results (see also Chapter 4).

First, it is not always necessary or in fact best to use the smallest possible incident electron beam. If the structure to be analysed is larger than the smallest spatial area that can be selectively excited in the specimen, it is wise to enlarge the beam to cover the entire structure, possibly using intentional beam astigmatism to shape the beam to fit the area. However, the increase in beam size should be done using the first condenser lens rather than the final lens, which should be kept adjusted to produce a beam crossover at the specimen. This is, of course, much easier to visualize with the TEM than the SEM.

The reason for this use of the lenses is to produce the greatest possible beam current at the specimen, and also to excite all of the areas containing the elements of interest. Increasing the beam diameter allows the beam current to be increased greatly, since the current is proportional to the diameter raised to the 8/3 power. The X-ray intensity is directly proportional to the current.

In nearly all cases the accelerating voltage should be fairly high for thin section analysis, in the range of about 10 times E_c for the elements of interest. This will efficiently excite the elements and produce a good X-ray signal, and will actually cause less specimen damage than lower voltages. Also, it provides adequate transmission imaging of sections. Most specimen damage is caused by either heating or electrical charging, and hence can be minimized by keeping the specimen in good contact with the grid. The use of collodion or other films, and coating the specimen with 3–10 nm of carbon, is very helpful for this purpose. The support film or coating should be on the bottom of the specimen as placed in the microscope so that it does not contribute either to electron scattering or X-ray absorption. It is also important to select a grid material that does not interfere with any of the elements of interest, and to keep the point of analysis well away from the grid to reduce the flux of X-rays from the grid (both characteristic and Bremsstrahlung) caused by scattered electrons. In TEMs the objective lens aperture should also be removed during analysis.

The usual analysing procedure is to collect a spectrum and then subtract the contribution of the support grid and/or film. Computer processing may then be desired to remove the Bremsstrahlung or strip any peak interferences that may be present, to show better the elemental peaks of interest. If quantitative results are desired the total counts for each element can be used either with suitable standards, or with previously established calibration curves to obtain absolute elemental mass or concentration, or the elemental concentration ratios can be calculated without standards.

2.6.3 X-RAY AREA MAPS

This method, while common in the materials field, suffers two key limitations for biological specimens:

(1) The 'background' of dots from non-characteristic (Bremsstrahlung) X-rays that have the same energy as the elemental X-rays can obscure the presence of dots from the desired element. These X-rays are produced by electron interactions in the sample, and have all energies up to the beam voltage. Since the elemental concentrations in most biological specimens are low, there is a poor 'peak-to-background' ratio between elemental and non-characteristic X-rays. This problem is worsened by the poor inherent peak sharpness of the energy-dispersive (ED) X-ray spectrometer as compared to the wavelength-dispersive (WD) one, but the former is still usually required because of its much higher efficiency.

The result is that the human eye, which does not easily recognize variations in dot density less than 3 or 4 to 1, cannot extract from the X-ray map much useful information on elemental distribution. If the specimen is thin, the low total count rate produced will in addition either produce images with very few dots that further mask the real variations because of statistical fluctuations in dot density, or will require excessively long times to produce an acceptable image which can lead to possible specimen damage;

(2) Most biological specimens have local density fluctuations, and the preparation process with its removal of water and introduction of other compounds – some containing heavy metals – adds to these fluctuations. Indeed, the transmitted electron image depends for its contrast on these density variations. The product of non-characteristic or background X-rays is roughly proportional to density, and the X-ray map will thus show corresponding variation in brightness or dot density independent of and superimposed on any real concentration variation. Since the typical variations in concentration are smaller than the density fluctuations, particularly in stained sections, the user can be badly misled.

The best solution is to record two X-ray maps each for the same length of time: one for the element, and one for an energy (wavelength) close by that does not correspond to an element. The variations in the second, or background, map are due to density variations primarily and must be visually 'subtracted' from the first image to deduce the true variation in the element of interest.

2.6.4 POINT ANALYSIS – SPECTRUM ARTIFACTS

It is often best in analysing biological specimens to select features for point analysis based on the electron image and the researcher's informed judgment. The resulting spectrum still requires interpretation, however, and the most noticeable artifact is the statistical nature of X-rays which is significant because of the low total intensities the biologist is commonly forced to use. In either ED or WD spectra this produces random fluctuations that superimpose on the smoothly varying background and peaks, which can sometimes be visually misinterpreted as peaks themselves. The probability of misinterpretation is reduced by using greater analysing times (number of counts), making repeated measurements, and comparing the shape (especially the width) of suspected peaks to real ones. Spectrum smoothing routines are widely available to improve the appearance of spectra by reducing random statistics, but these can make a false peak look better, too, and must be used with caution.

False peaks in ED spectra can result from pulse pile-up (two X-rays entering the detector at the same time are measured as a single X-ray with the combined energy), or escape peaks. This is rarely troublesome at the low count rates and concentrations common in biological specimens. Additional spectrum background can be produced if high energy scattered electrons penetrate the very thin beryllium window, and deposit their energy in the ED detector. These artifacts are all recognizable, and should be detected in any case in a control spectrum.

2.6.5 PEAK OVERLAPS

Much attention has been focused on the problems of peak overlaps and interferences because of the poorer spectrum resolution of the ED analysers that have largely supplanted WD types because of their efficiency and ability to simultaneously detect all elements present. Elaborate computer stripping programmes are available to uncover buried peaks, and these may be quite useful for quantitative work. However, they are rarely, if ever, better for finding peaks and identifying them than simple common sense. ED spectral peaks have Gaussian shapes, and unless statistical fluctuations mask the shape (which equally confuses computer programmes), the researcher's eye will readily detect asymmetry and suspect the presence of a buried peak. Furthermore, most elements (especially the heavy ones whose L-lines interfere significantly) produce several peaks whose relative heights are predictable for a given accelerating voltage. At least one of these peaks is usually present without interference, and in any case the presence of all the peaks in their proper relative heights should be used as a criterion on the element's presence. Some published results show questionable interpretation of spectra that have peaks identified as certain elements when consideration of all of the peaks suggest a different conclusion.

The methods used in computer programmes that deal with energy-dispersive X-ray spectra can be broken down into several classes as outlined below:

(1) Distinguish peaks from background
- (a) require standard samples
  - (i) measure background on a similar control
  - (ii) calculate background from a dissimilar control;
- (b) require operator judgment
  - (i) interpolation using straight or curved lines;
- (c) require neither
  - (i) frequency filter on spectrum.

(2) Separate peak overlaps
- (a) require standard samples
  - (i) measure overlap factors
  - (ii) store active spectrum library;
- (b) require no standards
  - (i) generate and fit peaks using Gaussian or modified Gaussian shapes one peak at a time
  - (ii) all peaks simultaneously.

Each of these methods has been described and used in the literature (see for instance Russ, 1977) and can be very satisfactory in some applications. No single method is applicable to all cases. There is also a great difference in the degree of computerization required. However, as more and more systems are presently equipped with a computer, this distinction is becoming less important.

At the same time, the computer is useful in carrying out the various computations for a quantitative analysis. As more work is concerned with the amount of an element present, rather than simply its presence or absence, the importance of the computer for dealing with both the spectrum and subsequent calculations will continue to increase.

2.6.6 VARIATIONS WITH TIME

Many researchers have observed – with frustration – the apparent presence of elemental peaks in the first few seconds of analysis, only to find them gone after the analysis is complete. This is sometimes due to wishful thinking, of course, but in many real situations it results from damage to the specimen by the electron beam. Some elements, if not well-bound, are 'pushed' away from the analysed point by local heating which can be hundreds of degrees centigrade. In other cases, contamination deposited on the specimen absorbs the characteristic X-rays or generates additional background X-rays to mask the peak. It is also possible for the beam to shift position due to charging, or for the specimen to move because of charging, heating, or stage drift.

There is no sure prevention for these problems. Cooling the specimen and proper coating minimize heating and charging effects, but in all cases it is wise to observe the specimen again after analysis. A very dark contamination spot or a severely etched appearance (a transmitted electron image will often be improved in contrast after this etching and mass loss) indicates that the results should not be trusted. If the contamination spot is streaked or has a comet-tail, it indicates specimen motion. Many researchers photograph the specimen with its contamination spots after analysis to indicate the locations. Remember, though, that the contamination usually builds up around, rather than on, the point of greatest temperature, and does not show the size of the excited region.

2.7 Summary

Microanalysis of biological tissue on a cellular or subcellular level can be obtained using the X-rays generated in the sample by the electron beam in an SEM or TEM. The advantages of this method over other histochemical techniques lie in the simplicity of elemental identification, the selectivity, and the ability to quantitate the results.

The mathematical formulae used for electron-excited microanalyses of bulk inorganic materials can be extended with minor modifications to biological tissue. The large subsurface volume excited in bulk tissue makes it difficult, however, to analyse selectively features imaged in the SEM. It is, therefore, preferable in most cases to use thin sections in either the TEM or STEM.

Either a TEM fitted with a scanning attachment or the SEM with a transmission detector and energy- or wavelength-dispersive X-ray detector can be used. The combination of energy-dispersive analysis for efficiency and simultaneous detection of all elements, and the TEM with its more flexible electron optics, provides the greatest flexibility and sensitivity.

Specimen preparation by rapid freezing and sectioning followed either by examining the frozen specimens or subsequent drying is necessary for the localization of mobile ions, but other more tightly bound elements can be studied after conventional fixation and embedding. Staining chemicals of the conventional type should be avoided, but the use of more selective stains can give selective elemental precipitates at reaction sites.

The areas of application of this technique include three principal areas: (1) the study of natural elemental concentrations; (2) the localization of pathological elements; and (3) the identification of sites of specific chemical activity using markers.

2.8 Bibliography

2.8.1 GENERAL

ECHLIN, P. & GALLE, P. (1975) Ed. *Biological Microanalysis.* Paris: Société Française de Microscopie Électronique.

HALL, T., ECHLIN, P. & KAUFMANN, R. (1974) Eds. *Microprobe Analysis as applied to Cells and Tissues.* London: Academic Press.

2.8.2 SAMPLE PREPARATION METHODS

Freezing

APPLETON, T. C. (1974) A cryostat approach to ultrathin 'dry' frozen sections for electron microscopy: a morphological and X-ray analytical study. *Journal of Microscopy,* **100**, 49–74.

FUCHS, W. & LINDEMANN, B. (1975) Electron beam X-ray microanalysis of frozen biological bulk specimen below 130° K; 1. Instrumentation and specimen preparation. *Journal de Microscopie et de Biologie Cellulaire,* **22**, 227–232.

MARSHALL, J. (1972) Ionic analysis of frozen sections in the electron microscope. *Micron,* **3**, 99–100.

SAUBERMANN, A. J. & ECHLIN, P. (1975) The preparation, examination, and analysis of frozen hydrated tissue sections by scanning transmission electron microscopy and X-ray microanalysis. *Journal of Microscopy,* **105**, 155–191.

Other

MORGAN, A. J., DAVIES, T. W. & ERASMUS, D. A. (1975) Changes in the concentration and distribution of elements during electron microscope preparative procedures. *Micron*, **6**, 11–23.

PANESSA, B. J. & RUSS, J. C. (1975) Techniques for practical biological microanalysis. In *Proceedings of the 8th Annual SEM Symposium*, 251–258.

2.8.3 STANDARDS

CHANDLER, J. A. (1976) A method for preparing absolute standards for quantitative calibration and measurement of section thickness with X-ray microanalysis of biological ultrathin specimens in EMMA. *Journal of Microscopy*, **106**, 291–302.

MORGAN, A. J., DAVIES, T. W. & ERASMUS, D. A. (1975) Analysis of droplets from iso-atomic solutions as a means of calibrating a transmission electron analytical microscope (TEAM). *Journal of Microscopy*, **104**, 271–280.

ROWSE, J. B., JEPSON, W. B., BAILEY, A. T., CLIMPSON, N. A. & SOPER, P. M. (1974) Composite elemental standards for quantitative electron microscope microprobe analysis. *Journal of Physics* E: *Scientific Instruments*, **7**, 512–514.

SPURR, A. (1975) Choice and preparation of standards for X-ray microanalysis of biological materials with special reference to macrocyclic polyether complexes. *Journal de Microscopie et de Biologie Cellulaire*, **22**, 287–302.

2.8.4 APPLICATIONS

Qualitative

'Natural' elements

BLAKEMORE, R. (1975) Magnetotactic Bacteria. *Science*, **190**, 377–379.

DAVIES, T. W. & ERASMUS, D. A. (1973) Cryoultramicrotomy and X-ray microanalysis in the transmission electron microscope. *Science Tools LKB.*, **20**, 9–13.

DEMPSEY, E. W., JARVIS, J. U. M. & PURKERSON, M. L. (1974) The location of sulphur in spermatozoa by energy-dispersive X-ray analysis and scanning electron microscopy. *Proceedings of the 7th Annual SEM Symposium*, 631–638.

DORGE, A., GEHRING, K., HAGEL, W. & THURAU, K. (1974) Intracellular Na^+ K^+ concentration of frog skin at different states of Na transport. In *Microanalysis as applied to cells and tissues*, ed. Hall, T. A., Echlin, P. & Kaufmann, R. London: Academic Press.

MARSHALL, A. T. & WRIGHT, A. (1973) Detection of diffusible ions in insect osmoregulatory systems by electron probe X-ray microanalysis using SEM and a cryoscopic technique. *Micron*, **4**, 31–45.

PALLAGHY, C. K. (1973) Electron probe microanalysis of potassium and chloride in freeze-substituted leaf sections of *Zea mays*. *Australian Journal of Biological Sciences*, **26**, 1015–1034.

YAROM, R., PETERS, P. D., HALL, T. A., KEDEM, J. & ROGEL, S. (1974) Studies with EMMA 4 on changes in the intracellular concentration and distribution of calcium in heart muscle of the dog in different steady states. *Micron*, **5**, 11–20.

'Unusual' elements

DEFILIPPIS, L. F. & PALLAGHY, C. K. (1975) Localization of zinc and mercury in plant cells. *Micron*, **6**, 111–120.

FOWLER, B. A. & GOYER, R. A. (1975) Bismuth localization within nuclear inclusions by X-ray microanalysis. Effects of accelerating voltage. *Journal of Histochemistry and Cytochemistry*, **23**, 722–726.

MIZUHIRA, V. (1976) Elemental analysis of biological specimens by electron probe X-ray microanalysis. *Acta Histochemica Cytochemica*, **9**, 69–87.

'Tag' elements

BRAATZ, R. & KOMNICK, H. (1976) Vacuolar calcium segregation in relaxed myxomycete protoplasm as revealed by combined electrolyte histochemistry and energy dispersive analysis of X-rays. *Science and Industry*, No. 6, 9–11.

ERASMUS, D. A. (1974) The application of X-ray analysis in the transmission electron microscope to a study of drug distribution in the parasite *Schistosoma mansoni* (Platyhelminthes). *Journal of Microscopy*, **102**, 59–69.

Quantitative

BEAMAN, D. R. & FILE, D. M. (1976) Quantitative determination of asbestos fibre concentrations. *Analytical Chemistry*, **48**, 101–110.

HENDRIKS, A. (1975) Microanalysis of thin sections in TEAM. *Edax Editor*, **5**, 13–22.

ZS.-NAGY & PIERI, C. (1976) Choice of standards for quantitative X-ray microanalysis of biological bulk specimens. *Edax Editor*, **6**, 28–30.

POOLEY, F. D. (1975) The identification of asbestos dust with an electron microscope microprobe analyser. *Annals of Occupational Hygiene*, **18**, 181–186.

RUSS, J. C. (1973) Microanalysis of thin sections, coatings and rough surfaces. *Proceedings of the 6th Annual SEM Symposium, 11TR1, Chicago*, 113.

2.8.4 PRINCIPLES

BEAMAN, D. R. & ISASI, J. A. (1972) Electron beam microanalysis. *American Society for Testing and Materials, Special Technical Publication*, 506.

CASTAING, R. (1951) *Thesis*, University of Paris, ONERA Publ., No. 55.

HALL, T. A. & WERBA, P. (1968) The measurement of total mass per unit area and elemental weight fractions along line scans in thin specimens. *Proceedings of the 5th International Congress on X-ray Optics and Microanalysis*, 93.

JACOBS, M. H. & BABOROVSKA, J. (1972) Quantitative microanalysis of thin foils with a combined electron microscope-microanalyser. *Proceedings of the 5th European Congress on Electron Microscopy*, 136–137.

MARTIN, P. M. & POOLE, D. M. (1971) Electron-probe microanalysis: the relationship between intensity ratio and concentration. *Metallurgical Reviews*, **150**, 19–46.

PHILIBERT, J. (1963) Electron probe microanalysis: a new method for calculating the absorption correction. *ASTM Special Technical Publication*, **399**.

REED, S. J. B. (1965) Characteristic fluorescence corrections in electron-probe microanalysis. *British Journal of Applied Physics*, **16**, 913–926.

REED, S. J. B. (1975) *Electron Microprobe Analysis*. Cambridge: Cambridge University Press.

RUSS, J. C. (1970) Energy dispersion X-ray analysis: X-ray and electron probe analysis. *ASTM Special Technical Publication*, **485**, 217–231.

RUSS, J. C. (1974) Quantitative microanalysis with minimum pure element standards. *Proceedings of the 5th Annual Conference of the Electron Probe Analysis Society*, 22a–c.

RUSS, J. C. (1977) Processing of energy-dispersive X-ray spectra. *X-ray Spectrometry*, **6**, 37.

THOMAS, P. M. (1963) Outline of a method for correcting for atomic number effects in electron probe microanalysis. *British Journal of Applied Physics*, **14**, 397–398.

YAKOWITZ, R. L., MYKLEBUST, R. L. & HEINRICH, K. F. J. (1973) 'Frame' – An on-line correction procedure for quantitative electron probe microanalysis. *U.S. Department of Commerce, National Bureau of Standards*, Technical note 796.

3

J. A. CHANDLER

The application of X-ray microanalysis in TEM to the study of ultrathin biological specimens – a review

3.1 Introduction

3.1.1 THIN SPECIMEN ANALYSIS

X-ray microanalysis has been used in biological research for over twenty years, but until the late 1960s its application to the analysis of ultrathin specimens was rather limited. A major disadvantage in the analysis of bulk material is the lack of spatial resolution due to diffusion of electrons entering the specimen surface (Fig. 3.1). In soft biological tissue this may be several micrometres for incident electrons of energy greater than 10 keV (Hall, 1971). When ultrathin specimens are analysed the lateral diffusion is very small (Chandler, 1973; Thurston & Russ, 1971) and observation of the transmitted image allows an almost exact localization of the subcellular region being analysed (Fig. 3.2). In addition, the use of ultrathin specimens with transmission imaging allows the tissue ultrastructure to be correlated with this localized analysis to a degree beyond that possible with bulk surface observation.

A third, and less obvious, advantage of thin specimen analysis arises from the fact that background radiation (white radiation) is reduced compared with that from bulk specimens. Since the detection of trace elements ($< 1\%$) depends critically on the ratio of peak-to-background X-ray counts, this reduction of background, due to the reduced amount of total matrix material being analysed in the chosen region of the specimen, enhances the sensitivity of the technique (Chandler, 1977). This also means however, that the total mass of the selected elements to be detected in the microvolume is extremely small and great attention must be paid to instrumental parameters affecting the detection of such low levels.

For these three reasons the analysis of ultrathin specimens by transmission electron microscopy holds unique promise in the study of biological tissue. A wealth of literature is available describing the use of conventional electron probe microanalysis of bulk material. This chapter, however, reviews the progress which has been made in the last 15–20 years in the development of ultrathin specimen analysis and, by way of application, demonstrates the extremely wide range of problems the technique is capable of solving.

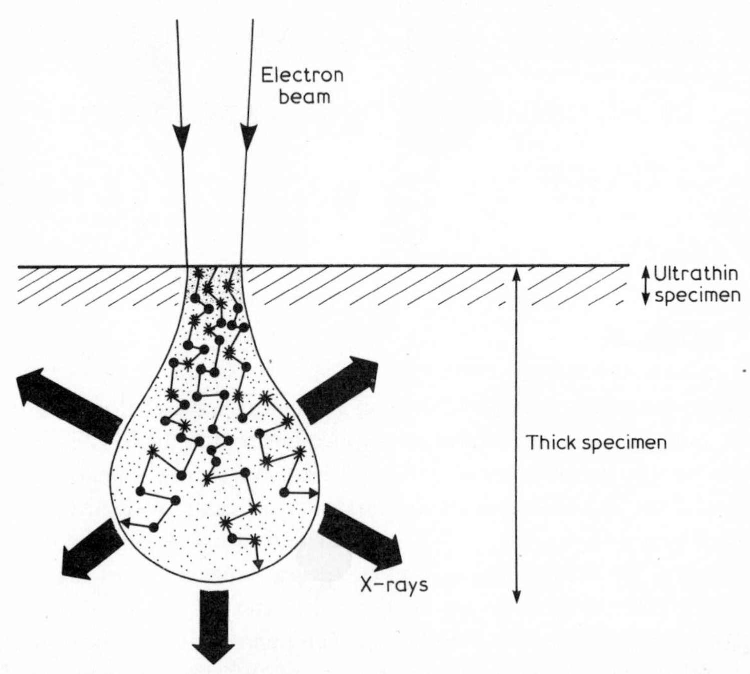

Fig. 3.1 Diffusion of electrons entering the surface of a thick specimen. Both elastic (dark circles) and inelastic (asterisks) collisions occur. X-rays are produced from a volume greater than the original beam dimension, but choosing a thin specimen (hatched region) limits the electron diffusion.

3.1.2 BIOLOGICAL ANALYSIS – LITERATURE

Since 1960, less than 300 publications have appeared in the literature describing applications of ultrathin specimen analysis (see Bibliography). The explosion of publications in the last 3–4 years (Fig. 3.3) has arisen due to the recent availability of analytical instruments, both as attachments for existing microscopes, and as integral instruments built specifically for microanalysis. As research workers' knowledge of, and confidence in, the technique increases so one may expect a progressively rapid rise in the number of publications describing biological applications. A number of factors will, however, limit the increasing popularity of this technique. Not least of these are the inherent difficulties of specimen preparation. These problems are discussed in Chapters 4 and 5 in this book but special mention is made here about certain methods used specifically for ultrathin specimen analysis.

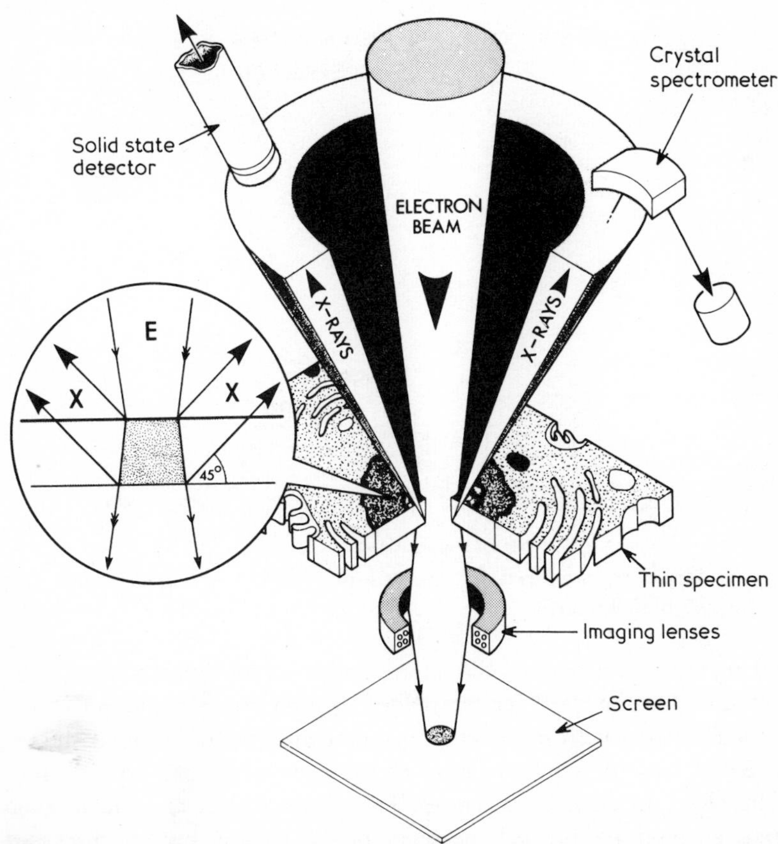

Fig. 3.2 X-ray analysis of an ultrathin specimen (e.g. of a cell nucleus). The electron beam is focused into the selected region of the specimen for analysis, the image of this area being formed on the screen. X-rays are collected by crystal spectrometers or solid state detectors. Inset: showing the diffusion of the electron beam passing through the specimen. A typical X-ray take-off angle is 45°. E, Electron beam; X, emitted X-rays.

Another limitation is the full appreciation which must be made of the true quantitative nature of X-ray analysis. A large number of early publications were concerned with very simple, and in some cases rather naive, qualitative analyses of very limited regions of biological samples. Most workers now realize the need for a proper method of specimen sampling in order that a thoroughly sound interpretation may be made of subcellular analyses compared with ultrastructural studies. It is no longer acceptable for workers to expect to spend two hours at an instrument, obtain analytical data from half a dozen organelles, and rush into print with an instant biological interpretation. Studies which are to have any worth will initially require detailed and

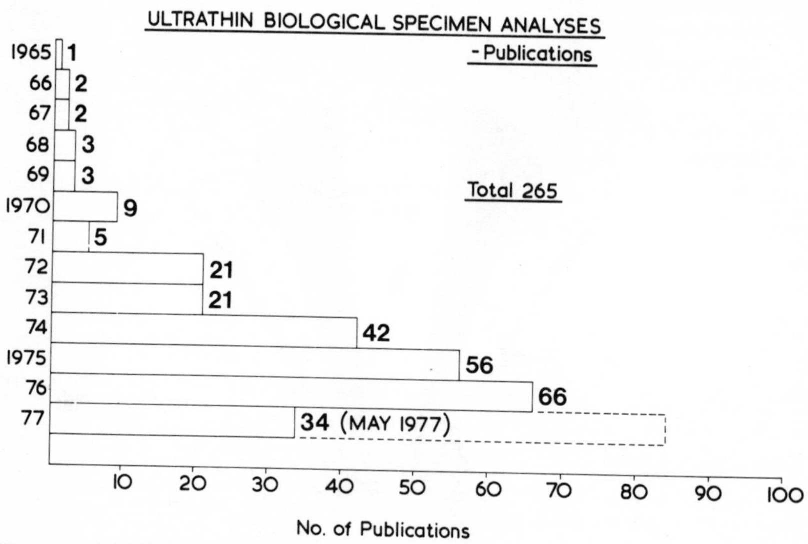

Fig. 3.3 Publications produced for analysis of ultrathin specimens since 1965 (as listed in the Bibliography).

painstaking analyses to obtain valid interpretations of biological analysis. The main reasons for the need for this added caution are to be found in the problems associated with specimen preparation and electron irradiation (i.e. the potential loss, or contamination of the specimen), and the inherent variation which occurs in most tissues. The effects of electron irradiation on biological analyses are beyond the scope of this chapter but are discussed elsewhere, both in this book and by Chandler (1977). In the section of this chapter devoted to specific applications, discussion is included on the precautions taken to accommodate these difficulties.

3.2 Instrumentation

Whereas an account of instrumentation used in biological X-ray microanalysis is also given elsewhere in this book (Chapter 2), it is important here to differentiate between the microscopes used for the analysis of bulk and thin material. The applications which are discussed in this chapter are restricted to those of ultrathin specimens, i.e. of thickness less than 0.25 μm for examination in transmission electron microscopy. For thick specimens there

Table 3.1 Instruments used for ultrathin specimen analysis

	Publications (to May 1977)
Electron microscope microanalyser (EMMA)	109
Transmission electron analytical microscope (TEAM)	102
Scanning transmission electron microscope (STEM)	45
Transmission electron probe analyser (TEPA)	15

are many hundreds of publications on biological analysis, mostly describing investigations with the electron probe microanalyser (EPMA). For ultrathin specimens, however, analysis has been performed with four major types of instrument. These are listed in Table 3.1.

EMMA

The electron microscope microanalyser (EMMA) was the first of the true transmission electron microscopes to be manufactured with capabilities for full quantitative X-ray analysis. As an integral instrument it did not compromise specimen facilities or high resolution imaging for the sake of sensitive analysis. Even though the number of EMMA instruments used regularly for biological analysis is only five worldwide, these still claim the greatest number of publications. EMMA instruments have the unique advantage of possessing twin wavelength dispersive crystal spectrometers in combination with energy-dispersive detectors to allow extremely high sensitivity of detection for trace elements with excellent X-ray wavelength resolution (Chandler, 1977). The mode of operation of the EMMA instrument is with a static probe, and although probe diameters of 20 nm are possible, the available beam current is rarely found sufficient for biological analysis so that probes of 100 nm are more frequently used. This electron probe focusing is achieved by the use of a third condenser lens (minilens) placed immediately above the specimen.

TEAM

Recent years have also witnessed a tremendous growth in the number of transmission electron microscopes having X-ray detectors as attachments, both as retrospective additions to existing designs and as new, purpose built instruments (TEAM, transmission electron analytical microscope; described separately from EMMA for the purpose of this chapter). In the main the X-ray detectors are of the solid state energy-dispersive type, although some early designs of wavelength dispersive spectrometer were employed. When these attachments first appeared in the early 1970s little attention was paid by manufacturers to the design of the specimen region or the interface for obtaining maximum sensitivity. More recently, however, such important factors have been taken into consideration and the sensitivity of the instrument is approaching, and in some designs is equal to, that of the purpose built EMMA design (Shuman, Somlyo & Somlyo, 1976). As with the EMMA instrument the normal mode of analysis is by a static probe. However, transmission microscopes can be fitted with attachments for scanning the beam and for producing very fine probes (down to 10 nm). Some analyses have been reported for ultrathin biological specimens using such small diameter probes (Wood & Harling, 1974; Parducz & Joó, 1976).

STEM

The scanning electron microscope (SEM) is generally used for imaging the surfaces of bulk specimens by the collection of secondary electrons. In some cases, however, it is possible to examine ultrathin specimens in the transmission mode by placing an electron detector beneath the specimen stage (scanning transmission electron microscope, STEM). One of the main advantages of this system is to be found in the examination of thicker samples (1–2 μm) because of the absence of chromatic abberation in the image (see Chapter 2). Some X-ray analyses have been performed in these instruments where the examination of specimens in the range 200–500 nm gives higher X-ray signals than ultrathin (100 nm) sections, yet permits reasonable ultrastructural resolution. In some of these instruments high gun brightness is possible using field emission sources. Thus small diameter probes can be used with high current densities for analysis. The STEM type of instrument (either in an SEM or a TEM with scanning attachment) may image the specimen by scanning the electron beam, but the analysis is usually performed with a static probe. Again, the most common form of X-ray detector used with the STEM is the solid state detector.

TEPA

One of the earliest instruments to be used to analyse thin specimens in transmission was the Cameca probe. This is basically an electron probe microanalyser operated in the scanning mode but with facilities for inserting transmission imaging lenses beneath the specimen to convert it to a TEM. When operated in this way a transmission image can be observed during simultaneous analysis with a static probe. The Cameca (TEPA, transmission electron probe analyser) is equipped with both crystal spectrometers and solid state detector similar to the EMMA, but performs with an inferior electron image resolution. Its main function is the examination and analysis of bulk specimens.

Of these four types of analytical instruments (EMMA, TEAM, STEM, TEPA), the majority of analyses of thin biological specimens have been performed on the EMMA and TEAM. If the need for very fine electron probes becomes greater then the STEM type of instrument may increase in importance.

There is no doubt that the development of solid state detectors (SSD) has played a vital role in the increased popularity of X-ray analysis in the last few years. There are some limitations in their usefulness when compared with crystal spectrometers, particularly with regard to energy resolution and mass detection sensitivity (Chapter 2). It is interesting, however, that of 265 publications appearing to-date on X-ray analysis of ultrathin specimens (see Bibliography), only 10 % have involved the use of crystal spectrometers. This is almost certainly due to the comparative ease of use of the solid state detector compared with the crystal spectrometer. The latter type of detector seems only to have been employed in cases where overlapping lines in the X-ray spectrum precluded the use of the SSD, or when extremely low elemental concentrations required high peak-to-background ratios. It seems likely that solid state detectors will continue to dominate the X-ray analysis of biological specimens and that continuous improvements will occur in energy resolution, mass detection sensitivity, and data handling (Shuman, Somlyo & Somlyo, 1976). Work by Shuman and Somlyo (1976) has indicated that increased sensitivity may be obtained using large detector sizes (30 mm^2) at short detector–specimen distances ($\simeq$ 10 mm) and with specimens maintained at reduced temperatures (−110 °C). With a 60 nm diameter probe these authors have detected iron in single ferritin molecules (10^{-19} g) in a TEAM instrument. Shuman, Somlyo and Somlyo (1976) have indicated that the theoretical limit of detection with optimization of such an arrangement is 10^{-22} g, or 2 atoms of iron. Such a predicted limit of detectability depends on instrumental and operational factors as well as on techniques of specimen preparation.

3·3 Specimen preparation

Undoubtedly, the greatest obstacle to performing satisfactory X-ray analysis of biological tissue is that of specimen preparation. The conventional techniques of fixation, dehydration, embedding, etc., are so traumatic that in many cases one is left with little but resin and stain to examine, most of the specimen having been left behind somewhere *en route* to the microscope. Whereas this may be sufficient for electron microscopy, the demands of X-ray analysis are considerably greater. The preparative procedure must provide a specimen with satisfactory ultrastructural detail while ensuring that the elements to be analysed have been neither lost nor displaced from their original *in vivo* sites (see Chapter 4). In almost all instances this is both impossible to perform and impossible to check. However, some procedures are adopted which, theoretically, are likely to minimize these artefacts.

Table 3.2 Specimen preparation methods for ultrathin sample analysis (to May 1977)

	Frequency
(1) Conventional fixation and embedding (sections)	149
(2) Frozen, freeze dried (sections)	36
(3) Pyroantimonate (sections)	31
(4) Air dry (suspensions)	18
(5) Silver acetate/lactate (sections)	12
(6) Freeze substitution (sections)	11
(7) Oxalate, Na or K (sections)	10
(8) Staining reactions (sections)	10
(9) Freeze dry and embedding (sections)	9
(10) Enzyme reactions (sections)	7
(11) Fixed, frozen, freeze dried (sections)	5
(12) Replicate (from tissue or histology sections)	5
(13) Ashing (from tissue or histology sections)	5
(14) Critical point dry (suspensions)	2
(15) Freeze dry (suspensions)	1
(16) Frozen sections or fixed and embedded tissue	1
(17) Oxine chelation (sections)	1
	311

The range of specimen preparation techniques that have been employed for subsequent X-ray microanalysis of thin specimens is shown in Table 3.2, together with their frequency of use. There are a number of factors to be considered when choosing a suitable preparation procedure. Electrolytes such as sodium, potassium, and chlorine are easily displaced by aqueous solutions and the usual recourse is to use frozen sections for these elements. Calcium, however, seems to have been reasonably well analysed in specimens prepared by other means. The loss or displacement of any element in tissue depends critically on the degree to which it is bound within the specimen. For tightly bound elements (often heavy metals), aqueous solutions may often be successfully employed, while histochemical techniques have been widely used for both diffusible and non-diffusible elements. The choice of preparation method must also take into account the possible loss or displacement of organic material to which the elements may be bound.

A number of authors have published investigations on elemental loss from specimens during various preparation procedures. Radioactive labelling (De Filippis & Pallaghy, 1975; Mehard & Volcani, 1975) has been used, as well as atomic absorption spectrophotometry (Chandler, 1975; Morgan, Davies & Erasmus, 1975), flame photometry and conductimetric titration (Harvey, Hall & Flowers, 1976), for monitoring losses during fixation, dehydration, embedding, and even sectioning and flotation. When detailed analyses are performed it is usually necessary to adopt a preparative procedure for the optimum retention of particular elements with some method for determining the integrity of the specimen composition. This may form a major part of the analytical investigation.

The most common methods used for ultrathin specimen preparation are listed in Table 3.2 and are described below.

3.3.1 CONVENTIONAL FIXATION AND EMBEDDING

In most cases the tissue is fixed in an aldehyde, usually glutaraldehyde, or sometimes in alcohol. Post-fixation with osmium tetroxide has been successfully used but adds a further risk to the possible loss of elements from the tissue. Alcohol or acetone dehydration may cause extraction, as may intermediate solvents, while displacement could occur during infiltration with the embedding resin. After surviving these steps, elements may still be lost during the flotation of sections on aqueous baths. Nevertheless, despite these hazards, some 50% of all analyses have been performed using virtually conventional electron microscope preparation techniques (see Bibliography).

3.3.2 HISTOCHEMICAL TECHNIQUES

Table 3.2 lists the histochemical methods which have been employed both to minimize losses during fixation and to produce visible precipitates in the specimen. The methods depend upon the availability of the elements to be precipitated, i.e. on their degrees of binding within the specimen. References to applications of these techniques are listed in the Bibliography.

The most widely used histochemical technique is that using potassium pyroantimonate. In this reaction the soluble potassium salt forms insoluble complexes with a number of different elements. Potassium is often reincorporated on to the precipitate (Tisher, Weavers & Cirksena, 1972), which is insoluble in alcohol, and phosphate ions have also been found to cause deposits (Chandler, 1977). Calcium is the most frequently detected element in combination with antimony in the precipitates but problems arise over the difficulty in separating the X-ray lines from these two elements with the solid state detector (Chandler, 1977). There is no apparent stoichiometry with the antimonate method. Many different combinations occur with sodium, magnesium, calcium, iron, manganese, zinc, cadmium, and potassium and the degree of precipitation does not necessarily indicate the concentration of any one cation. The method is also strongly dependent on the pH of the solution. It is usually used in combination with osmium tetroxide fixative.

The oxalate technique depends upon the addition of sodium or potassium oxalate to the fixative (usually an aldehyde) to form insoluble calcium oxalate complexes in the tissue. The method has been mostly used for the detection of calcium in muscle (see Bibliography).

In the method employing silver acetate and silver lactate, tissue chlorine causes the formation of dense silver chloride deposits. Between 0.5 and 1% silver acetate in 1% osmium tetroxide is employed as the fixative, the procedure taking place under safe-light conditions and then being followed by washing with nitric acid to remove unspecific silver precipitates. Dehydration and embedding procedures are performed as usual (Lauchli, Stelzer, Guggenheim & Henning, 1974). The histochemical technique has been used for the analysis of chlorine in gastric mucosa, muscle, and plant tissue.

A promising technique for the precipitation of a wide range of elements for subsequent analysis is that using oxine (8-hydroxyquinoline). This organic chelating agent does not have the disadvantages of the pyroantimonate or silver acetate methods in that elements are precipitated *in situ* without being associated with any other heavy element (such as antimony or silver). Thus there is no interference from extra X-ray lines or from elevated X-ray background radiation. Oxine is a strong chelating agent of many elements (e.g. Al, Ca, Zn, Cd, Ag, Ba, Co, Cu, Fe) and, as such, will produce non-specific reactions in tissue. The method has been used in combination with aldehydes (Mizuhira & Kimura, 1973) for the localization of cadmium in liver cells. The reaction may, like the pyroantimonate technique, be critically dependent on pH, method of fixation, etc. and it would be advisable to perform preliminary *in vitro* tests with salts of varying molarity to determine the efficiency of precipitation for the elements to be analysed in the tissue.

A number of other histochemical techniques have been used for thin specimen analysis. An obvious application is the analysis of final reaction products in enzyme histochemical reactions. For example, Dierkes (1977) analysed lead deposits for the localization of esterase in cellular slime mould, while Lewis (1973) has used copper deposits for the determination of subcellular cholinesterase activity. As an alternative to heavy metal deposits in tissue, the use of azo dyes for localization of acid phosphatase has been suggested by Bowen, Ryder and Downing (1976). In this reaction bromine was detected by X-ray microanalysis as part of the final reaction product even though it produced no electron-dense deposits (see Chapter 6).

A number of important factors must be borne in mind when employing precipitating reactions for subsequent analysis (Chandler, 1977; Lechene & Warner, 1977):

(1) All precipitation is an artefact. Ions do not actually exist in discrete particles in cells but must diffuse along a concentration gradient to form the 'insoluble' precipitate.

(2) The degree of precipitation in tissues depends on access of the salt to intracellular locations, pH, buffers, composition and ionic strength of the medium, and competition between ions.

(3) Insolubility may not be absolute when the precipitate forms a small fraction of the whole bathing medium.
(4) Stoichiometry often does not occur in the formation of precipitates so that, for example, the amount of antimony in a subcellular region of tissue treated with pyroantimonate will not necessarily relate to the amount of calcium present. Other elements may also be present within the same precipitate.

There is no absolute method of checking the efficiency of the precipitating agent in retaining elements in their *in vivo* sites. In some cases cryo-ultramicrotomy (Chandler & Battersby, 1976*b*) (see below) has been used to compare frozen material with histochemically treated tissue. A combination of biochemical, morphological, and X-ray analytical data has often proved the most reliable approach to an interpretation of subcellular elemental composition.

Further examples of histochemical techniques and labelling methods are discussed in the section devoted to applications.

3.3.3 FREEZE SUBSTITUTION

In order to overcome the elemental losses due to aqueous fixatives, a number of investigators have employed freeze substitution as a means of replacing tissue water with resin. One of the first to use the technique successfully for the retention of electrolytes in thin sections of soft tissue was Spurr (1972*a*, 1972*b*). A number of other workers have used the method in both animal and plant tissues. For example, Forrest and Marshall (1976) detected potassium, phosphorus, sulphur, and chlorine in nerve cells prepared by freeze substitution. Van Zyl, Forrest, Hocking and Pallaghy (1976) also used freeze substitution in the demonstration of sodium in nerve cells, while Mehard and Volcani (1974; 1976*a*, 1976*b*) have indicated the retention of magnesium, silicon, sulphur, chlorine, phosphorus, and aluminium after similar procedures. The method originally adopted by Spurr (1972*a*, 1972*b*), and subsequently modified in various ways, involves first quenching the tissue in isopentane cooled with liquid nitrogen. The tissue is then transferred to the freeze substitution fluid, diethyl ether, containing a saturated solution of benzamide at the temperature (−80 °C) of dry ice. The benzamide is used to produce sodium salts insoluble in ether. After a few days the tissue is transferred to vinyl cyclohexene dioxide at −20 °C for a few hours to allow subsequent penetration of resin. Infiltration and embedding takes place at room temperature with a low viscosity epoxy resin (Spurr, 1969). Surprisingly, sections of the tissue cut on to water baths are reported to retain electrolyte content. Alternative methods of freeze substitution for analysis are given by Harvey, Hall and Flowers (1976), Pallaghy (1973), Van Zyl, Forrest and Hocking (1976), and others. The author (unpublished data) has used the Spurr method successfully for retention of some elements in human sperm cells but with loss of sodium and potassium. Lechene and Warner (1977) have suggested that *in vitro* tests with salt solutions should be performed prior to treating the tissue to determine if the desired elements remain insoluble.

3.3.4 FREEZE DRYING AND EMBEDDING

A simple method for the replacement of water in the tissue by resin, without contact with aqueous solutions, involves quenching small pieces of the tissue in liquid nitrogen and freeze drying them for a period of several hours. The tissue is then embedded by allowing it to be immersed in unpolymerized resin under vacuum. When the vacuum is released and a positive pressure ensues, the resin penetrates to the centre of the tissue pieces and is subsequently polymerized. The technique has been used for the analysis of copper, chromium, and arsenic in wood (Chou, Chandler & Preston, 1973), calcium, and phosphorus in blood platelets (Skaer, Peters & Emmines, 1974) and in tendon (Höhling, Barckhaus, Krefting & Schreiber, 1976), and calcium, phosphorus, and sulphur in teeth (Höhling, Nicholson, Schreiber, Zessack & Boyde, 1972; Höhling & Nicholson, 1975; Nicholson, Ashton, Höhling, Quint, Schreiber, Ashton & Boyde, 1977). Owing to the inevitable collapse of the tissue during freeze drying the ultrastructure of soft tissue is not preserved as well as by conventional means, but for harder materials, or where less requirements are placed on cellular structure, the method provides some improvement in retention of elements.

For the examination of whole cell suspensions or cultures, the method of Garfield and Somlyo (1975) may be appropriate. Whole cells spread on to a coated grid were frozen in liquid nitrogen. They were then freeze dried in a vacuum coating unit, and examined directly in a TEAM instrument. Satisfactory retention was reported for sodium, sulphur, chlorine, potassium, phosphorus, and calcium in mitochondrial granules of these whole cells.

3.3.5 CRYO-ULTRAMICROTOMY

Possibly, the most likely way of preserving the tissue composition close to its *in vivo* state is to rapidly freeze it. Small pieces of tissue have been frozen in cryo-protectants within liquid nitrogen, or in liquid nitrogen slush, or against copper blocks held at liquid nitrogen temperatures. Rapid freezing is essential to avoid ice crystal formation in the tissue. Ultrathin sections of the frozen material are then cut at temperatures between −80 °C and −150 °C. To date, analysis has been reported only for ultrathin sections which have been freeze dried, but some attempts are being made to examine thicker material kept frozen hydrated (Gupta, Hall, Maddrell & Moreton, 1976). This requires the use of a cold stage in the electron microscope or electron probe microanalyser. Shuman and Somlyo (1976) have analysed single ferritin particles maintained at −100 °C in a TEAM instrument, but difficulties of specimen stability during analysis need to be overcome before the technique is generally viable for routine analysis of thin sections at high image resolution.

Freeze drying is usually accomplished by placing the sections, while still frozen, on grids in a vacuum chamber for a few hours. The ultrastructural information obtained is far inferior with frozen ultrathin sections than that achieved with alternative methods, and it is frequently very difficult to find enough sections with good enough image detail to perform statistical analyses. No one has yet recommended cryo-ultramicrotomy as a routine method of preparation for analyses where many subcellular regions are to be analysed. However, frozen sections form the second most frequent method of specimen preparation for analysis (see Table 3.2). The method has been successfully used in the examination of sodium, phosphorus, sulphur, chlorine, potassium, calcium, and zinc in sperm cells (Chandler & Battersby, 1976*b*), magnesium, phosphorus, sulphur, chlorine, potassium, calcium, silicon, and sodium in muscle (Sjostrom & Thornell, 1975), and sodium, chlorine, and potassium in kidney (Trump, Berezesky, Chang & Bulger, 1976), as well as in other tissues.

A few workers (Weavers, 1972; Sjostrom, 1975; Mizuhira, 1976*a*, 1976*b*) have also cut and analysed ultrathin sections of fixed and frozen tissue. This does not overcome the problems of losses during aqueous fixation but obviates the need for organic solvents and resins.

Possible artefacts arise with freeze-dried material due to the inevitable movement of ions as the ice–vapour interface moves across the tissue. Electrostatic forces may even cause some displacement and ice crystal formation could result in some relocation of ions even during the freezing process. For tissues with relatively small intercellular spaces (animal rather than plant material), such ionic movements may be quite minimal, however, compared to the spatial resolution of analysis.

Only two reports have been found (Clarke, Salsbury & Willoughby, 1970; Salsbury & Clarke, 1972) of analysis of tissue prepared by critical point drying. Such a method is unlikely to be of any great value in analysis of ultrathin specimens, except perhaps for whole cell suspensions, the technique being more readily applied to bulk material analysis.

3.3.6 AIR DRYING

Possibly the most simple and rapid method of preparing cell or organelle suspensions for analysis is by air drying. It has been shown (Chandler & Battersby, 1976*b*) that the process does not cause translocation of ions in human sperm cells. The method has also been successfully applied to the analysis of whole spasmonemes (Routledge, Amos, Gupta, Hall & Weiss Fogh, 1975), blood cells (Skaer, Peters & Emmines, 1976; Skaer & Peters, 1975; Takaya, 1975*a*), as well as to subcellular organelles such as nuclei (Takaya, 1975*d*), secretory granules of the pituitary (Takaya, 1975*b*), zymogen granules of the pancreas (Takaya, 1975*c*), and chromosomes (Chandler, 1974).

Allowance must be made during the air drying process for the possibility that debris from the medium (e.g. in the case of sperm cells, from the semen) may contaminate the cells or organelles. In most cases, washing is unacceptable because of the risks of ionic loss.

3.3.7 ASHING AND REPLICATION

For the analysis of particles or residual elements in tissue without need of the preservation of ultrastructure, the techniques of ashing and replication have been used. In this way, the tissue is incinerated while mounted on the grid, or the residue may be dispersed onto a grid after the ashing process, or the ash may be dissolved in acid and the solution sprayed onto grids (Davies & Morgan, 1976). Tissues of the heart (Davies & Morgan, 1976), and ovary (Henderson & Griffiths, 1975) have been examined in this way using the TEAM and EMMA instruments.

Replication of tissue has also been employed for the analysis of particulate inclusions (Henderson & Griffiths, 1975). The tissue, fixed or unfixed, or in the form of a histological section on a slide, is immersed in cellulose acetate, and then coated with polyvinyl acetate (PVA). When the PVA is stripped away from the tissue, particles are removed and transferred to the microscope for analysis by means of a carbon replica evaporated on to the PVA sheet. Foreign particles in stomach tumours (Henderson, Evans, Davies & Griffiths, 1975), and ovaries (Henderson & Griffiths, 1975) have been analysed using this method.

A wide range of methods has thus been employed with the aim of performing satisfactory subcellular analysis in ultrathin sections. No method can guarantee to provide a faithful representation of the *in vivo* situation, and for any meaningful interpretation of X-ray data the operator needs to be fully aware of the artefacts inherent in each preparative technique. In the applications of the EMMA and TEAM instruments described below, some of these possible artefacts are discussed together with the particular attention which has been paid to the analytical operation in order to achieve maximum sensitivity and reproducibility.

3.3.8 STANDARDS

The correct choice of standards is very important when performing quantitative microanalysis on thin sections. A number of these are listed by Chandler

(1977), the most promising standards for ultrathin section analysis being those involving the addition of elements to resin. Spurr (1974) first described the technique of introducing macrocyclic polyether complexes with alkali elements into epoxy resins. This was modified by Chandler (1976) for using potassium in Araldite as an absolute calibration standard. Jessen, Peters and Hall (1974) described the use of sulphur in Araldite by using varying amounts of the natural components. Possibly the most promising standards are those devised by Roomans and Gaal (1977). Four different kinds of compounds are used in epoxy resin or low viscosity resin: (1) phenyl compounds containing Group Vb elements; (2) cyclopentadienyl derivatives; (3) a pentanedione derivative (acetylacetonate); (4) complexes of metals with dialkyldithiocarbamates. Various elements can be added in concentrations up to 1% in epoxy resin or up to 2% in low viscosity resin.

Because of the similarity of the standard to normal embedded tissue sections they provide more reliable quantitative data in thin section form than bulk standards. Provided the X-ray detector is suitably calibrated for relative detection efficiency between elements (Chandler, 1976), a single standard of one element allows a complete absolute calibration over the whole elemental range.

3.4 Applications of thin specimen analysis

Broadly, the application of X-ray analysis to biological tissue may be classified into four types:

(1) Naturally occurring elements (endogenous electrolytes, metals, etc.).
(2) Accidentally introduced matter (foreign particles, toxic substances).
(3) Deliberately introduced foreign material (tracers, labels, drugs, competitive elements, etc.).
(4) Histochemical reaction products (localization of molecular groups, etc.).

All of these categories have been described in the literature (see Bibliography) for animal, plant, and protozoan tissues. In Table 3.3 is shown the range of elements that have been detected for all the tissue types analysed. It may be seen that calcium and phosphorus are by far the most frequently analysed elements. Animal tissues account for 80% of all analyses performed and of these muscle, kidney, liver, and blood cells have most often been examined. In over 80% of the publications listed in the Bibliography no quantitative analysis was performed. Instead, most workers, while investigating the presence of one or two elements in tissue, also reported their observations on the presence of other elements as shown by the X-ray energy spectrum. It is difficult to estimate the importance of purely qualitative analyses when they are based on a relatively small number of assays, especially in view of specimen preparation and electron irradiation effects, together with

Table 3.3 Frequency of analysis of elements, species and tissues in ultrathin specimens (to May 1977)

Elements		*Species*	
Ca	144	Animal	218
P	127	Plant	36
S	91	Protozoa	13
Cl	82		
K	67	*Tissues*	
Fe	45		
Mg	42	Muscle	37
Si	36	Kidney	22
Zn	34	Liver	16
Na	34	Blood	16
Cu	25	Leaf	14
Au	18	Sperm	10
Ag	16	Synovium	10
Al	12	Pancreas	10
Sb	12	Nerve	8
Pb	11	Lung	8
Others < 10		Teeth	8
		Prostate	7
		Others	7

biological variation which occurs in most soft tissues. However, with increased availability and use of proper standards and computerized programmes, it is expected that more applications will be of a quantitative nature in the future.

Examples are given below of each of the four classifications previously indicated.

3.4.1 NATURALLY OCCURRING ELEMENTS

Muscle

By far the greatest number of analyses of ultrathin specimens have been on the detection of endogenous elements in tissue. Many investigators have applied X-ray analysis to the investigation of calcium in muscle. Yarom (1977) has extensively studied muscle prepared in a variety of ways and has localized the element to the different subcellular zones during various physiological states. In order to separate X-ray lines from calcium and antimony, crystal spectrometers were employed for the analysis of muscle prepared by the pyroantimonate method (Plate 3.1; Yarom & Chandler, 1974). In order to obtain sufficient X-ray counts to be statistically meaningful, this analysis was performed with a counting time of 20 seconds and an electron beam current of 40 nA in an EMMA instrument. These conditions are typical for the analysis of ultrathin specimens in transmission at accelerating voltages of

80 kV. Higher count rates are produced with higher beam currents but the electron probe diameter increases with the beam current (Chandler, 1977).

Sjostrom and Thornell (1975) also examined muscle but prepared by cryo-ultramicrotomy (Plate 3.2) of fixed and unfixed tissue and compared with embedded material analysed in a TEAM. Ultrathin frozen sections were cut at −100 °C and freeze dried. Analysis of sodium, silicon, phosphorus, sulphur, chlorine, potassium and calcium was performed on carbon coated sections at ambient temperature. Fixation of the tissue before cryo-sectioning was found to cause loss of ions from the fibres, while dehydration followed by plastic embedding resulted in almost total loss of the ionic pattern of the muscle fibres compared with frozen tissue.

Yarom, Peters, Scripps and Rogel (1974) examined muscle prepared by fixation in glutaraldehyde, osmium tetroxide, or potassium pyroantimonate. Although calcium was retained in the tissue by each method, the use of glutaraldehyde alone was found to cause large elemental loss except in the nucleus where it was considered to be more tightly bound. The pyroantimonate method was found to retain calcium in its subcellular localization more satisfactorily than osmium tetroxide fixation. Calcium was also found to be removed from pyroantimonate-treated tissue sections after staining with aqueous lead or uranium solutions. In another report Yarom, Maunder, Scripps, Hall and Dubowitz (1975) used concentrations of glutaraldehyde increasing to 50% before direct embedding in Epon. This was found to retain calcium, zinc, copper and other elements for analysis in both muscle and blood cells.

The potassium pyroantimonate method has also been used in an attempt to localize calcium, magnesium, sodium, and potassium in kidney tissue (Tisher, Weavers & Cirksena, 1972). In this preparation the tissue was first perfused *in vivo* with a 2% solution of pyroantimonate and then fixed with glutaraldehyde containing the antimonate. Omission of osmium tetroxide as a post fixative produced a large amount of coarse and fine precipitation both intracellularly and extracellularly in the kidney sections. Extended buffer rinsing of the fixed tissue was found to remove the precipitation. Analysis of the precipitates was performed in an EMMA with an electron beam current of 0.7–1.0 μA at 60–100 kV for periods of 100 seconds using both crystal spectrometers and solid state detectors. Calcium, sodium, potassium, and magnesium were all found to be associated with the precipitates in the nuclei, microvilli, and lateral plasma membrane of the cells in the proximal convoluted tubule. Sodium was present in lower quantities than expected from other investigations. It was suggested that the sodium ions may not have been fully retained by precipitation with the pyroantimonate due to both competition with other ions and unstable bonding. Phosphate ions in the buffer may have inhibited sodium precipitation or, most likely, the efficiency of detection for sodium with the instrumental arrangement employed was too low.

The efficiency of detection for each element depends on the energy of the X-rays emitted. For elements of atomic number less than 16 (sulphur) the efficiency falls off very rapidly due to both lower X-ray yield and increased X-ray absorption in windows, etc. (Chandler, 1977). Thus in any analysis attempting to quantify relative concentrations of elements across tissue, account must be taken of the varying efficiency with which X-rays are produced and detected (Chandler, 1976).

Sperm cells

Human sperm cells have been analysed by the author both as a model for X-ray studies and during investigation of the effect of heavy metals on sperm motility. Initially it was shown (Chandler & Battersby, 1976*b*) that air dried whole sperm cells did not differ in their subcellular elemental composition from ultrathin sections of frozen and freeze-dried sperm cells. Drops (5 μl) of human semen were air dried on to carbon coated aluminium or nylon grids and analysed in the acrosome, nucleus, and midpiece regions for sodium, silicon, phosphorus, sulphur, chlorine, potassium, calcium, copper, and zinc. Compared with similar regions in frozen sections (cut at −85 °C) from the same semen samples (Plate 3.3), no differences were found for any of the elemental concentrations except calcium in the acrosome (Fig. 3.4). Analytical conditions employed for both types of preparation in the EMMA instrument involved an accelerating voltage of 80 kV, a beam current of 0.05 μA, and an analysis time of 100 seconds per cell region. Quantitation was carried out with the Hall (1971) method using a resin standard loaded with potassium (Chandler, 1976). The comparative elemental concentrations for the two preparation methods as shown in Fig. 3.4 were calculated for the three regions in 20 cells. Previous investigations (Battersby, 1976) had indicated that this was a suitable number of cells to analyse to represent the sample as a whole. Even so, large ($\doteq$ 30%) standard deviations of the mean were obtained for most elements. For any quantitative investigation to be truly representative of a biological sample sufficient number of analyses must be performed to take into account the very wide variation which occurs in most biological systems.

Sperm cells were also treated with the pyroantimonate method (Chandler & Battersby, 1976*a*) and analysed for calcium and zinc using EMMA. Fixation was accomplished by centrifugation of the semen followed by resuspension of the pellet in 2% potassium pyroantimonate in osmium tetroxide. Alcohol dehydration was followed by embedding in resin. Ultrathin sections exhibited precipitation mainly in the midpiece (Plate 3.4). Analysis of the three subcellular regions as before, showed zinc and calcium to be faithfully retained when compared with frozen or air dried specimens. In addition, the formation or otherwise of antimonate precipitates with these elements gave some indication of their relative binding within the cells. Absolute concentrations, calculated as before with resin standards, indicated that the elements were also retained at levels expected from the other analytical methods.

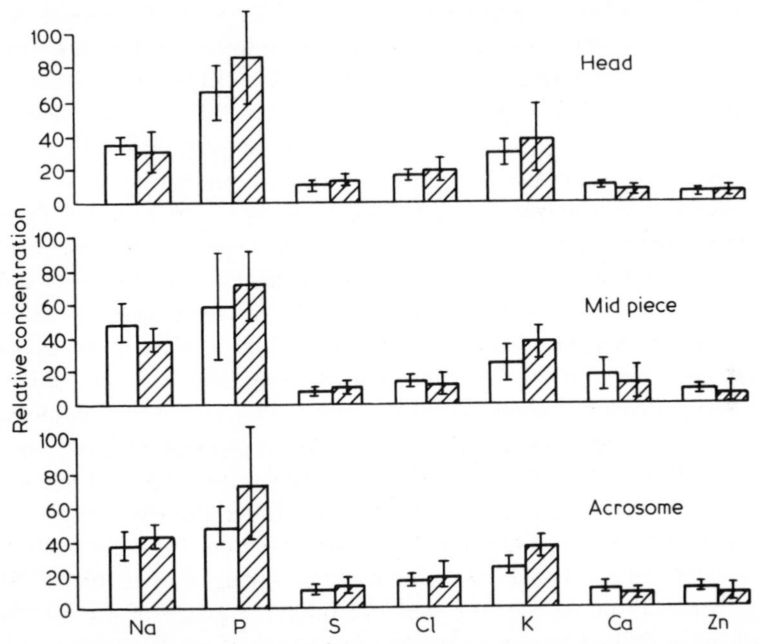

Fig. 3.4 Comparative analysis of air dried (clear bars) and frozen sections (hatched bars) of human sperm cells. Vertical indicators represent one standard deviation of the mean.

The pyroantimonate method thus seems to be a promising technique for the analytical study of a number of elements over a wide range of tissue types, although regard should be given to the precautions mentioned previously (Section 3.3.2). The variable precipitation found with different uses of the fixative indicates that visual interpretation without X-ray analysis is insufficient to estimate local elemental concentrations.

One of the major problems associated with the use of frozen sections for analysis is the extremely poor structural detail present in the preparation. The method is usually justified by claiming that image quality needs only to be good enough to allow recognition of organelles for analysis. Very few 'good quality' sections have been presented in the literature for TEAM studies and the trend seems to be towards using slightly thicker sections viewed by the STEM imaging system since the latter may provide rather better contrast as well as allowing cold stages to be used with more convenience.

Kidney

An examination has been reported by Trump, Berezesky, Chang and Bulger (1976) who performed analyses on sections of frozen kidney in an electron probe microanalyser fitted with a transmitted electron detector (equivalent to STEM). Small pieces of kidney were first frozen by being pressed against a copper block cooled to liquid nitrogen temperatures. This avoided the 'gas-layer' effect of immersion in liquid nitrogen reported to slow down the cooling rate and form large crystals in the tissue. Sections were cut to 100 nm thickness at −71 °C and freeze dried within the cryo-chamber on microscope grids. After being warmed to ambient temperature in dry nitrogen, they were carbon coated and analysed with a beam current of 10^{-10} A and an accelerating voltage of 25 kV. The X-ray data was collected from areas of cytoplasm in a number of cellular regions using a counting time of 100–600 seconds. Image resolution was just sufficient to allow nuclear and cytoplasmic regions to be identified. The electron beam was scanned in a raster mode within the cytoplasm during the analysis to allow mean cytoplasmic elemental concentrations to be determined. Using standards of sodium chloride and potassium chloride in frozen sections of bovine serum albumin, absolute concentrations of sodium, potassium, and chlorine were determined in the cytoplasmic regions of the kidney with the Hall method (Hall, Anderson & Appleton, 1973). The analysis indicated that intracellular electrolyte concentrations could be related to physiological states induced experimentally in the tissue. The STEM image was inferior to that obtained by normal TEM imaging, probably because of the inherent resolving power of the STEM arrangement as limited by the probe diameter.

It seems clear from these applications that freezing techniques are the most promising way of retaining mobile electrolytes in tissue for subsequent analysis (see Chapters 4 and 5), but the progress made in the last 10 years in producing worthwhile ultrastructural information from such material has been very disappointing.

Paramecium

Plattner and Fuchs (1975) used the oxalate method of fixation in order to precipitate calcium in *Paramecium*. The organisms were immersed in 1 % O_sO_4 containing dipotassium oxalate. In addition specimens were fixed in glutaraldehyde alone. Prior to fixation they were incubated in media containing calcium in varying molar concentrations or with ionophores applied before calcium incubation. Unstained ultrathin sections were prepared for analysis from routinely embedded material and examined in a TEAM instrument fitted with STEM facilities. This allowed the formation of a very small electron probe for analysis. The electron beam was scanned in a raster

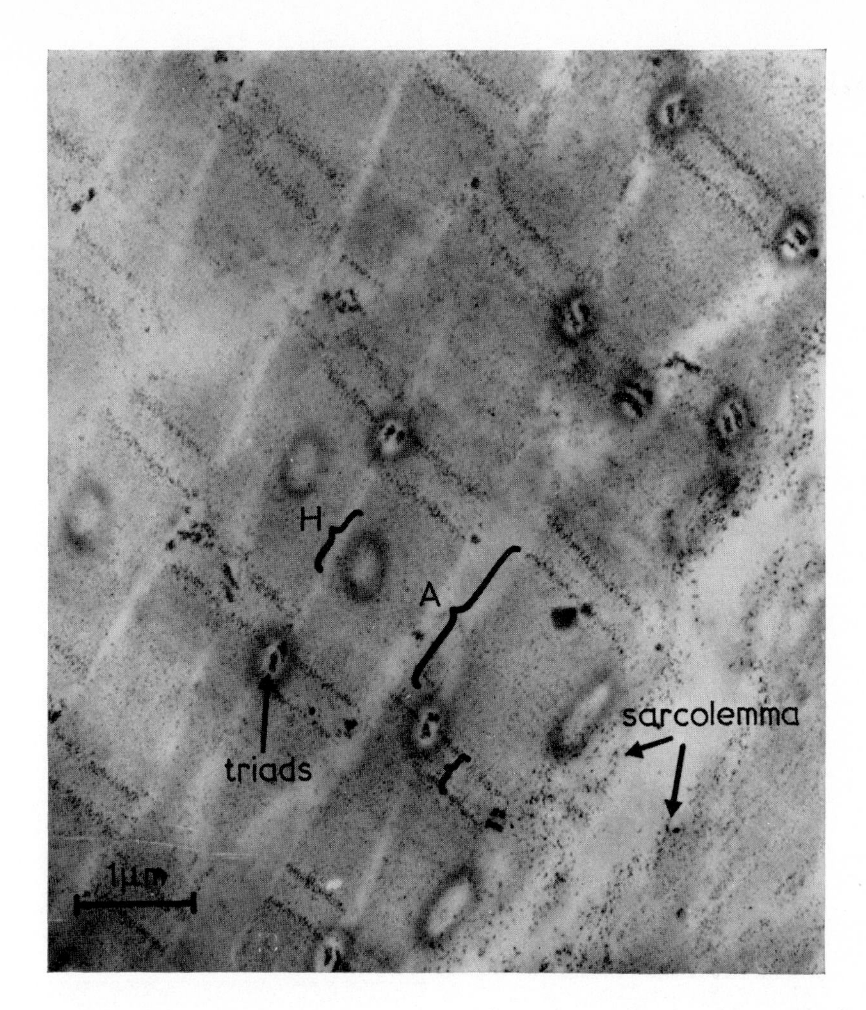

Plate 3.1 Frog muscle prepared by the potassium pyroantimonate method for X-ray analysis. Calcium is localized in the regions shown, and is approximately related to the density of precipitation. The dark circles are contamination marks left by the electron beam.

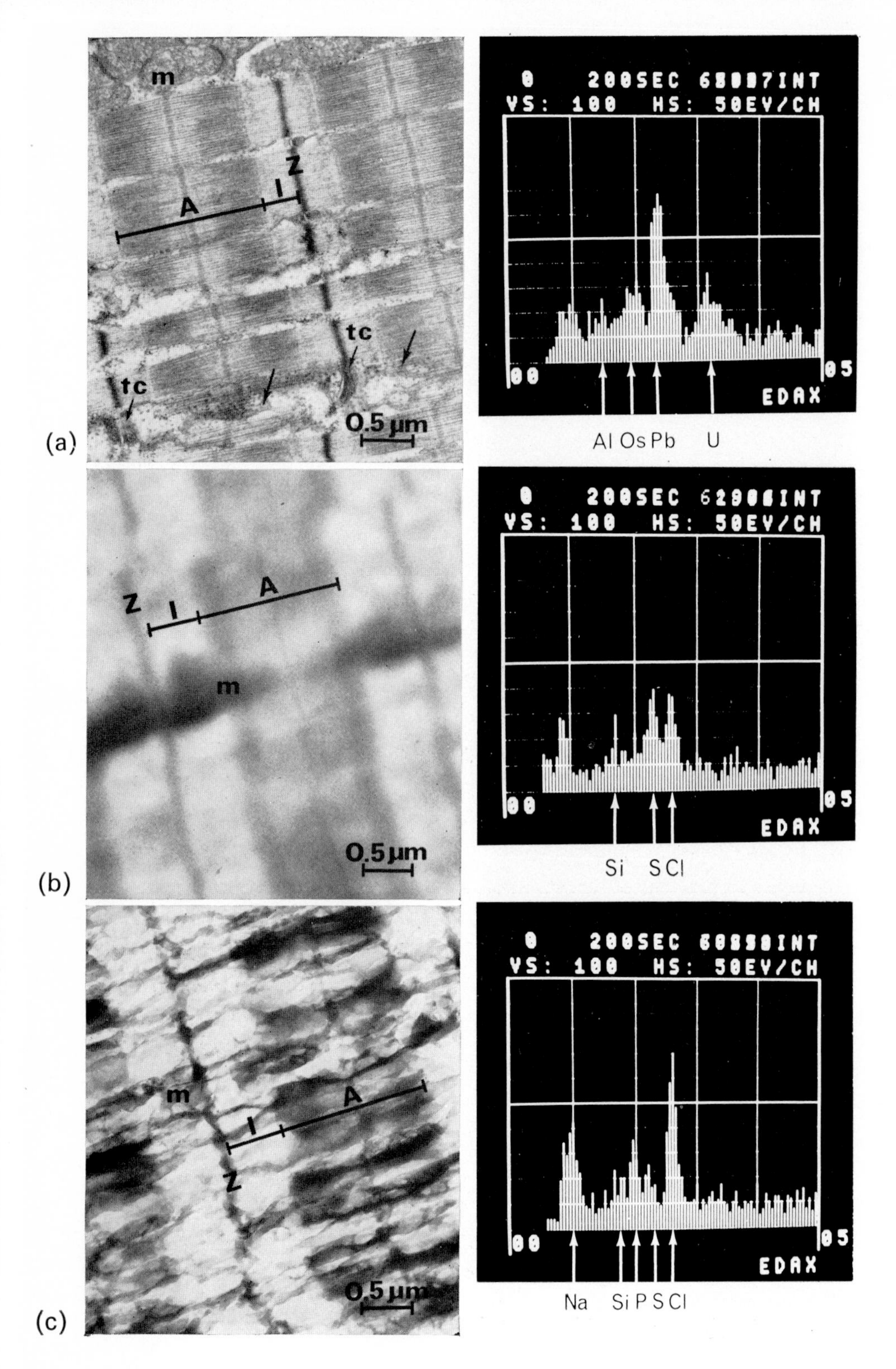

Plate 3.2 Frog skeletal muscle prepared by various methods for analysis of thin sections: (*a*) conventional fixation and resin embedding; (*b*) fixed and frozen, sections cut on to DMSO, rinsed and air dried; (*c*) fixed and frozen, frozen sections cut dry. Regions shown are A, A-band; I, I-band; Z, Z-band; tc, terminal cisternae; m, mitochondria. (Courtesy of Dr M. Sjostrom, and Chapman and Hall, Publishers.)

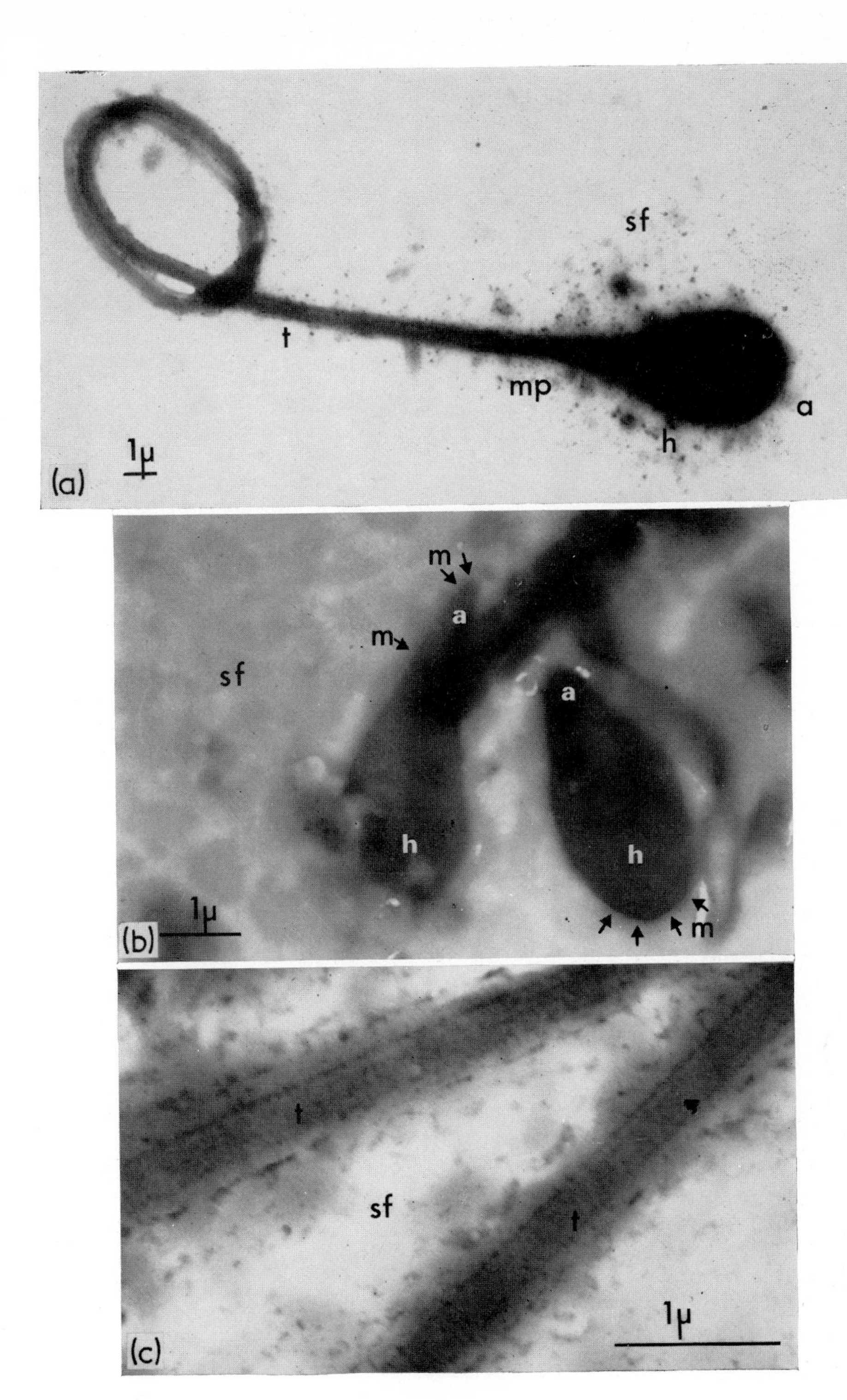

Plate 3.3 Human sperm cells prepared for analysis by: (*a*) air drying; (*b*) and (*c*) cryo-ultramicrotomy. Analysis was performed on acrosomal, nuclear, and midpiece regions for both preparations. t, tail; mp, midpiece; h, head; a, acrosome; sf, seminal fluid (dried); m, membrane (arrowed, just visible in frozen sections).

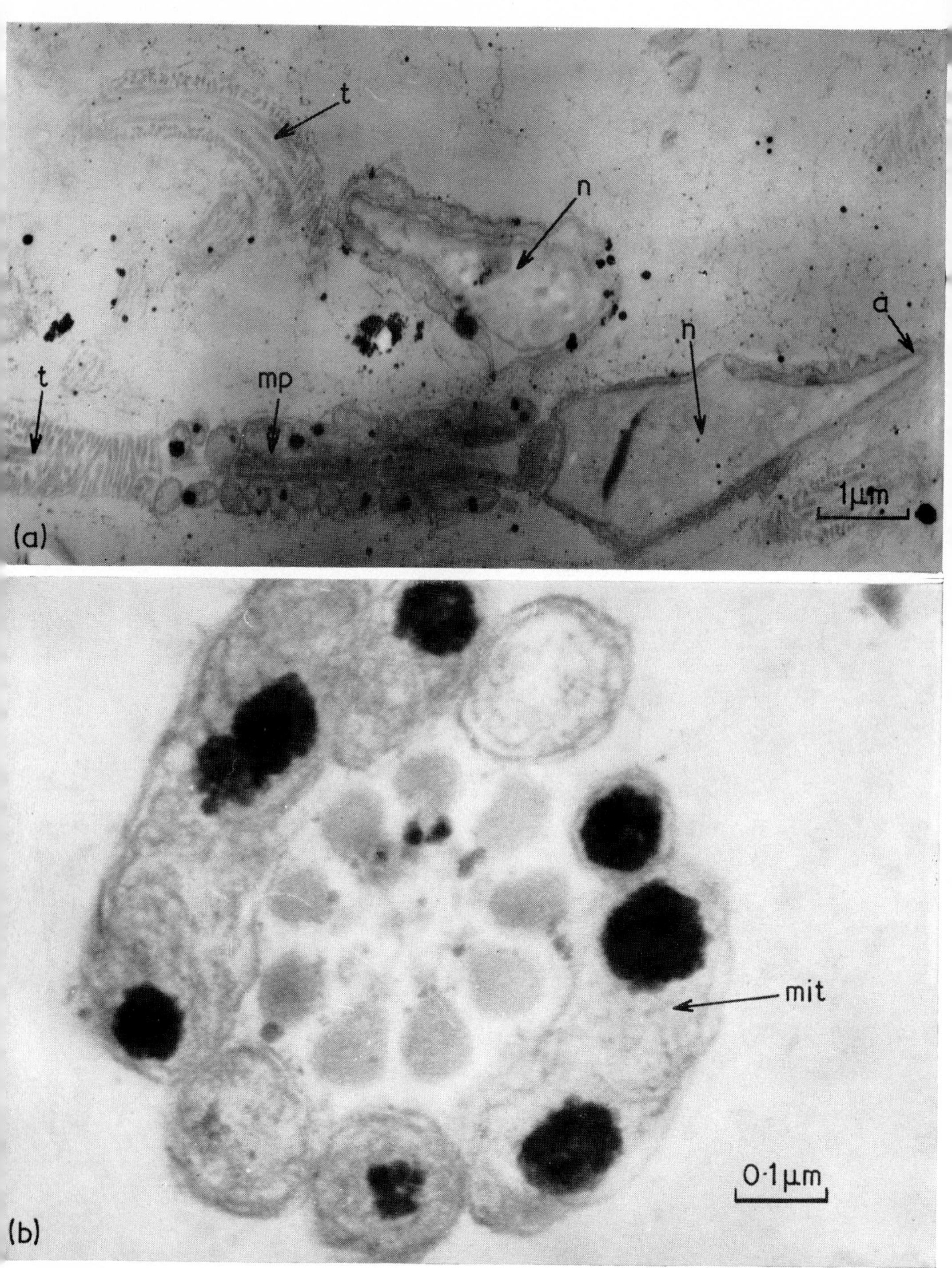

Plate 3.4 Human sperm cells treated with potassium pyroantimonate. Precipitation occurs mostly in the midpiece mitochondria. Analysis indicates calcium and zinc. (*a*) Longitudinal section; (*b*) transverse section of midpiece. mp, midpiece; t, tail; n, nucleus; a, acrosome; mit, mitochondria.

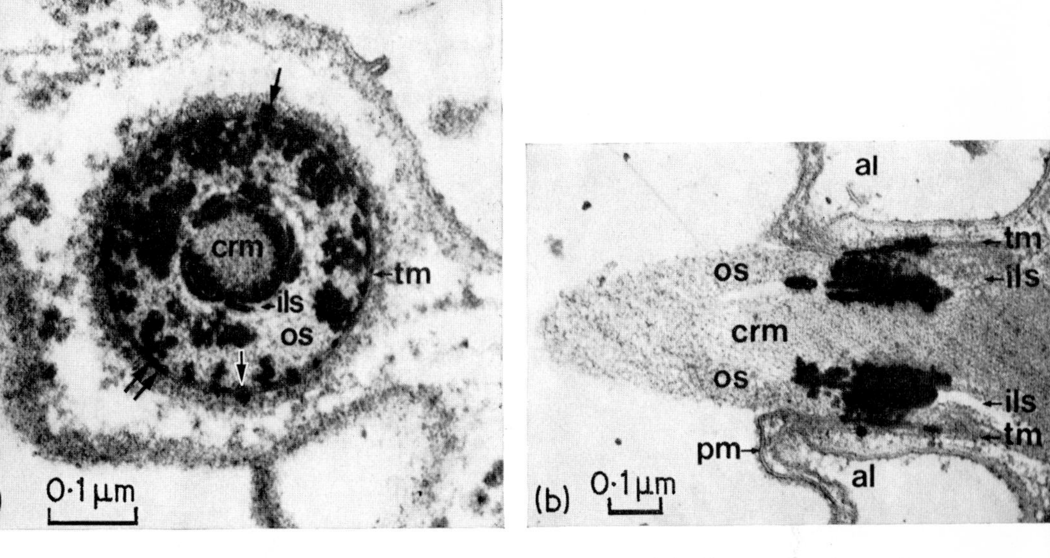

Plate 3.5 Trichocysts of *Paramecium* showing precipitation, analysed as calcium, in transverse (*a*) and longitudinal sections (*b*). Treatment with ionophores (*b*) causes exocytosis from the trichocyst tip. crm, crystalline matrix; ils, inner lamellar sheath; os, outer sheath; tm, trichocyst membrane; al, alveoli; pm, plasma membrane. (Courtesy of Dr Plattner and Springer–Verlag, Publishers.)

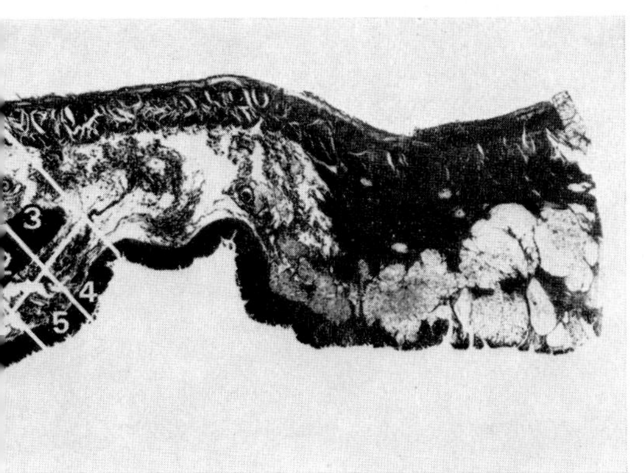

ate 3.6 Histological slide of stomach mour. Areas for subsequent replication are dicated. (×7)

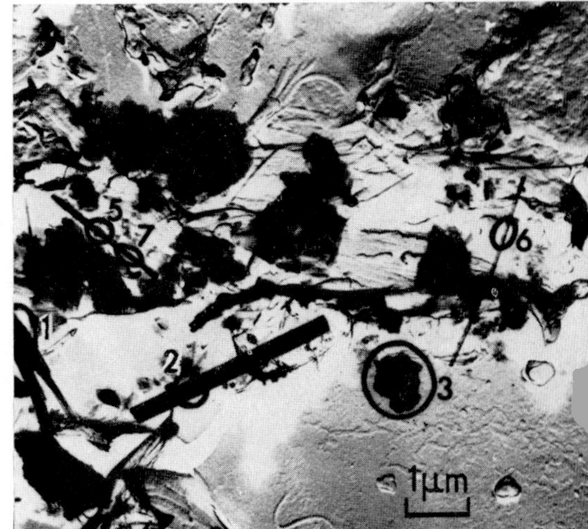

Plate 3.7 Replica of ashed residue from stomach tumour showing mineral particles subsequently analysed in EMMA.

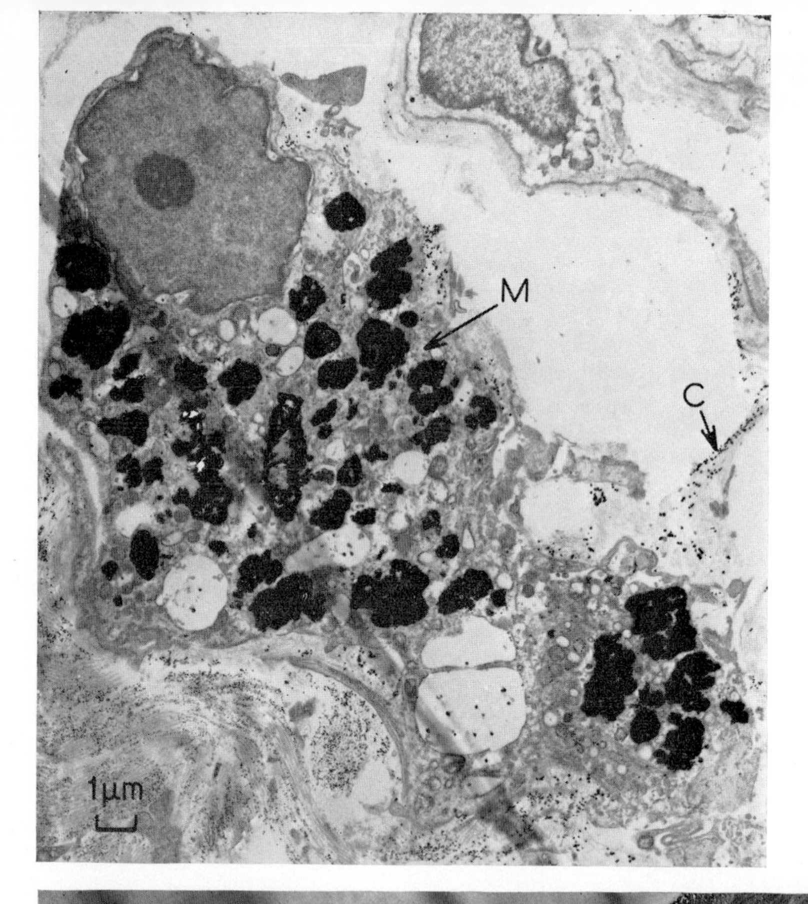
M
C
1µm

S3
lumen
S2
M1
1 µ

Plate 3.8 Electron-dense material in a macrophage and within collagen of conventionally prepared mouth biopsy. Analysis indicates the presence of Hg, Ag, and Sn in the particles, similar to that found in dental amalgam. M, macrophage; C, collagen. (Courtesy of Dr J. D. Harrison and Longman's Group, Publishers.)

Plate 3.9 Ultrathin section of wood microtubular double wall. Wood was treated with copper–chrome–arsenic preservative and analysed in the regions shown by the circles for Cu, Cr, As. ML, Middle lamella.

Plate 3.10 Unstained sections of (*a* and *b*) knee and (*c* and *d*) synovium after treatment with gold analgesia. Gold was detected by analysis with EMMA in lysosomes. N, Nucleus. (Courtesy of Dr Yarom and Springer–Verlag, Publishers.)

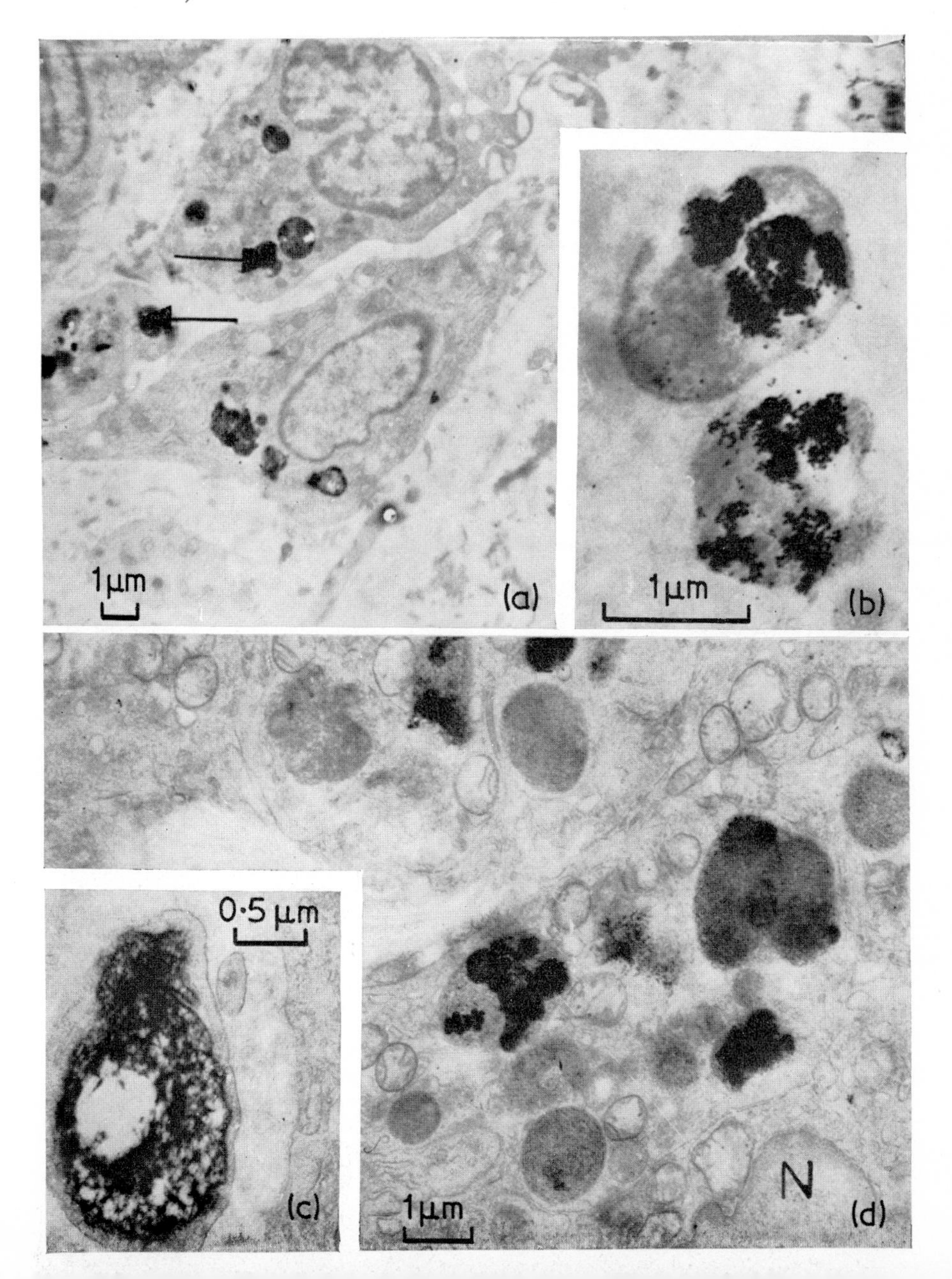

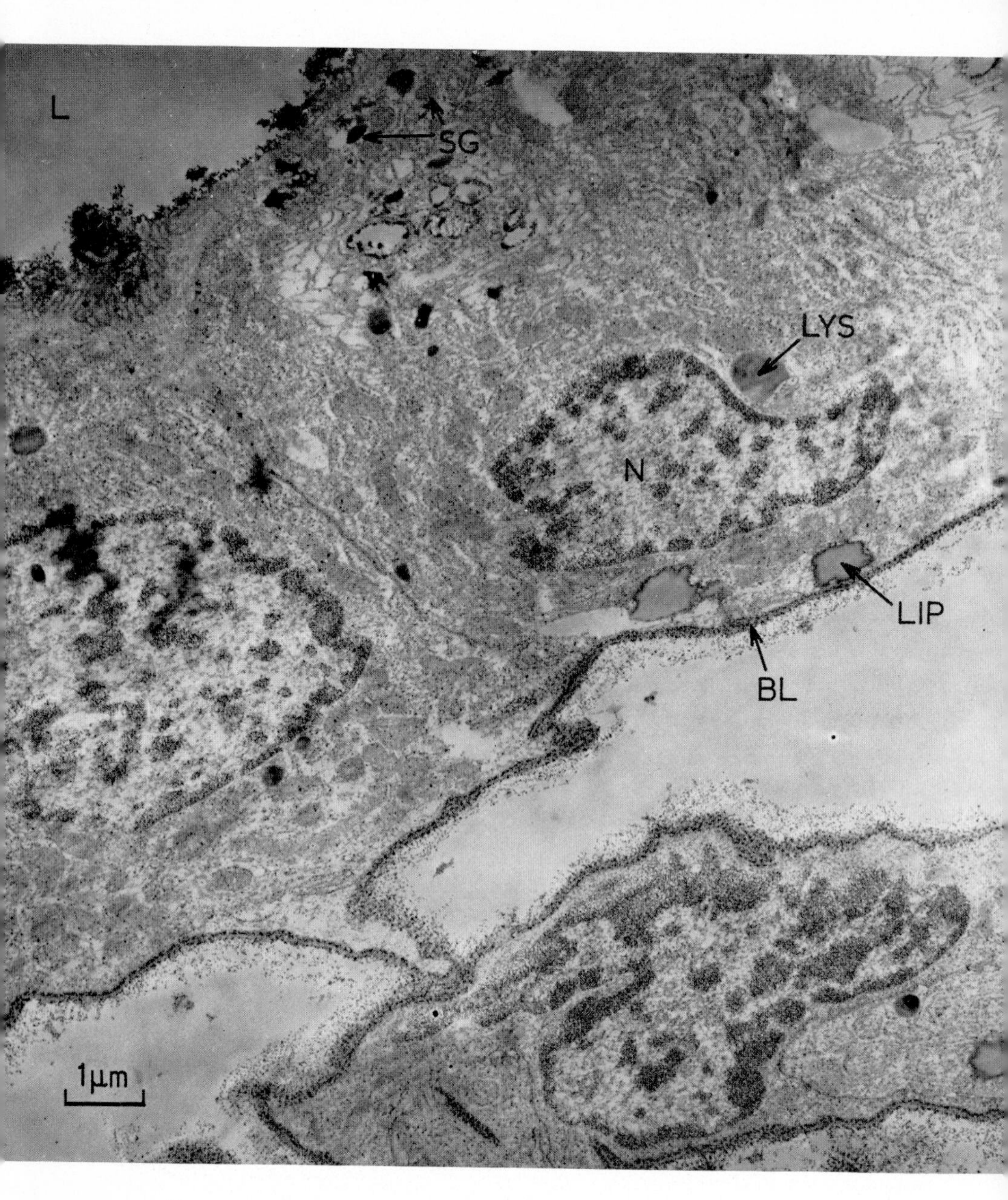

Plate 3.11 Rat prostate epithelium treated with potassium pyroantimonate for analysis of subcellular zinc and cadmium. Heavy precipitation occurs along the basement membrane and in the nuclei. N, nucleus; LYS, lysosomes; BL, basal lamina; LIP, lipid; SG, secretory granules; L, lumen.

measuring 125×125 nm over the individual areas of the section to be analysed for periods of 200 seconds per analysis. In this arrangement the X-ray detector (SSD) could be placed just 2 cm from the specimen.

Calcium enrichment at certain subcellular sites was indicated by the formation of electron-dense deposits (Plate 3.5*a*) and was confirmed by X-ray analysis. Both methods of fixation produced similar electron-dense deposits, but the oxalate technique was seen to produce increased precipitation after the cultures had been exposed to elevated Ca^{++} concentrations prior to fixation. The analysis indicated that the electron-dense deposits were rich in calcium and phosphorus and that exposure of the paramecia to ionophores prior to incubation with calcium produced massive exocytosis in the trichocysts (Plate 3.5*b*), as did incubation with certain molar concentrations of calcium. In this study, quantitative analyses were not performed but sites of calcium enrichment were also analysed for phosphorus, sulphur, and magnesium. A comparison of various cellular regions showed that calcium levels in the cytoplasm were not elevated after calcium incubation, and that calcium enrichment occurred within discrete sites. The experiments suggested that Ca^{++} influx and deposition must take place rapidly within individual trichocysts of the paramecia and not synchronously over the trichocyst population of a cell. Because the deposition pattern was similar with both methods of preparation it may be suggested that binding of calcium at these sites was quite strong. The value of the oxalate technique may be more readily seen, as with the pyroantimonate method, in tissues where the ions are less tightly bound to the tissue (see Section 3.3.2).

3.4.2 ACCIDENTALLY INTRODUCED MATERIAL

Most applications in this category take the form of a study of toxic poisoning of both plant and animal tissues by both organic and inorganic contaminants. A number of applications have been described for the analysis of ingested materials of human tissues.

Mineral inclusions

Henderson, Evans, Davies and Griffiths (1975) have described a variety of minerals present in stomach tumours of Japanese males. Preparation of tissue for this type of examination is less critical, since elemental loss from the particles is unlikely to occur, but still precise as very small amounts of material are handled during the procedure.

Sections (10 μm thick) were cut from paraffin embedded blocks of tissue that had been fixed for histological examination. After being cleared and dehydrated the tissue was examined in a photographic enlarger (Plate 3.6) and regions of the section were selected for subsequent processing. The areas indicated in Plate 3.6 were replicated as described in Section 3.3.7 and the replicas were mounted on electron microscope grids for subsequent analysis in EMMA.

Alternatively, similar paraffin-embedded 10 μm thick sections were mounted on glass slides, cleared, dehydrated, and then incinerated at 500 °C in an oven. A replica was then made on the glass slide to remove the residue and again mounted on a microscope grid for examination.

Plate 3.7 shows extracted particles from the tissue. These were analysed using an accelerating voltage of 80 kV and an electron beam current of 0.05 μA. Care had to be taken to avoid losing water of crystallization from some crystals during the analysis and they were coated with carbon before inspection. Both platey clay minerals (kaolin, talc) as well as fibrous particles (asbestos) were found, together with diatomaceous material and various silicates. Classification of the minerals was achieved by comparison with certain standard minerals of known origin and by quantitation of the ratios of elements within individual particles as given by the X-ray detector (SSD). For this type of analysis the SSD had a great advantage over the crystal spectrometer in being able to exhibit simultaneously all elements within the analysed sample. For some minerals, where certain elemental ratios are almost identical, identification may depend on the presence or absence of other elements as shown in the X-ray spectrum.

Metal absorption

Harrison, Rowley and Peters (1977) have investigated the accidental insertion of amalgam filling material into the soft tissues of the mouth during dental procedures. Excision biopsies were fixed conventionally in glutaraldehyde and osmium tetroxide and embedded in resin. Ultrathin sections were cut (with difficulty) and examined unstained in an EMMA instrument with both crystal spectrometers and solid state detectors.

The tattoo material was seen, in the electron microscope, to be in the form of discrete electron-dense particles usually associated with the surface of collagen fibrils. In some regions the particles were found within the basement membrane of muscle cells of blood vessels, and in others they were associated with the basement membrane of mucosal epithelium, with striped muscle fibres, and in macrophages (Plate 3.8). Analysis of the particles indicated the presence of mercury, silver, tin, iron, selenium, chlorine, zinc, copper, and silicon. It was necessary to use the crystal spectrometers to distinguish the SKα line from HgMα. All the elements found could be ascribed to the absorption of amalgam material during surgery, silver and tin being most commonly detected. The analysis indicated that there was both a qualitative and quantitative redistribution of elements from the original dental amalgam introduced into the soft tissues due to variable corrosive release from the site of injection and to different binding affinities to subcellular sites for each element. This type of analysis requires no special techniques of specimen preparation and, due to the relatively large quantities of elements present in the tissue, rather less caution over detection sensitivity.

A similar study was undertaken by Harry and Triparthi (1970) who examined deposits of copper in the cornea from a pathological case of Wilson's disease. Copper is absorbed by the liver, brain, and eye from the alimentary tract and gives rise to the Kayser–Fleischer ring – a brown smokey halo in the cornea of the eye. In this examination, tissue was prepared simply by conventional fixation and embedding, and ultrathin sections were examined in an EMMA instrument. Electron-dense deposits within the Descemet's membrane of the cornea were found to contain copper. Again no special techniques were required for the analysis of this tissue.

3.4.3 DELIBERATELY INTRODUCED ELEMENTS

Many experimental investigations have employed X-ray microanalysis to trace the passage of elements, either free or bound to molecular groups, through animal and plant tissues.

Wood preservation

The distribution of copper, chromium, and arsenic in samples of wood treated with a fungicide was examined by Chou, Chandler and Preston (1973). Pieces of Scots pine wood were impregnated with a copper–chromium–arsenic (CCA) preservative and infected with a fungus. Sections were prepared for analysis by first freeze drying small samples in a tissue drier, and then vacuum embedding them with Epon. Sections of thickness 100–150 nm were cut with a diamond knife and were carbon coated on microscope grids for analysis in an EMMA instrument (Plate 3.9). Quantitation was performed by comparison with ultrathin standards of resin containing dispersions of oxides from copper, chromium, and arsenic for the subcellular concentrations of the three elements in the wood sample. The Hall method was employed to determine elemental distribution and it was seen that the toxic elements of the preservative were located within the cell wall. The lumen surface of the wood was also protected by a coating of the CCA, an important finding in view of the location of the fungal hyphae at this site after infection.

Analgesic gold injections

A number of workers (see Bibliography) have investigated the fate of colloidal gold injected into tissues for the treatment of rheumatoid arthritis and various joint conditions. In the study of Yarom, Hall, Stein, Robin and Makin (1973), synovial biopsies from patients suffering with chronic effusions were fixed conventionally in glutaraldehyde and osmium tetroxide. Using EMMA, and with an accelerating voltage of 40–80 kV, a beam current of 70–250 nA, and a probe diameter of 200–300 nm, ultrathin sections were analysed for subcellular gold. The ultrathin specimen analysis was compared with that obtained from thick (2 μm) sections of similar biopsy material. For the thick section analysis a conventional electron probe microanalyser was used and the sections were obtained from a standard histological paraffin embedded block, having been cleared in xylol and carbon coated on microscope grids. X-ray analysis of the synovia from treated patients showed abundant gold deposits within phagosomes and secondary lysosomes (Plates 3.10a, 3.10b, 3.10c and 3.10d).

In a similar study with human biopsies, Ghadially, Oryschak and Mitchell (1976) examined material from patients treated with sodium aurothiomalate. For analysis, the tissue was fixed with glutaraldehyde and osmium tetroxide as before, and ultrathin specimens (100–150 nm) were examined in a STEM configuration of an SEM. With a static electron probe, analyses were performed on subcellular regions for periods of 200 seconds at a time. Compared with unaffected tissue, arthritic material exhibited increases in intracytoplasmic filaments and lysosomes in cells of the synovial membrane. In tissue treated with gold injections the lysosomes contained highly electron-dense material in the form of fine particles, spicules, and curled formations. These lysosomes contained high concentrations of gold, but the metal was not found to be associated with any other element. The authors suggested that the element occurred as the metallic form within the lysosomes and performed its analgesic function by an inactivation of lysosomal enzymes.

Cadmium in prostate

Timms, Chandler, Morton and Groom (1977) have investigated the uptake of cadmium by the prostate in a study of the possible involvement of this metal in prostatic disease. Rats were given single subcutaneous injections of cadmium chloride and were then sacrificed after 20 days. Using the pyroantimonate technique, pieces of prostate were fixed in 1% osmium tetroxide containing 2% potassium pyroantimonate and then immediately dehydrated in absolute alcohol. Ultrathin sections were cut and mounted on aluminium grids for analysis in EMMA (Plate 3.11). Both zinc and cadmium were analysed in the stromal and epithelial portions of the alveoli comprising the prostate gland, by examining epithelial nuclei, nucleoli, secretory granules, cytoplasm, microvilli, lysosomes, lumen, and connective tissue. The normal zinc distribution (Chandler, Timms & Morton, 1977) was found to be markedly altered by cadmium administration, the latter element appearing to mimic zinc in its subcellular localization (Fig. 3.5) and causing epithelial cell necrosis and basal cell hyperplasia.

Analysis of the prostate sections was performed with an electron beam

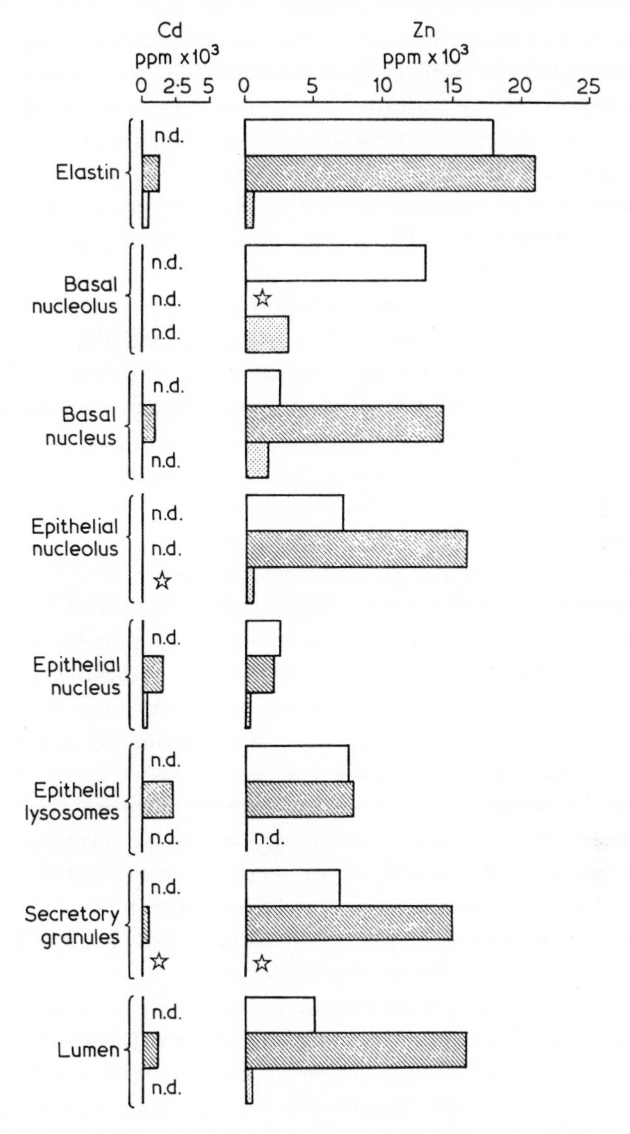

Fig. 3.5 Distribution of zinc and cadmium in rat prostate during cadmium administration *in vivo*. Open bars = control untreated tissue; hatched bars = 1 day after cadmium treatment; stippled bars = 20 days after cadmium treatment. Analysed with EMMA. n.d., Not detectable; star = none seen.

current of 0.1 μA for periods of 40 seconds per point analysis. Because of the very low level of elements present the crystal spectrometers were employed for detection of the characteristic lines for both zinc and cadmium while the solid state detector was used to monitor the white radiation. Absolute quantitation was again performed by the method of Chandler (1976). In order to permit the detection of the trace levels of cadmium the sections were mounted on aluminium grids and supported in a beryllium specimen holder (Chandler, 1973). This ensured that no characteristic X-ray lines interfered with the analysis and that the X-ray background due to scattered electron interaction was kept very low.

Using these techniques and calibrating with resin standards (Chandler, 1976), a sensitivity of 50 ppm has been established with the EMMA instrument for local point analyses of many elements. Such sensitivity, now also possible with the TEAM or STEM configurations (Shuman & Somlyo, 1976), is necessary for the progress of X-ray analysis in the further study of endogenous elements at physiological levels.

3.4.4 HISTOCHEMICAL APPLICATIONS

Although relatively few of the published papers on X-ray analysis deal with histochemical applications, this is possibly the field of work which offers most promise. In addition to the detection of elements in tissue as described above, histochemical procedures enable the subcellular localization and quantitation of many organic groups and molecular configurations to be determined by the use of appropriate labelling and staining techniques (see Chapter 6).

An important approach to this has been indicated by Okagaki and Clark (1977) in extending the use of conventional histological stains to the electron microscope. Many biological stains used in routine histological procedures contain heavy metals, which in thin sections can be localized subcellularly. Using a STEM arrangement in the TEAM instrument these authors have developed a technique for making thin sections from paraffin embedded tissue containing stains with the marker elements.

Ultrathin (100–400 nm) sections were cut from the paraffinized block in a cryostat at −30 °C, mounted on grids, deparaffinized, dried, and stained on the grid. The 'Fontana–Masson' silver reaction introduced silver into the tissue (epidermis) where melanocytes were present; Mayer's Mucicarmine reaction was used to stain secretory cells of the endocervix and the secretory granules were subsequently found to contain aluminium; Weigert's Iron Haematoxylin was used to localize iron in nucleic acids of the cervical cells; and Eosin Y staining of uterine squamous epithelium was indicated by the presence of bromine in the cytoplasm. Many other biological stains which contain marker elements may be used in a similar fashion.

Kirkham, Goodman and Chappell (1975) have localized neural tracts in insect tissue by the detection of cobalt after selective staining of neurones with cobalt chloride. Injection of the cobalt salt into the axons and subsequent development of the tissue with ammonium sulphide resulted in the deposition of cobalt sulphide in the neural processes. Cobalt was present in concentrations readily detectable in the TEAM with an energy-dispersive analyser. Other selective stains such as copper, iron, or nickel were suggested as alternatives to cobalt for the labelling process.

Immunocytochemistry has become a well established technique for subcellular localization of antigens in tissue using the light and electron microscope. The technique of X-ray microanalysis may well be applied to the cytochemically stained sections for quantitative estimation of antigenic distribution. For example, a common technique involves reacting the tissue (say rat) with specific antisera from another species (say rabbit–anti-rat). This is subsequently reacted with a second antibody (sheep–anti-rabbit) that has been labelled with a suitable marker. One well-known method uses the enzyme horseradish peroxidase as the marker. With this enzyme firmly bound to the original rat antigen (for example a protein hormone) the tissue sections are incubated with a substrate such as diaminobenzidene (DAB). The DAB is split to produce a final reaction product which is visible at both the light-optical and electron-optical level. Osmium has been used as a subsequent heavy metal stain due to its ready conversion to osmium black by the DAB. X-ray analysis of the subcellular osmium may then provide a quantitative estimation of the distribution of the original tissue antigen. The localization and quantitation of prolactin in rat tissues has been studied in this way using an EMMA instrument (Sibley & Chandler, to be published). Other substrates already having heavy metals or detectable elements may be employed as alternative markers for antigenic localization.

Labelling has been employed by a number of workers to trace the passage of compounds through cellular pathways after external administration. For example, Erasmus (1974) examined the distribution of the antimony-containing drug Astiban in tissues of the parasite *Schistosoma mansoni* maintained in mice. After intraperitoneal injection of the drug the parasites were recovered and conventionally fixed and embedded. The presence of antimony revealed the localization of the drug in the ovary and vitelline cells.

Chromium has been used as a marker for intracellular biogenic amines. Wood (1974, 1975) incorporated chromium into the nervous system of adrenomedullary tissue from cats by fixation with glutaraldehyde containing potassium dichromate. Using the TEAM, positive analysis for chromium, which labelled the biogenic amines of the tissue, was found within norepinephrine cells whereas no chromium was detected in the epinephrine cells. It was concluded that the chromium precipitates within the cytoplasm were incorporated into the reactive sites of the amines.

Other labelling techniques may prove to be equally promising for subcellular analysis. Iodination is routinely used for assaying protein hormones, using the radioisotope firmly bound to the protein antibody. The sensitivity of X-ray analytical techniques described here should make it possible to localize such proteins at the subcellular level. Alternatively, direct labelling of proteins with iodine, as has been used for autoradiographic studies, may allow direct determination of protein receptor sites at the cells after suitable incubations.

Other techniques of labelling have been reviewed by Chandler (1975, 1977) and are discussed elsewhere in this book. Also covered in Chapter 6 are the applications of X-ray analysis to enzyme cytochemistry.

3.5 Conclusions

Analysis of ultrathin specimens is now limited not so much by instrumental considerations but by an incomplete knowledge of specimen preparation effects on subcellular elemental localization. The minimum detectable limit of many analytical systems is now in the region of 10^{-18} g with some evidence of 10^{-19} g being possible. In terms of local mass fractions this is approximately equal to 50 ppm within an ultrathin section, or for an electrolyte such as calcium, equal to almost 1 milliequivalent in the region of analysis. Whether or not this detectable level represents the true concentration of the *in vivo* tissue is very difficult to say. In each case the analysis should be checked with biochemical data and corroborated with evidence from associated techniques.

Theoretically, cryo-ultramicrotomy offers the best chance of retaining tissue near to its *in vivo* state. Progress with this technique has been disappointingly slow, however, and it is difficult to envisage any further development that may revolutionize specimen preparation in the way that the use of epoxy resins changed the face of normal transmission electron microscopy. In any event, the research worker will still have to make certain assumptions about the nature of his ultrathin specimen during the analysis, assumptions which must be based on correlative techniques to ascertain the integrity of the prepared sample.

Future publications on ultrathin specimen analysis must provide evidence of adequate sampling and statistical accuracy in quantitation. Biological variation makes this a necessity and the highly sophisticated and expensive techniques offered by X-ray analysis will be wasted if data has not been presented in such a responsible manner. With manufacturers providing instrumentation which is becoming more and more easy to use, it is important for all research workers to appreciate the need for caution in the interpretation of data which can be obtained by the push of a button.

Much progress has been made in the last 10–15 years since the inception of thin specimen X-ray microanalysis. The next 5 years will be successfully spent if the technique can be rid of the black magic which surrounds it and it becomes established as a reliable scientific tool.

3.6 Acknowledgements

The author is grateful to the Tenovus Organization for general financial support. Original photographs for illustration of some applications were kindly provided by Dr R. Yarom, Hebrew University, Haddassah Medical School, Jerusalem; Dr J. Harrison, Department of Oral Pathology, Kings College Hospital Dental School, London; Dr Plattner, Institut für Zellbiologie, Universität München, Federal Republic of Germany; and Dr M. Sjostrom, University of Umeå, Sweden.

3.7 Bibliography

For full citation of references designated by a dagger (†), and for details of specimen preparation procedures, readers are requested to refer to the Footnotes at the end of the Bibliography.

References which are textually cited are designated by an asterisk (*).

Elements in parentheses, e.g. (Sb), refer to those deliberately introduced into tissue as stains.

References/Bibliography	*Instrument*	*Specimen preparation*	*Tissue/ organism*	*Elements*	*Species*
Ahmed, A. 1975 Calcification of human breast carcinomas: ultrastructural observations. *Journal of Pathology*, **117**, 247–251.	EMMA	1	Breast	P, Ca	A
Ali, S. Y., Wisby, A. & Craig-Gray, J. 1977 Preparation of thin cryo-sections for electron probe analysis of calcifying cartilage. *Proceedings of the Royal Microscopical Society*, **12**, 71.	TEAM	8	Cartilage	P, Ca	A
Ap Gwyn, I., Evans, P. M., Jones, B. M. & Chandler, J. A. 1977 Shape changes and reduced calcium levels in the surface membrane regions of human platelets exposed to manganese *in vitro*. *Cytobios*, **16**, 97–106.	EMMA	1	Blood	Ca, Mn	A
Appleton, T. C. 1972 Dry ultrathin frozen sections for electron microscopy and X-ray microanalysis: the cryostat approach. *Micron*, **3**, 101–105.	EMMA	8	Pancreas	Na, P, K, Ca	A
Appleton, T. C. 1973 X-ray microanalysis of diffusible electrolytes in ultrathin frozen sections. *Journal of Physiology*, **233**, 15–17b.	EMMA	8	Spleen	Na, Mg, Al, Si, P, S	A

References/Bibliography	*Instrument*	*Specimen preparation*	*Tissue/ organism*	*Elements*	*Species*
Appleton, T. C. 1974*a* A cryostat approach to ultrathin dry frozen sections for electron microscopy: a morphological and X-ray analytical study. *Journal of Microscopy*, **100**, 49–74.	EMMA	8	Pancreas	Na, Mg, Si, P, S, Cl, K, Ca	A
Appleton, T. C. 1974*b* Cryo-ultramicrotomy: possible applications in cytochemistry. *EM and Cytochem*,† 229–241.	EMMA	8	Skin	Na, Mg, Si, P, S, Cl, K, Ca	A
Appleton, T. C. & Newell, P. F. 1977 X-ray microanalysis of freeze-dried ultrathin frozen section of a regulating epithelium from the snail *Otala*. *Nature*, **266**, 854–855.	TEAM/ EMMA	8	Snail	Si, P, S, Cl, K, Ca, Fe, Zn	A
Ashraf, M. & Bloor, C. M. 1976 X-ray microanalysis of mitochondrial deposits in ischaemic myocardium. *Virchows Archiv B. Cell Pathology*, **22**, 287–298.	STEM	1	Muscle	P, Ca	A
Ashraf, M., Sybers, H. D. & Bloor, C. M. 1976 X-ray microanalysis of ischaemic myocardium. *Experimental and Molecular Pathology*, **24**, 435–440.	STEM	3	Muscle	P, Ca	A
Bacaner, M., Broadhurst, J. Hutchinson, T. & Lilley, J. 1973 Scanning transmission electron microscope studies of deep frozen unfixed muscle correlated with spatial localization of intracellular elements. *Proceedings of the National Academy of Sciences*, **70**, 3423–3427.	STEM	8	Muscle	P, S, Cl, K, Ca	A
Baccetti, B., Pallini, V. & Burrini, A. G. 1976 The accessory fibres of the sperm tail, II. Their role in binding zinc in mammals and cephalopods. *Journal of Ultrastructure Research*, **54**, 261–275.	TEAM	1	Sperm	P, S, Cu, Zn	A
Baker, J. R. & Appleton, T. C. 1976 A technique for electron microscope autoradiography and X-ray microanalysis of diffusible substances using freeze dried fresh frozen sections. *Journal of Microscopy*, **108**, 307–315.	TEAM	8	Kidney	Si, P, S, Cl, K, Ca, Mn, Fe	A
*Battersby, S. 1976 Electron microscope microanalysis of trace metals in human sperm. *M.Sc. thesis*, University of Wales.	EMMA	2, 11	Sperm	Na, Mg, P, S, Cl, K, Ca, Cu, Zn, Cd	A

References/Bibliography	*Instrument*	*Specimen preparation*	*Tissue/ organism*	*Elements*	*Species*
Battersby, S. & Chandler, J. A. 1977 Correlation between elemental composition and motility of human spermatozoa. *Fertility and Sterility*, **28**, 557–561.	EMMA	11	Sperm	Na, Mg, P, S, Cl, K, Ca, Cu, Zn	A
Berridge, M. J., Oschman, J. L. & Wall, B. A. 1975 Intracellular calcium reservoirs in *Calliphora* salivary glands. *Ca transport*,† 131–138.	EMMA	11	Sperm	Na, Mg, P, S, Cl, K, Ca, Cu, Zn	A
Boquist, L. 1975 Ultrastructural study of calcium-containing precipitation in human parathyroid glands. *Virchows Archiv A. Pathology and Anatomy*, **368**, 99–108.	STEM	2	Para- thyroid	Ca	A
Boquist, L. 1977*a* Ultrastructural changes in the parathyroids of mongolian gerbils induced experimentally *in vitro*. *Acta Pathologica et Microbiologica Scandinavica*, **85**, 203–218.	STEM	2	Para- thyroid	Ca	A
Boquist, L. 1977*b* Electron-dense plaques (calcium binding sites) in the plasma membrane of the endocrine cells in rodent pancreatic islets and parathyroid glands.	STEM	1	Pancreas, para- thyroid	Ca	A
Boquist, L. & Lundgren, E. 1975 Effects of variations in calcium concentration on parathyroid morphology *in vitro*. *Laboratory Investigation*, **33**, 638–647.	STEM	2	Para- thyroid	Ca (Sb)	A
*Bowen, I. D., Ryder, T. A. & Downing, N. L. 1976 An X-ray microanalytical azo-dye technique for the localization of acid phosphatase activity. *Histochemistry*, **49**, 43–50.	TEAM	9 (15)	Liver	(Br)	A
Braatz, R., & Komnick, H. 1973 Vascular calcium segregation in relaxed myxomycete protoplasm as revealed by combined electrolyte histochemistry and energy dispersive X-ray analysis. *Cytobiologie*, **8**, 158–163.	TEAM	3	Fungus	Ca	P
Buja, L. M., Dees, J. H., Harling, D. F. & Willerson, I. 1976 An electron microscope study of mitochondrial inclusions in canine myocardial infarcts. *Journal of Histochemistry and Cytochemistry*, **24**, 508–516.	TEAM	1	Muscle	Ca	A
Chandler, J. A. 1971 Analytical electron microscope. *American Laboratory*, **3**, 50–57.	EMMA	1	Cornea, liver, wood,	Cr, Fe, Cu, As, Au blood	A, P
*Chandler, J. A. 1973 Recent development in analytical electron microscopy. *Journal of Microscopy*, **98**, 359–378.					

References/Bibliography	*Instrument*	*Specimen preparation*	*Tissue/ organism*	*Elements*	*Species*
*Chandler, J. A. 1974 X-ray microanalysis of human chromosomes. *Lancet*, **7859**, 687.	EMMA	11	Chromosomes	Mg, P, S, Ca, Fe, Zn	A
*Chandler, J. A. 1975 Electron probe X-ray microanalysis in cytochemistry. *Techniques Biochem*,† **2**, 308–437.					
*Chandler, J. A. 1976 A method for preparing absolute standards for quantitative calibration and measurement of section thickness with X-ray microanalysis of biological ultrathin specimens in EMMA. *Journal of Microscopy*, **106**, 291–302.					
*Chandler, J. A. 1977 *X-ray Microanalysis in the Electron Microscope*. North Holland Publishing Co.					
*Chandler, J. A. & Battersby, S. 1976*a* X-ray microanalysis of zinc and calcium in ultrathin sections of human sperm cells using the pyroantimonate technique. *Journal of Histochemistry and Cytochemistry*, **24**, 740–748.	EMMA	2	Sperm	Ca, Zn (Sb)	A
*Chandler, J. A. & Battersby, S. 1976*b* X-ray microanalysis of ultrathin frozen and freeze dried sections of human sperm cells. *Journal of Microscopy*, **107**, 55–65.	EMMA	8	Sperm	Na, P, S, Cl, K, Ca, Zn	A
Chandler, J. A. & Chou, C. K. 1970*a* The distribution of preservatives in wood and uptake by fungal hyphae, examined in the analytical electron microscope.	EMMA	1	Wood	Cr, Cu, As	P
Chandler, J. A. & Chou, C. K. 1970*b* Subcellular elemental analysis with the analytical electron microscope. *Proc. 28th EMSA*.†	EMMA	7	Wood fungus	S, Cr, Cu, As	P
Chandler, J. A., Harper, M. E., Blundell, G. K. & Morton, M. S. 1975 Examination of the subcellular distribution of zinc in rat prostate after castration using EMMA. *Journal of Endocrinology*, **65**, 34b.	EMMA	2	Prostate	Zn	A
Chandler, J. A., Harper, M. E. & Griffiths, K. Studies on subcellular zinc distribution in relation to hormone levels in rat prostatic tissue using the electron microscope microanalyser EMMA. *Journal of Endocrinology*, **61**, Iv–Ivi.	EMMA	2	Prostate	Zn	A
Chandler, J. A., Sinowatz, F., Timms, B. G. & Pierrepoint, C. G. 1977 The subcellular distribution of zinc in dog prostate studied by X-ray microanalysis. *Cell and Tissue Research*, **185**, 89–103	EMMA	2	Prostate	Zn	A

References/Bibliography	Instrument	Specimen preparation	Tissue/ organism	Elements	Species
Chandler, J. A. & Timms, B. G. 1977 The effect of testosterone and cadmium on the rat lateral prostate in organ culture. *Virchows Archiv B. Cell Pathology* (in press).	EMMA	2	Prostate	Zn, Cd	A
*Chandler, J. A., Timms, B. G. & Morton, M. S. 1977 Subcellular distribution of zinc in rat prostate studied by X-ray microanalysis. I. Normal prostate. *Histochemical Journal*, **9**, 103–120.	EMMA	2	Prostate	Ca, Zn	A
*Chou, C. K., Chandler, J. A. & Preston, R. A. 1973 Microdistribution of metal elements in wood impregnated with a Cu–Cr–As preservative as determined by analytical electron microscopy. *Wood Science and Technology*, **7**, 151–160.	EMMA	7	Wood	Cr, Cu, As	P
*Clarke, J. A., Salsbury, A. J. & Willoughby, D. A. 1970 Application of electron probe microanalysis and electron microscopy to the transfer of antigenic material. *Nature*, **277**, 69–71.	EMMA	10	Blood	(I)	A
Clemente, F. & Meldolesi, J. 1975 Calcium in pancreatic secretion, I. Subcellular distribution of calcium and magnesium in the exocrine pancreas of the guinea pig. *Journal of Cell Biology*, **45**, 88–102.	TEPA	2	Pancreas	Mg, K, Ca (Sb)	A
Crocker, P. R., Dieppe, P. A., Tyler, G., Chapman, S. K. & Willoughby, D. A. 1977 The identification of particulate matter in biological tissues and fluids. *Journal of Pathology*, **121**, 37–40.	TEAM	1	Synovium	P, Ca	A
Davies, T. W. & Erasmus, D. A. 1973 Cryo-ultramicrotomy and X-ray microanalysis in the transmission electron microscope. *Science Tools*, **20**, 9–13.	TEAM	8	Liver parasite	Na, P, S, Cl, K, (Sb)	A
*Davies, T. W. & Morgan, A. J. 1976 The application of X-ray analysis in the transmission electron analytical microscope (TEAM) to the quantitative bulk analysis of biological microsamples. *Journal of Microscopy*, **107**, 47–54.	TEAM	13	Heart	Na, Mg, P, S, K, Ca	A
*De Filippis, L. F. & Pallaghy, C. K. 1975 Localization of zinc and mercury in plant cells. *Micron*, **6**, 111–120.	STEM	1, 6	Leaf, root	Zn, Hg	P
Dempsey, E. W., Agate, F. J., Lee, M. & Purkerson, M. L. 1973 Analysis of submicroscopic structures by their emitted X-rays. *Journal of Histochemistry and Cytochemistry*, **21**, 580–586.	STEM	1	Kidney	P, S, Cl, Ag	A

References/Bibliography	Instrument	Specimen preparation	Tissue/ organism	Elements	Species
*Dierkes, U. 1977 On the problem of histochemical demonstration of esterase activity by the thiolacetic acid method in the cellular slime mold *Physerum comfortum*. *Microscopica Acta*, **79**, 23–28.	TEAM	1 (15)	Slime mold	Si, S, (Pb)	P
Dore, J. L. & Vernon Roberts, B. 1976 A method for the selective demonstration of gold in tissue sections. *Medical Laboratory Science*, **33**, 209–213.	EMMA	1	Kidney	Au (Ag)	A
Duckett, J. G. & Chescoe, D. 1976 A combined ultrastructural and X-ray microanalytical study of spermatogenesis in the bryophyte *Phaeoceros laevis* (L) Prosk. *Cytobiologie* **13**, 322–340.	TEAM	1	*Phaeoceros laevis*	P, S, Ca	P
Duckett, S. & White, R. 1974 Cerebral lipofuschinesis induced with tellurium: electron-dispersive X-ray spectrophotometry analysis. *Brain Research*, **73**, 205–214.	TEAM	1	Brain	Te	A
Eley, B. M., Garrett, J. R. & Harrison, J. D. 1976 Analytical ultrastructural studies on implanted dental amalgam in pigs. *Histochemical Journal*, **8**, 647–650.	TEAM	1	Teeth	S, Cu, Ag, Sn, Hg	A
*Erasmus, D. A. 1974 The application of X-ray analysis in the transmission electron microscope study of drug distribution in the parasite *Schistosoma mansoni*. *Journal of Microscopy*, **102**, 56–59.	TEAM	1, 8	Parasite	P, S, Cl, K, Ca, (Sb)	A
Ettiene, E. 1972 Subcellular localization of calcium repositories in plasmodia of the acellular slime mold *Physarum polycephalum*. *Journal of Cell Biology*, **54**, 179–184.	TEPA	1, 8, (3)	Slime mold	K, Ca	P
Feris, J., Coleman, J. R., Davis, & Morehouse, B. 1976 Electron microprobe and analysis of particles. *Archives of Environmental Health*, **31**, 113–115.	STEM	1	Lung	Ti	A
Fisher, G., Kaneshiro, E. S. & Peters, P. D. 1976 Divalent cation affinity sites in *Paramecium aurelia*. *Journal of Cell Biology*, **69**, 429–442.	EMMA	1	Paramecium	P, S, Cl, Ca, (Mn)	Pr
*Forrest, O. G. & Marshall, A. T. 1976 Comparative X-ray microanalysis of frozen hydrated and freeze-substituted specimens. *6th Europ. Cong.*,† 218–220.	TEAM	6	Nerve cord	P, S, Cl, K	A
Fowler, B. A. & Goyer, R. A. 1975 Bismuth localization within nuclear inclusions by X-ray microanalysis: effects of accelerating voltage. *Journal of Histochemistry and Cytochemistry*, **23**, 722–726.	TEAM	1	Kidney	Bi	A

References/Bibliography	*Instrument*	*Specimen preparation*	*Tissue/ organism*	*Elements*	*Species*
Fowler, B. A. & Parker, P. 1973 Observations on the preparation of epoxy embedded tissue samples for energy dispersive X-ray analysis. *Stain Technology*, **48**, 333–335.	TEAM	1	Kidney	Hg	A
Fujita, H. 1975 X-ray microanalysis on the thyroid follicle of the Crayfish, *Eptetretus burgeri*, and Lamprey, *Lampreta japonica. Histochemistry*, **43**, 283–290.	TEAM	1	Thyroid	S, Cl, I	A
Gabor, A., Janossy, S., Mustardy, L. A. & Faludi-Daniel, A. 1977 X-ray microanalytical study of Mn and Fe compartmentation in maize chloroplasts. *Acta Histochemica*, **58**, 317–323.	TEAM	1	Leaf	Mn, Fe	P
Galle, P. 1966 Microanalyse des inclusions minérales du rein. *3rd Nephron*†, 306–319.	TEPA	1	Kidney	P, Ca, U	A
Galle, P. 1969 Cytochimie élémentaire sur coupes ultrafines de poumons. *Le Poumon*, **3**, 307–317.	TEPA	1, 13	Lung	Mg, Al, Si, P, Ca, Fe, Ti, Ba	A
Galle, P. 1974 Rôle des lysosomes et des mitochondries dans le phénomènes de concentration et d'élimination d'élements minéreux par le rein. *Journal de Microscopie et Cellulaire Biologie*, **19**, 17–24.	TEPA	1	Kidney	P, Au, U	A
Galle, P., Berry, J. P. & Stuve, J. 1968 New applications of electron micro-probe analysis in renal pathology. *3rd EMPA*,† paper 41.	TEPA	1	Kidney	Mg, P, S, Ca, Au	A
Galle, P. & Morel-Maroger, L. 1965 Les lésions rénales du saturnisme humain et expérimental. *Nephron*, **2**, 273–286.	TEPA	1	Kidney	Fe, Pb	A
Gambetti, P. & Erulkar, S. E. 1974 Calcium and phosphorus containing structures in ependymal glial cells of the spinal cord of the frog. *Anatomical Record*, **178**, 359.	TEAM	1	Spinal cord	P, Ca	A
Gambetti, P., Erulkar, S. E., Somlyo, A. P. & Gonates, N. K. 1975. Calcium containing structures. *Journal of Cell Biology*, **64**, 322–330.	TEAM	1	Spinal cord	P, Ca	A
Garfield, R. E. & Somlyo, A. P. 1975 Electron probe analysis and ultrastructure of cultured, freeze-dried vascular smooth muscle. *33rd EMSA*,† 588–589.	TEAM	8	Muscle	Na, P, S, Cl, K, Ca, Sr	A
Gay, J. L. 1972 X-ray microanalysis in development of the fungus *Saprolegnia. Micron*, **3**, 139–143.	EMMA	1, 8	Fungus	P, S	P

References/Bibliography	Instrument	Specimen preparation	Tissue/organism	Elements	Species
Gebhart, W., Ebner, H. & Erlach, E. 1976 Comparative X-ray microanalysis of Epon embedded and frozen sections of human epidermis. *6th Eur. Cong.*,† 226–228.	TEAM	1, 8	Skin	P, S, Cl, K, Ca	A
Ghadially, F. N., Ailsby, R. L. & Yong, N. K. 1976 Ultrastructure of the haemophilic synovium membrane and electron probe X-ray analysis of haemosiderin. *Journal of Pathology*, **120**, 201–208.	STEM	1	Synovium	P, Fe	A
Ghadially, F. N., Lalonde, J. M. A. & Oryschak, A. F. 1976 Electron probe X-ray microanalysis of siderosomes in the rabbit haemathrotic synovial membrane. *Virchows Archiv B. Cell Pathology*, **22**, 135–142.	STEM	1	Synovium	P, Fe	A
Ghadially, F. N., Oryschak, A. F., Ailsby, R. L. & Mehta, P. N. 1974 Electron probe X-ray microanalysis of siderosomes in haemarthrotic articular cartilage. *Virchows Archiv B. Cell Pathology*, **16**, 43–49.	STEM	1	Cartilage	P, K, Fe	A
Ghadially, F. N., Oryschak, A. F. & Mitchell, D. M. 1976 Ultrastructural changes produced in rheumatoid synovial membrane by chrysotherapy. *Annals of Rheumatic Diseases*, **35**, 67–72.	STEM	1	Synovium	P, Ca, Au	A
Ghadially, F. N. & Yong, N. K. 1976 Ultrastructural and X-ray analytical studies on intranuclear bismuth inclusions. *Virchows Archiv B. Cell Pathology* **21**, 45–55.	STEM	1	Kidney	S, Bi	A
Gonzalez-Angulo, A. & Azner-Ramos, R. 1976 Ultrastructural studies on the endometrium of women wearing T-Cu 200 IUDs by means of transmission and scanning EM and X-ray dispersive analysis. *American Journal of Obstetetrics and Gynaecology*, **125**, 170–178.	TEAM	1	Endo-metrium	Cu	A
Griffiths, K., Henderson, W. J., Chandler, J. A. & Joslin, C. A. F. 1973 Ovarian cancer: some new analytical approaches. *Postgraduate Medical Journal*, **49**, 69–72.	EMMA	12	Ovary	Mg, Si	A
Grimaud, J. A., Czyba, J. C. & Guillot, N. 1977 Energy-dispersive X-ray spectrometry of human spermatozoa in electron microscopy. *Comptes Rendus des Séances de la Société de Biologie*, **170**, 1233–1226.	TEAM	1	Sperm	Zn	A

References/Bibliography	Instrument	Specimen preparation	Tissue/ organism	Elements	Species
*Gupta, B. L., Hall, T. A., Maddrell, S. H. P. & Moreton, R. B. 1976 Distribution of ions in a fluid-transporting epithelium determined by electron probe X-ray microanalysis. *Nature*, **264**, 284–287.					
Gustafson, R. A. & Thurston, E. L. 1974 Calcium deposition in the myxomycete *Didymium squamulosum*. *Mycologia*, **66**, 397–412.	STEM	1	Fungus	P, Ca	P
Hales, C. N., Luzio, J. P., Chandler, J. A. & Herman, L. 1974 Localization of calcium in the smooth endoplasmic reticulum of rat isolated fat cells. *Journal of Cell Science*, **15**, 1–15.	EMMA	2	Fat cells	Ca (Sb)	A
*Hall, T. A. 1971 The microprobe assay of chemical elements. *Physical Techniques*,† 148–275.					
*Hall, T. A., Anderson, H. C. & Appleton, T. C. 1973 The use of thin specimens for X-ray microanalysis in biology. *Journal of Microscopy*, **99**, 177–182.	EMMA	1	Cartilage	Ca	A
Hall, T. A., Hohling, H. J. & Bonnucci, E. 1971 Electron probe X-ray analysis of osmiophilic globules as possible sites of early mineralization in cartilage. *Nature*, **231**, 535–536.	EMMA	1	Cartilage	P, Ca	A
Hall, T. A., Peters, P. D. & Scripps, M. C. 1974 Recent microprobe studies with an EMMA 4 analytical microscope. *Microprobe Analysis*,† 385–408.	EMMA	1, 2, 4	Synovium, muscle, nerve, kidney	Mg, Al, Si, P, S, Cl, K, Ca, Ti, V, Fe, Br, Ag, Ba, Au	A
Harper, M. E., Danutra, V., Chandler, J. A. & Griffiths, K. 1976 The effects of 2-bromo-α-ergocryptine (CB154) administration on the hormone levels, organ weights, prostatic morphology, and zinc concentrations in the male rat. *Acta Endocrinologica*, **83**, 211–224.	EMMA	2	Prostate	Zn	A
*Harrison, J. D., Rowley, P. S. A. & Peters, P. D. 1977 Amalgam tattoos; light and electron microscopy and electron probe microanalysis. *Journal of Pathology*, **121**, 83–92.	EMMA	1	Mouth epithelium	S, Cl, Fe, Cu, Zn, Se, Ag, Hg	A
*Harry, J. & Triparthi, R. 1970 Kayser–Fleischer ring, a pathological study. *British Journal of Ophthalmology*, **54**, 794–800.	EMMA	1	Cornea	Cu	A
*Harvey, D. M. R., Hall, J. L. & Flowers, T. J. 1976 The use of freeze substitution in the preparation of plant					

References/Bibliography	Instrument	Specimen preparation	Tissue/organism	Elements	Species
tissue for ion localization studies. *Journal of Microscopy*, **107**, 189–198.					
Henderson, W. J., Blundell, G., Richards, R., Hext, P. M., Volcani, B. E. & Griffiths, K. 1975 Ingestion of talc particles by cultured lung fibroblasts. *Environmental Research*, **9**, 173–178.	EMMA	1	Lung	Mg, Si	A
Henderson, W. J., Chandler, J. A., Blundell, G. K., Griffiths, C. & Davies, J. 1973 The application of analytical electron microscopy to the study of diseased biological tissue. *Journal of Microscopy*, **99**, 183–193.	EMMA	1, 12	Ovary, stomach	Mg, Al, Si, K, Fe	A
*Henderson, W. J., Evans, D. M. D., Davies, J. D. & Griffiths, K. 1975 Analysis of particles in stomach tumours from Japanese males. *Environmental Research*, **9**, 240–249.	EMMA	12	Stomach	Mg, Al, Si, P, S, Cl, K, Ca, Ti, Fe	A
*Henderson, W. J. & Griffiths, K. 1975 Identification of talc particles in ovarian tissue. *Gyn. Malig.*,† 225–240.	EMMA	12, 13	Ovary, cervix, stomach	Mg, Al, Si, Fe	A
Henderson, W. J., Melville Jones, C., Barr, W. T. & Griffiths, K. 1975 Identification of talc on surgeons' gloves and in tissue from starch granulomas. *British Journal of Surgery*, **62**, 944.	EMMA	12	Stomach, peritoneum, appendix, gall bladder	Mg, Al, Si, S, Ca, Fe	A
Herman, L., Sato, T. & Hales, C. N. 1973 The electron microscopic localization of cations to pancreatic islets of Langerhans and their possible role in insulin secretion. *Journal of Ultrastructure Research*, **42**, 298–311.	EMMA	2	Pancreas	Na, Mg, K, Ca, Mn, (Sb)	A
Herman, L., Sato, T. & Weavers, B. A. 1971 An investigation of the pyroantimonate reaction for sodium localization using the analytical electron microscope EMMA 4. *29th EMSA*†	EMMA	2	Pancreas	Na, Mg, K, Ca, Mn, (Sb)	A
Heumann, H. G. 1976 The subcellular localization of calcium in vertebrate smooth muscle: calcium concentration and calcium accumulating structures in muscle cells. *Cell and Tissue Research*, **169**, 221–231.	TEAM	3	Muscle	Ca	A
Hillman, D. E. & Llinas, R. 1974 Calcium containing electron dense structures in the axons of the squid giant synapse. *Journal of Cell Biology*, **61**, 146–155.	TEAM	1	Nerve	Cl, P, Ca	A
Hodson, S. & Marshall, J. 1970 Tissue sodium and potassium: direct detection in the electron microscope. *Experientia*, **26**, 1283–1284.	EMMA	8	Cornea	Na, K	A

References/Bibliography	*Instrument*	*Specimen preparation*	*Tissue/ organism*	*Elements*	*Species*
Höhling, H. J., Schopfar, H., Hohling, R. A. & Hall, T. A. 1970 The organic matrix of developing tibia and femur and macromolecular deliberations. *Naturwissenschaften*, **57**, 357–358.	EMMA	7	Bone	P, Ca	A
Höhling, H. J., Hall, T. A., Boothroyde, B., Cooke, C. J., Duncumb, P., Fearnhead, R. W. & Fitton-Jackson, S. 1968 Electron probe studies of prestages of apatite formation in bone and aorta. *Tiss. Calc.*,† 323–328.	EMMA	1	Bone	P, S, Ca	A
Höhling, H. J., Hall, T. A., Boyde, A. & Von Rosensteil, A. P. 1968 Combined electron probe and electron diffraction analysis of prestages and early stages of dentine formation in rat incisors. *Calcified Tissue Research*, **2**, Supp. 5.	EMMA	1	Teeth	P, S, Ca	A
Höhling, H. J., Hall, T. A., Mutschelknauss, R. & Nien, J. 1969 Kombinierte Elektronenbeugungs und electron probe microanalysis untersuchungen an Fruhstadien der Zahnsteinbilding. *Zeitschrift fur Naturforschung*, **24**, 58–60.	EMMA	1	Teeth	P, Ca	A
Höhling, H. J. 1972 Quantitative electron microscopy and electron probe microanalysis for studying hard tissue formation. *4th H & C*,† 89–90.	TEPA	7	Teeth	P, Ca	A
*Höhling, H. J., Barckhaus, R. H., Krefting, E. T. & Schreiber, J. 1976 Electron microscope microprobe analysis of mineralized collagen fibrils and collagenous regions in turkey leg tendon. *Cell and Tissue Research*, **175**, 345–350.	TEAM	7	Tendon	P, Ca	A
*Höhling, H. J. & Nicholson, W. A. P. 1975 Electron probe analysis in hard tissue research. *Journal de Microscopie et Biologie Cellulaire*, **22**, 185–192.	TEPA	7	Teeth	P, S, Ca	A
*Höhling, H. T., Nicholson, W. A. P., Schreiber, J., Zessack, U. & Boyde, A. 1972 The distribution of some elements in predentine and dentine of rat incisors. *Naturwissenschaften*, **59**, 423–424.	TEPA	7	Teeth	P, S, Ca	A
Höhling, H. J., Steffens, H. & Heucke, F. 1972 Investigation of mineralization in hard tissues with protein-polysaccharides of collagen as main constituents of the matrix. *Cell and Tissue Research*, **134**, 283–296.	TEPA	1	Teeth, bone	P, Ca	A
Howell, S. L., Montague, W. & Tyhurst, M. 1976 Calcium distribution	EMMA	8	Pancreas	Ca	A

References/Bibliography	Instrument	Specimen preparation	Tissue/organism	Elements	Species
in islets of Langerhans: a study of calcium concentrations and calcium accumulation in B cell organelles. *Journal of Cell Science*, **19**, 395–409.					
Ishitani, R., Miyakawa, A. & Iwamoto, I. 1977 Elemental analysis of autoradiographic grains by X-ray microanalysis. *Experientia*, **33**, 440–442.	TEAM	1 (16)	Cerebral cortex	Si, Cl, Ag, Pb	A
Jensen, T. E. & Sicko Goad, L. 1977 The use of X-ray energy-dispersive analysis for the identification of polyphosphate bodies. *Journal of Cell Biology*, **63**, 155a.	STEM	1	*Plectonema borganum*	P, Ca	Pr
*Jessen, H., Peters, P. D. & Hall, T. A. 1974 Sulphur in different types of keratohyeline granules. A quantitative assay by X-ray microanalysis. *Journal of Cell Science*, **15**, 359–377.	EMMA	1	Esophagus	S	A
Joó, F. & Toth, I. 1975 X-ray micro-analysis of a peripheral nerve of the rat. *Balaton 75*,† BT 4.	TEAM	8	Nerve	Na, P, Cl, K	A
Kendell, M. D. 1975 EMMA 4 analysis of iron in cells of the thymic cortex of a bird (*Quelea quelea*). *Philosophical Transactions of the Royal Society, B*, **273**, 79–84.	EMMA	1	Thymus	Fe	A
Kianura, M., Seveus, L. & Maramorosch, K. 1975 Ferritin in insect vectors of the Maize Streak disease agent. Electron microscopy and electron microscope analysis. *Journal of Ultrastructure Research*, **53**, 366–377.	TEAM	1	*Cicadulina mbila*	Fe	A
*Kirkham, J. B., Goodman, L. J. & Chappel, R. L. 1975 Identification of cobalt in processes of stained neurones using energy spectra in the electron microscope. *Brain Research*, **85**, 33–37.	TEAM	1 (16)	Brain	(S, Co)	A
Knowles, J. C., Weaver, B. A. & Cooper, E. H. 1972 Accumulation of calcium in the mitochondrial dense bodies of mice. *Experimental Cell Research*, **73**, 230–233.	EMMA	1	Bladder wall	P, Ca	A
Krstic, R. & Golaz, J. 1977 Ultra-structural and X-ray microprobe comparison of gerbil and human pineal acervuli. *Experientia*, **33**, 507–508.	TEAM	1	Pineal bodies	P, Ca, Sr	A
Lacy, D. & Pettitt, A. J. 1972 Biological applications of combined transmission electron microscopy and X-ray microanalysis with special reference to studies of the mammalian testis. *Micron*, **3**, 115–129.	EMMA	1	Testis	Si, P, S	A

References/Bibliography	*Instrument*	*Specimen preparation*	*Tissue/ organism*	*Elements*	*Species*
*Lauchli, A., Stelzer, R., Guggenheim, R. & Henning, L. 1974 Precipitation techniques as a means for intracellular ion localization by use of electron probe analysis. *Microprobe Analysis,*† 107–118.	EMMA	1			
*Lechene, C. P. & Warner, R. R. 1977 Ultramicroanalysis: X-ray spectrometry by electron probe excitation. *Annual Review of Biophysics and Bioengineering* (in press).					
Lever, J. D., Santer, R. M., Lu, K. S. & Presley, R. 1977 Electron probe X-ray microanalysis of small granulated cells in rat sympathetic ganglia after sequential aldehyde and dichromate treatment. *Journal of Histochemistry and Cytochemistry,* **25**, 275–279.	TEAM	1 (16)	Ganglion	Cr	A
Lewis, P. R. 1972 The application of an analytical electron microscope to enzyme studies of brain tissue. *5th Eur. Cong.,*† 93	EMMA	1 (15)	Brain	(S, Cu)	A
*Lewis, P. R. 1973 The use of EMMA to measure cholinestrase activity. *Journal of Physiology,* **233**, 12–13p.	EMMA	1 (15)	Brain	(S, Cu)	A
Lim, D. J. 1972 Applications of the scanning electron microscope mounted energy-dispersive X-ray analyser to ear research. *EDAX,*† 63–77.	STEM	1, 8	Ear	P, S, Cl, K, Ca, Hg	A
Maatta, K. & Arstila, A. U. 1975 Pulmonary deposits of titanium dioxide in cytological and lung biopsy specimens. *Laboratory Investigation,* **33**, 342–346.	TEAM	1	Lung sputum	Si, Ti	A
Makita, T. 1975 X-ray microanalysis of mitochrondrial dense bodies in the ileal goblet cells of muscular dystrophic mice. *Mic. Canada,*† 106–107.	TEAM/ STEM	1	Ileum	S, Cl (Ru)	A
Makita, T., Cho, F. & Honjo, S. 1974 X-ray microanalysis of elements in the frozen–thawed spermatozoa of Cynomolgus Monkey. *5th Int. Prim.,** 134–140.	TEAM	1	Sperm	P, S, Cl, Ca	A
Makita, T. & Kiwaki, S. 1974 Detection of elements in the lumen and the follicle cells of the thyroid gland of the horse by X-ray microanalysis. *Archivum Histologicum Japonicum,* **37**, 143–148.	STEM	1	Thyroid	P, S, Cl, I	A
Makita, T., Marimoto, M. & Kiwaki, S. 1974 The formation and continuity of secretion granules in the splenic lobe of the pigeon pancreas as	STEM	1	Pancreas	Zn	A

References/Bibliography	Instrument	Specimen preparation	Tissue/ organism	Elements	Species
revealed by freeze etching, micro-X-ray analysis, and cytochemistry. *Histochemical Journal*, **6**, 185–198.					
Makita, T., Matsubara, K. & Kiwaki, S. 1975 Intramitochondrial dense bodies in the goblet cells of the ileum of muscular dystrophic mice. *Archivum Histologicum Japonicum*, **38**, 109–116.	TEAM/ STEM	1	Ileum	Mg, P, S, Cl, Fe, (Ru)	A
Makita, T. & Yamamoto, M. 1974 X-ray microanalysis of the components of the avian cardiac muscle. *8th Int. EM*,† 96–97.	STEM	1	Muscle	P, S, Cl	A
Martin, J. H., Carson, F. L. & Race, G. J. 1974 Calcium containing platelet granules. *Journal of Cell Biology*, **60**, 775–777.	EMMA	1	Blood	P, Ca	A
Maunder, C. A., Dubowitz, V., Hall, T. A. & Yarom, R. 1976 X-ray microanalysis of human muscle biopsies. *6th Eur. Congr.*,† 229–231.	EMMA	1, 2	Muscle	P, Cl, Ca, Zn	A
Maynard, P. V., Elstein, M. & Chandler, J. A. 1975 The effect of copper on the distribution of elements in human spermatozoa. *Journal of Reproduction and Fertility*, **43**, 41–48.	EMMA	11	Sperm	Na, P, S, Cl, K, Ca, Cu, Zn	A
*Mehard, C. W. & Volcani, B. E. 1974 Electron probe X-ray microanalysis of subcellular silicon in rat tissues and the diatom. *Journal of Cell Biology*, **63**, 220a.	TEAM	6	Diatom, liver, spleen, kidney	Si	A, P
*Mehard, C. W. & Volcani, B. E. 1975 Evaluation of Si and Ge retention in rat tissues and diatoms during cell and organelle preparation for electron probe microanalysis. *Journal of Histochemistry and Cytochemistry*, **23**, 348–358.	TEAM	1, 6	Diatom, liver	Si, Ge	A, P
*Mehard, C. W. & Volcani, B. E. 1976*a* Silicon in rat liver organelles, electron probe microanalysis. *Cell and Tissue Research*, **166**, 255–263.	TEAM	6	Liver	Al, Si, P, S, Cl	A
*Mehard, C. W. & Volcani, B. E. 1976*b* Silicon containing granules of rat liver, kidney and spleen mitochondria. *Cell and Tissue Research*, **174**, 315–328.	TEAM	6	Kidney, liver, spleen	Mg, Si, P, S, Cl	A
Mizuhira, V. 1973 Demonstration of the elemental distribution in biological tissues by means of the electron microscope and electron probe X-ray analyser. *Acta Histochemica Cytochemica*, **6**, 44–52.	STEM/ TEAM	2, 4	Stomach	P, Cl, K, Ca (Ag, Sb)	A
*Mizuhira, V. 1976*a* Elemental analysis of biological specimens by electron probe X-ray microanalysis. *Acta Histochemica Cytochemica*, **9**, 69–87.	TEAM	9	Lung	Na, Mg, Al, Si, P, S, Cl, Ca, Fe	A

References/Bibliography	Instrument	Specimen preparation	Tissue/organism	Elements	Species
*Mizuhira, V. 1976*b* A review of the various usages of energy-dispersive X-ray analysis in biology. *11th MAS,*† 5A–5D.	TEAM	8	Liver, kidney	Na, Mg, Si, P, S, Cl, K, Ca, Fe, Cd	A
Mizuhira, V., Fataesaku, Y., Mimura, M., Fujioka, S. & Yotsumoto, H. 1974 Subcellular distribution of heavy metals by means of an analytical electron microscope. *8th Int. EM,*† 84–85.	TEAM	1	Liver, kidney	P, S, Cl, Fe, (Cd, Hg, Pb)	A
*Mizuhira, V. & Kimura, M. 1973 Localization of cadmium in the mouse cells induced by acute cadmium poisoning. *31st EMSA,*† 402–403.	TEAM	5	Liver	P, S, Cl, Fe, Cd	A
Mizuhira, V., Nakamura, H., Yotsumoto, H. & Namae, T. 1972 An application of the electron probe X-ray microanalyser to the biological sections. II. Chloride and some other elemental distributions in the gastric mucosal epithelium. *4th H & C*† 275–276.	STEM	4	Stomach	Na, P, Cl, K, Ca, Zn (Ag)	A
Morgan, A. J. & Bellamy, D. 1973 Microanalysis of the elastic fibres of rat aorta. *Age and Ageing,* **2**, 61.	TEAM	8	Aorta	P, S, Cl, K, Ca	A
*Morgan, A. J., Davies, T. W. & Erasmus, D. A. 1975 Changes in the concentration and distribution of elements during electron microscope preparation procedures. *Micron,* **6**, 11–23.	TEAM	1, 8	Aorta, tendon	P, S, K, Ca	A
Murphy, M. J. & Piscopo, J. C. 1976 Cellular iron in aplastic anaemic human bone marrow: a study by energy-dispersive analysis of X-rays. *Journal of Submicroscopic Cytology,* **8**, 269–276.	TEAM	1	Bone	Fe	A
Namae, T., Harling, D. & Cogswell, G. 1972 Use of an EDAX system on an electron microscope fitted with a scanning attachment. *EDAX,*† 79–86.	TEAM	1	Liver, bacteria, leaf	Al, Cl, Ca, Fe	A, P, Pr
Neumann, D. & Janossy, A. G. S. 1977 Effect of gibberellic acid on the ion ratios in a dwarf maize mutant (*Zea mays*). *Planata,* **134**, 151–153.	TEAM	6	Leaf	Cl, K, Ca, Fe	P
Newell, P. F. & Machin, J. 1976 Water regulation in aestivating snails. Ultrastructural and analytical evidence for an unusual cellular phenomenon. *Cell and Tissue Research,* **173**, 417–421.	EMMA	8	Snail	Si, P, S, Cl, Ca	A
*Nicholson, W. A. P., Ashton, B. A., Hohling, H. J., Quint, P., Schreiber, J. Ashton, I. K. & Boyde, A. 1977. Electron microprobe investigations into the process of hard tissue formation. *Cell and Tissue Research,* **177**, 331–345.	EMMA/TEAM	1, 7	Teeth	P, S, Ca	A

References/Bibliography	*Instrument*	*Specimen preparation*	*Tissue/ organism*	*Elements*	*Species*
Nott, J. A. & Parkes, K. R. 1975 Calcium accumulation and secretion in the serpulid polychaete *Spirorbis spirorbis* L. at settlement. *Journal of the Marine Biological Association*, **55**, 911–923.	TEAM/ STEM	3	Tubeworm	Ca	A
Ohkura, T., Iwatsuki, H. & Uehira, K. 1974 Electron probe microanalysis of the dog synovial membrane stained with Alcian blue 8GS. *Journal of Electron Microscopy*, **23**, 320.	TEAM	1 (16)	Synovium	Cu	A
*Okagaki, T. & Clark, B. 1977 X-ray microprobe analysis of biological stains using ultrathin paraffin embedded sections. *SEM 77*,† 153–158.	STEM	1 (16)	Cervix	Al, Fe, Cu, Br, Ag	A
Ophus, E. M. & Gullvag, B. M. 1974 Localization of lead within leaf cells of *Rhytidiadelphus squarrosus* (Hedw.) Wermst by means of transmission electron microscopy and X-ray microanalysis. *Cytobios*, **10**, 45–58.	TEAM	1	Leaf	P, Fe, Ni, Cu, Zn, Cd, Pb	P
Oryschack, A. F. & Ghadially, F. N. 1974 Aurosomes in rabbit articular cartilage. *Virchows Archiv B. Cell Pathology*, **17**, 159–168.	STEM	1	Cartilage	P, Au	A
Oryschack, A. F. & Ghadially, F. N. 1976 Evolution of aurosomes in rabbit synovial membrane. *Virchows Archiv B. Cell Pathology*, **20**, 29–39.	STEM	1	Synovium	P, Ca, Au	A
Oschman, J. L., Hall, T. A., Peters, P. D. & Wall, B. D. 1974 Association of calcium with membranes of squid giant axon ultrastructure and microprobe analysis. *Journal of Cell Biology*, **61**, 156–165.	EMMA	1	Nerve	P, Ca	A
*Pallaghy, C. K. 1973 Electron probe microanalysis of potassium and chloride in freeze-substituted leaf sections of *Zea mays*. *Australian Journal of Biological Sciences*, **26**, 1015–1034.	STEM	6	Leaf	Cl, K	P
*Parducz, A. & Joó, F. 1976 Visualization of stimulated nerve endings by preferential calcium accumulation of mitochondria. *Journal of Cell Biology*, **69**, 513–517.	TEAM	1	Nerve cord	Ca	A
Pariente, R., Barry, J. P., Cayrol, E. & Bronet, G. 1972 A study of pulmonary dust deposits using the electron microscope in conjunction with the electron probe microanalyser. *Thorax*, **27**, 80–82.	TEPA	1	Lung	Mg, Al, Si, Fe	A
Peters, P. D. 1975 Biological microanalysis using EMMA 4. *Journal*	EMMA	1, 4, 11	Blood, muscle,	P, Cl, Ca, Ag, Sn, Hg	A, P

References/Bibliography	Instrument	Specimen preparation	Tissue/organism	Elements	Species
de Microscopie et Biologie Cellulaire, **22**, 449–452.			leaf, seed		
Peters, P. D., Yarom, R., Dorman, A. & Hall, T. A. 1976 X-ray microanalysis of intracellular Zn: EMMA 4 examination of normal and injured muscle and myocardium. *Journal of Ultrastructure Research*, **57**, 121–131.	EMMA	1	Muscle	Zn	A
Plattner, H. 1976 Electron microscope energy-dispersive X-ray microanalysis of ciliary granule plaques. *6th Eur. Congr.*,† 221–222.	TEAM	3	*Paramecium*	P, S, Ca	Pr
*Plattner, H. & Fuchs, S. 1975 X-ray microanalysis of calcium binding sites in *Paramecium* with special reference to exocytosis. *Histochemistry*, **45**, 23–47.	TEAM	3	*Paramecium*	P, S, Ca	Pr
Plattner, H., Reichel, K. & Matt, H. 1977 Bivalent cation stimulated ATPase activity at preformed exocytosis sites in paramecium cells. Coincidence with membrane-intercalculated particle aggregates. *Nature* (in press).	TEAM	1 (16)	*Paramecium*	P, Pb	Pr
Podolsky, P. J., Hall, T. A. & Hatchett, S. L. 1970 Identification of oxalate precipitates in striated muscle fibres. *Journal of Cell Biology*, **44**, 699–702.	EMMA	3	Muscle	Ca	A
Popescu, L. M. & Diculescu, I. 1975 Calcium in smooth muscle sarcoplastic reticulum *in situ*. Conventional and X-ray analytical electron microscopy. *Journal of Cell Biology*, **67**, 911–915.	TEAM	3	Muscle	Ca	A
Remagen, W., Höhling, H. J., Hall, T. A. & Caesar, R. 1969 Electron microscope and microprobe observations on the cell sheath of stimulated osteocytes. *Calcified Tissue Research*, **4**, 60–65.	EMMA	1	Bone	P, Ca	A
Roomans, G. M. 1975 Calcium binding to the acrosomal membrane of human spermatozoa. *Experimental Cell Research*, **96**, 23–32.	TEAM	1	Sperm	P, S, Ca	A
Roomans, G. M. & Van Gaal, H. L. M. 1977 Organometallic and organometalloid compounds as standards for microprobe analysis of epoxy resin embedded tissue. *J. Microscopy*, **109**, 235–240.					
Roomans, G. M. & Seveus, L. A. 1976 Subcellular localization of diffusible ions in the yeast *Saccharomyces cerevisiae*: quantitative microprobe analysis of thin freeze-dried sections. *Journal of Cell Science*, **21**, 119–127.	TEAM	1	Yeast	Cl, K, Rb, Cs	P

References/Bibliography	Instrument	Specimen preparation	Tissue/ organism	Elements	Species
Roseda, A., Hacuja, L., Delgado, N. M., Merchant, H. & Pancardo, R. M. 1977 Elemental composition of subcellular structures of human spermatozoa. A study by energy-dispersive analysis by X-ray. *Life Sciences*, **20**, 647–656.	STEM	1	Sperm	Na, Mg, Si, P, S, Cl, Ca, Fe, Zn	A
*Routledge, L. M., Amos, W. B., Gupta, B. L., Hall, T. A. & Weiss Fogh, T. 1975 Microprobe measurements of calcium binding in the contractile spasmoneme of a corticellid. *Journal of Cell Science*, **19**, 195–201.	STEM	11	Spasmo-nemes	Ca	Pr
Ryder, T. A. & Bowen, I. D. 1974 The use of X-ray microanalysis to investigate problems encountered in enzyme cytochemistry. *Journal of Microscopy*, **101**, 143–151.	TEAM	1, 8, (15)	Planarian	P, Cl, Pb	A
Saetersdal, T. S., Myklebust, R., Jutsea, N., Berg, P. & Olsen, W. C. 1974 Ultrastructural localization of calcium in the pigeon papillary muscle as demonstrated by cytochemical studies and X-ray microanalysis. *Cell and Tissue Research*, **155**, 57–74.	STEM	2	Muscle	Ca	A
*Salsbury, A. J. & Clarke, J. A. 1972 Electron probe microanalysis in relation to immunology. *Micron*, **3**, 135–137.	EMMA	10	Blood	I	A
Sato, T., Herman, L., Chandler, J. A., Stracher, A. & Detweiler, T. C. 1975 Localization of a thrombin sensitive calcium pool in platelets. *Journal of Histochemistry and Cytochemistry*, **23**, 103–106.	EMMA	2	Blood	Ca	A
Schafer, P. W. & Chandler, J. A. 1970 Electron probe X-ray microanalysis of a normal centriole. *Science*, **170**, 1204–1205.	EMMA	1	Kidney	Si, P, S, Cl	A
Schenermann, D. W. & De Groot-Lasseel, M. H. A. 1976 Electron microscope cytochemical and X-ray microanalytical study on phagocytosis of alveolar macrophages. *6th Eur. Cong.*,† 215–217.	TEAM	1, 15	Lung	S, (Ba), Au, (Pb)	A
Scheuer, P. J., Thorpe, M. E. E. & Marriot, P. 1967 A method for the demonstration of copper under the electron microscope. *Journal of Histochemistry and Cytochemistry*, **15**, 300–301.	EMMA	1, (16)	Liver	(Ag)	A
Schippert, M. A. & Moll, S. H. 1967 Analytical applications of a combined	TEAM	1	Rotifer	Ca	A

References/Bibliography	Instrument	Specimen preparation	Tissue/organism	Elements	Species
electron microscope electron microanalyser. *Analytical Chemistry*, **39**, 867–876.					
Schroter, K., Lauchli, A. & Sievers, A. 1975 Mikroanalytische identifikation von Barimen sulfat-bristallen in den Statolilthen der Rhizoide von *Clara fragilis* Derv. *Planta*, **122**, 213–215.	TEPA	1	Chara-rhizoid	P, S, Cl, K, Ca, Ba	P
*Shuman, H. & Somlyo, A. P. 1976 Electron probe X-ray analysis of single ferritin molecules. *Proceedings of the National Academy of Sciences*, **73**, 1193–1195.	TEAM	11	Ferritin	Fe	A
*Shuman, H., Somlyo, A. V. & Somlyo, A. P. 1976 Quantitative electron probe microanalysis of biological thin sections, method and validity. *Ultramicroscopy*, **1**, 317–339.	TEAM	8	Blood, muscle	Cl, K	A
Siegesmund, K. A., Funaheshi, A. & Pintar, K. 1974 Identification of metals in lung from a patient with interstitial pneumonia. *Archives of Environmental Health*, **28**, 345–349.	STEM	1	Lung	Ni, Cr, Fe, Co	A
Silverberg, B. A. 1975 Ultrastructural localization of lead in *Stigeoclonium-tenue* (Chlorophyeae Ulotrichales) as demonstrated by cytochemistry and X-ray analysis. *Phycologia*, **14**, 265–274.	TEAM	1	Algae	Pb	Pr
Silverberg, B. A. 1976*a* Localization of selenium in bacterial cells using transmission electron microscopy and energy-dispersive X-ray analysis. *Archives of Microbiology*, **107**, 1–6.	TEAM	1	Bacteria	Se	Pr
Silverberg, B. A. 1976*b* Cadmium induced ultrastructural changes in mitochondria of freshwater green algae. *Phycologia*, **15**, 155–159.	TEAM	1	Algae	Cd	Pr
Silverberg, B. A., Stokes, P. M. & Ferstenberg, L. B. 1976 Intranuclear complexes in a copper tolerant green alga. *Journal of Cell Biology*, **69**, 210–214.	TEAM	1	Algae	Cu	Pr
*Sjostrom, M. 1975 X-ray micro-analysis in cell biology. Detection of intracellular elements in the normal muscle cell and in muscle pathology. *Journal de Microscopie*, **22**, 415–424.	TEAM	1, 8, 9	Muscle	Na, Mg, Si, P, S, Cl, K, Ca	A
Sjostrom, M., Johansson, R. & Thornell, L. E. 1973 Cryo-ultramicrotomy of muscles in defined functional states. *EM & Cytochem*,† 387–391.	TEAM	8	Muscle	Mg, Si, P, S, Cl, K, Ca	A

References/Bibliography	Instrument	Specimen preparation	Tissue/ organism	Elements	Species
*Sjostrom, M. & Thornell, L. E. 1975 Preparing sections of skeletal muscle for transmission electron analytical microscopy (TEAM) of diffusible elements. *Journal of Microscopy*, **103**, 101–112.	TEAM	1, 8	Muscle	Na, Mg, Si, P, S, Cl, K, Ca	A
Skaer, H., Ophus, E. & Gullvag, B. M. 1973 Lead accumulation within nuclei of moss leaf cells. *Nature*, **241**, 215–216.	EMMA	1	Leaf	Pb	P
Skaer, R. J. & Peters, P. D. 1975 The state of chlorine and potassium in human platelets and red cells. *Nature*, **257**, 719–720.	EMMA	11, 14	Blood	P, S, Cl, K, Ca	A
*Skaer, R. J., Peters, P. D. & Emmines, J. P. 1974 The localization of calcium and phosphorus in human platelets. *Journal of Cell Science*, **15**, 679–692.	EMMA	7	Leaf	Pb, Ca	A
*Skaer, R. J., Peters, P. D. & Emmines, J. P. 1976 Platelet dense bodies: a quantitative microprobe analysis. *Journal of Cell Science*, **20**, 441–457.	EMMA	11	Blood	P, Cl, Ca	A
Sohal, R. S., Peters, P. D. & Hall, T. A. 1976 Fine structure and X-ray microanalysis of mineralized concretions in the malphigian tubules of the housefly *Musca domestica. Tissue and Cell*, **8**, 447–458.	EMMA	1	Malpighian	P, S, Cl, K, Ca, Fe, Cu, Zn	A
Sohal, R. S., Peters, P. D. & Hall, T. A. 1977 Origin, structure, composition and age dependence of mineralized dense bodies (concretions) in the midgut epithelium of the adult housefly *Musca domestica. Tissue and Cell*, **9**, 87–102.	EMMA	1	Malpighian	P, S, Cl, Ca, Fe, Cu, Zn	A
Somlyo, A. V., Shuman, H. & Somlyo, A. P. 1976*a* Electron probe analysis of vertebrate smooth muscle: distribution of calcium and chlorine. *34th EMSA*,† 334–335.	TEAM	8	Muscle	Na, K, Cl, Ca	A
Somlyo, A. V., Shuman, H. & Somlyo, A. P. 1976*b* Chloride compartmentalization in striated muscle quantitative electron probe analysis. *Journal of Cell Biology*, **70**, 336a.	TEAM	8	Muscle	Na, Mg, P, Cl, K	A
Somlyo, A. V., Shuman, H. & Somlyo, A. P. 1977*a* Elemental distribution in striated muscle and the effects of hypertonicity: electron probe analysis of cryosections. *Journal of Cell Biology*, **74**, 828–857	TEAM	8	Muscle	Mg, P, S, Cl, K, Ca	A

References/Bibliography	Instrument	Specimen preparation	Tissue/organism	Elements	Species
Somlyo, A. V., Shuman, H. & Somlyo, A. P. 1977*b* Composition of the sarcoplasmic reticulum in *situ* by electron probe X-ray microanalysis. *Nature*, **268**, 556–558.	TEAM	8	Muscle	Na, Mg, P, S, Cl, K, Ca	A
Somlyo, A. V., Silcox, J. & Somlyo, A. P. 1975 Electron probe analysis and cryo-ultramicrotomy of cardiac muscle: mitochondrial granules. *33rd EMSA*,† 523–533.	TEAM	8	Muscle	P, Ca	A
Somlyo, A. P., Somlyo, A. V., Ashton, F. T. & Vallieres, J. 1976 Vertebrate smooth muscle: ultrastructure and function. *Cell Mot.*,† 165–183.	TEAM	8	Muscle	K, Ca	A
Somlyo, A. P., Somlyo, A. V., Devine, C. E., Peters, P. D. & Hall, T. A. 1974 Electron microscopy and electron probe analysis of mitochondria cation accumulations in smooth muscle. *Journal of Cell Biology*, **61**, 723–742.	EMMA	1	Muscle	Ca, Sr, Ba	A
Spencer, P. S. 1973 Effects of thallium salts on neuronal mitochondria in organotypic cord–ganglia–muscle combination cultures. *Journal of Cell Biology*, **58**, 79–95.	EMMA/TEAM	1	Spinal cord	Ti	A
Sprey, B., Gliem, G. & Janossy, A. G. S. 1976 Iron containing inclusions in chloroplasts of *Nicotiana clevelandii* × *Nicotiana glutinosa*. I. X-ray microanalysis and ultrastructure. *Zeitschrift fur Pflanzenphysiol*, **79**, 165–176.	TEAM	1	Leaf	P, Fe	P
Sprey, B., Gliem, G. & Janossy, A. G. S. 1977 Changes in the iron and phosphorus content of stroma inclusions during etioplast-chloroplast development in *Nicotiana*. *Zeitschrift fur Naturforschung*, **32c**, 138–139.	TEAM	1	Leaf	P, Fe	P
*Spurr, A. R. 1969 A low viscosity epoxy resin embedding medium for electron microscopy. *Journal of Ultrastructure Research*, **26**, 31–43.					
*Spurr, A. R. 1972*a* Freeze substitution additives for sodium and calcium retention in cells studied by X-ray analytical electron microscopy. *Botanical Gazette*, **133**, 263–270.	EMMA	6	Seed	Na, Ca	P
*Spurr, A. R. 1972*b* Freeze substitution systems in the retention of elements in tissues studied by X-ray analytical electron microscopy. *EPAX*,† 49–61.	EMMA	6	Root, leaf, seed	Na, Ca	P
*Spurr, A. R. 1974 Macrocyclic polyether complexes with alkali elements					

References/Bibliography	Instrument	Specimen preparation	Tissue/ organism	Elements	Species
in epoxy resin as standards for X-ray analysis of biological tissues. *Microprobe Analysis,*† 213–228.					
Stein, H., Yarom, T., Robin, G. C., Peters, P. D., Hall, T. A. & Makin, M. 1976 Spread of gold injected into the joints of healthy rabbits. *Journal of Bone and Joint Surgery*, **58**, 496–503.	EMMA	1	Synovium, liver, spleen, kidney, lymph	Au	A
Stoeckel, M. E., Hinderlang-Gertner, C., Dellman, H. D., Porte, A. & Stutinsky, F. 1975 Subcellular localization of calcium in the mouse hypophysis. *Cell and Tissue Research*, **157**, 307–322.	STEM	2	Pituitary	Ca	A
Stuve, J. & Galle, P. 1970 Role of mitochondria in the handling of gold by the kidney. *Journal of Cell Biology*, **44**, 667–676.	TEPA	1	Kidney	Au	A
Suftin, L. V., Holtrop, M. E. & Ogilvie, R. E. 1971 Microanalysis of individual mitochondrial granules with diameter less than 1000 Å. *Science*, **174**, 947–949.	STEM	1	Cartilage	P, Ca	A
Suzuki, Y., Aita, S., Hoshina, T. & Iwata, H. 1974 Identification of submicroscopic asbestos fibrils in tissue by analytical electron microscopy. *JEOL News*, **12e**, 2–4.	TEAM	1	Lung	Mg, Si, Fe	A
*Takaya, K. 1975*a* Electron probe microanalysis of the dense bodies of human blood platelets. *Archivum Histologicum Japonicum*, **37**, 335–341.	TEAM	11	Blood	P, Cl, Ca	A
*Takaya, K. 1975*b* Energy-dispersive X-ray microanalysis of neurosecretory granules of mouse pituitary using fresh air dried tissue spreads. *Cell and Tissue Research*, **159**, 227–232.	TEAM	11	Pituitary	P, S, Cl, K, Ca	A
*Takaya, K. 1975*c* Energy-dispersive X-ray microanalysis of zymogen granules of mouse pancreas using fresh air dried tissue spreads. *Archivum Histologicum Japonicum*, **37**, 387–393.	TEAM	11	Pancreas	P, S, Cl, K	A
*Takaya, K. 1975*d* Intranuclear silicon detection in a subcutaneous connective tissue cell by energy-dispersive X-ray microanalysis using fresh air dried spread. *Journal of Histochemistry and Cytochemistry*, **23**, 681–685.	TEAM	11	Skin	Si, P, S, Cl, K, Ca	A
Takaya, K. 1975*e* Electron microscopy of unstained fresh air dried tissue spreads – application to electron probe microanalysis. *10th Anat.*† 518.	TEAM	11	Blood	P, S, Cl, Ca	A

References/Bibliography	*Instrument*	*Specimen preparation*	*Tissue/ organism*	*Elements*	*Species*
Takaya, K. 1976 Electron microscopy of unstained fresh air dried spreads of mouse pancreas acinar cells and energy-dispersive X-ray microanalysis of zymogen granules. *Cell and Tissue Research*, **166**, 117–124.	TEAM	11	Pancreas	Mg, P, S, Cl, K, Ca	A
Tapp, R. L. 1975 X-ray microanalysis of the mid-gut epithelium of the fruit-fly *Drosophila melanogaster. Journal of Cell Science*, **17**, 449–455.	EMMA	1	Gut	S, Cu	A
Takaya, K. 1977*a* Electron microscopy of human melanosomes in unstained, fresh air-dried hair bulbs and their examination by electron probe microanalysis. *Cell and Tissue Research*, **178**, 169–173.	TEAM	11	Hair	Mg, P, S, Cl, K, Ca	A
Takaya, K. 1977*b* Dense particles in subcutaneous collagen fibres. *Experientia*, **32**, 162–163.	TEAM	11	Skin	Si, P, S, Cl, K, Ca	A
Takuma, S., Katagiri, S. & Ozasa, S. 1966 Electron probe microanalysis of horse dentin. *Journal of Electron Microscopy*, **15**, 86–89.	TEAM	1	Teeth	P, Ca	A
Thoroughgood, P. V. & Craig-Gray, J. 1975 Demineralization of bone matrix: observations from electron microscopy and electron probe analysis. *Calcified Tissue Research*, **19**, 17–26.	EMMA	1	Bone	P, Ca	A
Thurston, E. L. 1974 Morphology, fine structure and ontogeny of the stinging emergence of *Notica dioica. American Journal of Botany*, **61**, 809–817.	STEM	1	Leaf	Si	P
*Thurston, E. L. & Russ, J. C. 1971 Scanning and transmission electron microscopy and microanalysis of structured granules in *Fischerella ambigua. SEM 71*,† 511–516.	STEM	1	Algae	Na, Mg, Si, S, Cl, K	Pr
Thurston, E. L. & Gustafson, R. A 1972 X-ray microanalysis of mitochondrial inclusions in the myxomycete *Didymium squamulosum. EDAX*,† 33–39.	STEM	1	Slime mold	P, Ca	P
Thurston, E. L., Russ, J. C., Teigler, D. J. & Arnott, H. J. 1972 Microanalysis of the malpighian tubule in the silkworm *Bombyx mari. 30th EMSA*,† 376–377.	STEM	1	Silkworm	Na, P, S, Ca	A
*Timms, B. G., Chandler, J. A., Morton, M. S. & Groom, G. V. 1977 The effect of cadmium administration *in vivo* on plasma testosterone and the ultrastructure of rat lateral prostate. *Virchows Archiv B. Cell Pathology* **25**, 33–52.	EMMA	2	Prostate	Zn, Cd	A

References/Bibliography	*Instrument*	*Specimen preparation*	*Tissue/ organism*	*Elements*	*Species*
*Tisher, C. G., Weavers, B. A. & Cirksena, W. J. 1972 X-ray microanalysis of pyroantimonate complexes in rat kidney. *American Journal of Pathology*, **69**, 255–270.	EMMA	2	Kidney	Na, Mg, K, Ca, (Sb)	A
*Trump, B. F., Berezesky, I. K., Chang, S. H. & Bulger, R. E. 1976 Detection of ion shifts in proximal tubule cells of the rat kidney using X-ray microanalysis. *Virchows Archiv B. Cell Pathology*, **22**, 111–120.	STEM	8	Kidney	Na, K, Cl	A
Van Staden, J. & Comins, N. R. 1976 Energy-dispersive X-ray analysis of protein bodies in *Protea compacta* cotyledons. *Planta*, **130**, 219–222.	STEM	1	Seed	Mg, P, S, K, Ca	P
Van Steveninck, M. E., Van Steveninck, R. F. M., Peters, P. D. & Hall, T. A. 1976 X-ray microanalysis of antimony precipitates in barley roots. *Protoplasma*, **90**, 47–63.	EMMA	2	Root	P, Fe	P
Van Steveninck, R. F. M. 1976 Cytochemical evidence for ion transport through plasmodesmata. *Inter. Comm. Plants.*†	EMMA	4	Leaf, root	Cl (Ag)	P
Van Steveninck, R. F. M., Armstrong, W. D., Peters, P. D. & Hall, T. A. 1976 Ultrastructural localization of ions. III. Distribution of chloride in mesophyll cells of mangrove (*Aegiceras corniculatum* Blanco). *Australian Journal of Plant Physiology*, **3**, 367–376.	EMMA	4	Leaf	Cl (Ag)	P
Van Steveninck, R. F. M., Ballment, B., Peters, P. D. & Hall, T. A. 1976 Ultrastructural localization of ions. II. X-ray analytical verification of silver precipitation products and distribution of chloride in mesophyll cells of barley seedlings. *Australian Journal of Plant Physiology*, **3**, 359–365.	EMMA	4	Leaf	Cl (Ag)	P
Van Steveninck, R. F. M., Van Steveninck, M. E. & Chescoe, D. 1976 Intracellular binding of lanthanum in root tips of barley *Hordeum vulgare*. *Protoplasma*, **90**, 89–97.	TEAM	1 (16)	Root	La	P
Van Steveninck, R. F. M., Van Steveninck, M. E., Hall, T. A. & Peters, P. D. 1974*a* A chlorine-free embedding medium for use in X-ray analytical microscope localization of chlorine in biological tissues. *Histochemistry*, **38**, 173–180.	EMMA	4	Algae	Cl (Ag)	P

References/Bibliography	Instrument	Specimen preparation	Tissue/organism	Elements	Species
Van Steveninck, R. F. M., Van Steveninck, M. E., Hall, T. A. & Peters, P. D. 1974*b* X-ray microanalysis and distribution of halides in *Nitella translucens*. *8th Int. EM*,† 602–603.	EMMA	4	Algae	Cl (Ag)	P
Van Steveninck, R. F. M., Van Steveninck, M. E., Peters, P. D. & Hall, T. A. 1976 Ultrastructural localization of ions. IV. Localization of chloride and bromide in *Nitella translucens* and X-ray energy spectroscopy of silver precipitation products. *Journal of Experimental Botany*, **27**, 1291–1312.	EMMA	4	Algae	Cl, Br, (Ag)	P
*Van Zyl, J., Forrest, Q. G., Hocking, C. & Pallaghy, C. K. 1976 Freeze substitution of plant and animal tissue for the localization of water-soluble compounds by electron probe microanalysis. *Micron*, **7**, 213–224.	TEAM	6	Nerve	Na, P, S, Cl, K	A
Vernon-Roberts, B, Dore, J. L., Jessop, J. D. & Henderson, W. J. 1976 Selective concentration and localization of gold in macrophages of synovial and other tissues during and after chrysotherapy in rheumatoid patients. *Annals of Rheumatic Diseases*, **35**, 477–486.	EMMA	1	Synovium	Au	A
Walker, G. 1977 Copper granules in the barnacle *Balanus balanoides*. *Marine Biology*, **39**, 343–349.	STEM	1	Barnacle	S, K, Ca, Cu	A
Walker, G., Rainbow, P. S., Foster, P. & Crisp, D. J. 1975 Barnacles: possible indicators of zinc pollution? *Marine Biology*, **30**, 57–65.	STEM/TEAM	1	Barnacle	Mn, Fe, Zn	A
Walker, G., Rainbow, P. S., Foster P. & Holland, D. 1975 Zinc phosphate granules in tissue surrounding the midgut of the barnacle *Balanus balanoides*. *Marine Biology*, **33**, 161–166.	TEAM	1	Barnacle	Mg, P, K, Ca, Fe, Zn	A
Watanabe, I., Whittier, F. C., Moore, J. & Cuppage, F. E. 1976 Gold nephropathy–ultrastructure, fluorescence, and microanalytical studies of two patients. *Archives of Pathology and Laboratory Medicine*, **100**, 632–635.	TEAM	1	Kidney	Au	A
*Weavers, B. A. 1972 Ultrathin frozen sections for EMMA 4 analysis. *Micron*, **3**, 107–113.	EMMA	9	Gastric mucosa	Mg, Si, P, Ca, Fe, Zn	A
Weavers, B. 1973 Combined transmission electron microscopy and X-ray microanalysis of ultrathin frozen sections – an investigation to determine the normal elemental composition of mammalian tissue. *Journal of Microscopy*, **97**, 331–341.	EMMA	9	Liver, kidney, heart, stomach, duodenum	Mg, P, S, Ca, Fe, Zn	A

References/Bibliography	*Instrument*	*Specimen preparation*	*Tissue/ organism*	*Elements*	*Species*
*Wood, J. G. 1974 Positive identification of intracellular biogenic amine reaction products with electron microscope X-ray analysis. *Journal of Histochemistry and Cytochemistry*, **22**, 1060–1063.	TEAM	1, (16)	Brain	Cr	A
*Wood, J. G. 1975 Use of the analytical electron microscope (AEM) in cytochemical studies of the central nervous system. *Histochemistry*, **41**, 233–240.	TEAM	1, (16)	Nerve	Cr, Fe	A
*Wood, J. G. & Harling, D. 1974 Analytical electron microscopy of specific cytochemical reaction products. *JEOL*, **12e**, 11–15.	TEAM	1, (16)	Adrenal, nerve	Mo, Cr	A
*Yarom, R. 1977 X-ray microanalysis in muscle research. *Israel Journal of Medical Science*, **13**, 121–125.	EMMA	1	Muscle	P, S, Ca, Zn	A
Yarom, R., Behar, A., Yanko, L., Hall, T. A. & Peters, P. D. 1976. Gold tracer studies of muscle regeneration. *Journal of Neuropathology and Experimental Neurology*, **35**, 445–457.	EMMA	1	Muscle	Au	A
*Yarom, R. & Chandler, J. A. 1974 Electron probe microanalysis of skeletal muscle. *Journal of Histochemistry and Cytochemistry*, **22**, 149–154.	EMMA	2	Muscle	Na, Mg, K, Ca	A
Yarom, R., Hall, T. A. & Peters, P. D. 1975 Calcium in myonuclei: electron microprobe X-ray analysis. *Experientia*, **31**, 154–157.	EMMA	1, 2	Muscle	Ca (Sb)	A
Yarom, R., Hall, T. A. & Polliack, A. 1976 Electron microscopic X-ray microanalysis of normal and leukemic human lymphocytes. *Proceedings of the National Academy of Sciences*, **73**, 3690–3694.	EMMA	1	Blood	P, S, Cl, Ca, Cu, Zn	A
*Yarom, R., Hall, T. A., Stein, H., Robin, G. C. & Makin, M. 1973 Identification of localization of intra-articular colloidal gold: ultrastructure and electron microprobe examination of human biopsies. *Virchows Archiv B. Cell Pathology*, **15**, 11–22.	EMMA	1	Synovium	P, S, Ca, Fe, Au	A
Yarom, R., Maunder, C. A., Hall, T. A. & Dubowitz, V. 1975 X-ray microanalysis of mast cells in rat's muscle. *Experientia*, **31**, 1339–1340.	EMMA	1	Muscle	P, S, Ca, Zn	A
*Yarom, R., Maunder, C. A., Scripps, M., Hall, T. A. & Dubowitz, V. 1975 A simplified method of specimen preparation of X-ray microanalysis of	EMMA	17	Muscle, blood	Ca, Cu, Zn	A

References/Bibliography	*Instrument*	*Specimen preparation*	*Tissue/ organism*	*Elements*	*Species*
muscle and blood cells. *Histochemistry*, **45**, 49–59.					
Yarom, R., Peters, P. D. & Hall, T. A. 1974 Effect of glutaraldehyde and urea embedding on intracellular ion elements. *Journal of Ultrastructure Research*, **49**, 405–418.	EMMA	1, 2, 3, 4	Muscle	Cl, Ca	A
Yarom, R., Peters, P. D. & Hall, T. A. 1976 Preparation of tissues and cells for X-ray microanalysis. *6th Eur. Cong.*,† 232–234.	EMMA	1, 4	Muscle, blood	S, Cl, K, Ca, Zn, Au	A
Yarom, R., Peters, P. D., Hall, T. A., Kadem, J. & Rogel, S. 1974 Studies with EMMA-4 on changes in the intracellular concentration and distribution of calcium in heart muscles of the dog in different steady states. *Micron*, **5**, 11–20.	EMMA	2	Muscle	Ca	A
*Yarom, R., Peters, P. D., Scripps, M. & Rogel, S. 1974 Effects of specimen preparation on intracellular myocardial calcium. *Histochemistry*, **38**, 143–153.	EMMA	1, 2	Muscle	Cl, K, Ca (Sb)	A
Yarom, R., Stein, H., Dormann, A., Peters, P. D. & Hall, T. A. 1976 Aurothiomalate as an ultrastructural marker: electron microscopy and X-ray microanalysis of various tissues after *in vivo* gold injections. *Journal of Histochemistry and Cytochemistry*, **24**, 453–462.	EMMA	1	Kidney, liver, spleen, heart	Au	A
Yarom, R., Stein, H. & Peters, P. D. 1975 Nephrotoxic effect of parenteral and intra-articular gold ultrastructure and electron microprobe examination of clinical and experimental material. *Archives of Pathology*, **99**, 36–42.	EMMA	1	Kidney	Au	A
Yarom, R., Wisenberg, E., Peters, P. D. & Hall, T. A. 1977 Zinc distribution in injured myocardium. EMMA-4 examination of dog's heart after coronary ligation. *Virchows Archiv B. Cell Pathology*, **23**, 63–78.	EMMA	1, 2	Muscle	Zn	A

FOOTNOTES TO BIBLIOGRAPHY

(A) *Specimen preparation procedures*

1 Conventional fixation and embedding, generally using glutaraldehyde, or osmium tetroxide, or both, followed by alcohol dehydration and resin embedding.

2 Pyroantimonate. Generally fixed in osmium tetroxide (sometimes glutaraldehyde) containing 2% potassium pyroantimonate, followed by normal resin embedding.

3 Oxalate. Generally using sodium or potassium oxalate in glutaraldehyde fixative. Normal embedding procedure.

4 Silver acetate/lactate. Silver salt in osmium tetroxide fixative to produce insoluble silver chloride. Normal embedding procedure.
5 Oxine. Conventional fixative containing 8-hydroxyquinoline chelator. Normal embedding procedure.
6 Freeze substitution. Various techniques. Tissue is embedded and sectioned.
7 Freeze dry and embed in resin. Cut ultra-thin sections.
8 Cryo-ultramicrotomy. Freeze tissue, cut frozen sections on a dry knife and freeze dry for analysis.
9 Frozen fixed sections. Tissue is first fixed, then frozen, sectioned, freeze dried, and analysed without embedding.
10 Critical point drying. Unembedded cell suspensions.
11 Air dry. Unembedded tissue or cells are air dried on grids for analysis.
12 Replication. Unfixed tissue or histology sections are replicated to extract particles for analysis.
13 Ashing. Tissue is incinerated at high temperature for analysis of the residue.
14 Freeze dry. Cell suspensions or tissue is freeze dried and analysed without embedding.
15 Enzyme reaction, for frozen sections or fixed and embedded tissue.
16 Staining reactions. Introduction of heavy metals into sections or tissue for subsequent analysis. Frozen sections or fixed and embedded.
17 Conventional fixation with 50% glutaraldehyde. Immediately embedded in resin for sectioning.

(B) *Book references (marked in Bibliography with a dagger †)*

29th EMSA: *29th Annual Proceedings of the Electron Microscope Society of America*, 1971.
30th EMSA: *30th Annual Proceedings of the Electron Microscope Society of America*, 1972.
31st EMSA: *31st Annual Proceedings of the Electron Microscope Society of America*, 1973.
33rd EMSA: *33rd Annual Proceedings of the Electron Microscope Society of America*, 1975.
34th EMSA: *34th Annual Proceedings of the Electron Microscope Society of America*, 1976.
4th H and C: *Proceedings of the 4th International Congress on Histochemistry and Cytochemistry*, 1972.
5th H and C: *Proceedings of the 5th International Congress on Histochemistry and Cytochemistry*, 1976.
3rd Nephron: *Proceedings of the 3rd International Congress on Nephrology*, 1966.
4th Nephron: *Proceedings of the 4th International Congress on Nephrology*, 1967.
5th Eur. Cong.: *Proceedings of the 5th European Congress on Electron Microscopy*, 1972.
6th Eur. Cong.: *Proceedings of the 6th European Congress on Electron Microscopy*, 1976.
EM and Cytochem: *Proceedings of the 2nd International Symposium on Electron Microscopy and Cytochemistry*, 1973.
5th Int. Prim.: Contemporary Primatology. *Proceedings of the 5th International Congress on Primatology*, 1974.
7th Int. EM: *Proceedings of the 7th International Congress on Electron Microscopy*, 1970.
8th Int. EM: *Proceedings of the 8th International Congress on Electron Microscopy*, 1974.
10th Anat: *Proceedings of the 10th International Congress of Anatomists*, 1975.
EDAX: Thin section microanalysis. *Proceedings of a symposium November 8th 1972*, St. Louis, Missouri. Ed. Russ, J. C. & Panessa, B. J. Edax Laboratories, North Carolina, USA.
Tiss. Calc.: Les Tissues Calcifies. *Proceedings of the 5th European Symposium on Calcified Tissues*, 1968.
SEM 71: *Proceedings of the Scanning Electron Microscope Symposium*, 1971.

SEM 77:	*Proceedings of the Scanning Electron Microscope Symposium*, 1977.
11th MAS:	*Proceedings of the 11th Annual Conference of the Microbeam Analysis Society*, 1976.
Microprobe Analysis:	*Microprobe Analysis as Applied to Cells and Tissues*, ed. Hall, T., Echlin, P. & Kaufmann, R. London: Academic Press, 1974.
Gyn Malig:	*Gynaecological Malignancy, Clinical and Experimental Studies*, ed. Brush, M. & Taylor, R. W. London: Bailiere Tindall, 1975.
Ca transport:	*Calcium Transport in Contraction and Secretion*, ed. Carafoli, E. & Drabikowski, W. North Holland, 1975.
3rd EMPA:	*Proceedings of the 3rd Congress on Electron Microprobe Analysis*, 1968.
Cell Mot:	*Cell Motility*, ed. Goldman, R. D., Pollard, T. D. & Rosenbaum, J. L. Cold Spring Harbour Laboratory, 1976.
Inter Comm. Plants:	*Intercellular Communications in Plants: Studies on Plasmodesmata*, ed. Gunning, B. E. S. & Robards, A. W. Berlin: Springer-Verlag, 1976.
Mic. Canada:	*Proceedings of the Annual Meeting of the Microscopical Society of Canada*, **2**, 1975.
Balaton 75:	*Balaton Conference on Electron Microscopy*, Veszprem, Hungary, 1975.
Physical Techniques 2nd edn.:	*Physical Techniques in Biochemical Research*, ed. Oster, G. New York: Academic Press.
Techniques Biochem.:	*Techniques of Biochemical and Biophysical Morphology*. Vol. 2. Ed. Glick, D. & Rosenbaum, R. New York: John Wiley and Sons, Inc.

(C) *Species*

A:	animal
P:	plant
Pr:	Protozoa

4

A. J. MORGAN, T. W. DAVIES AND
D. A. ERASMUS

Specimen preparation

4.1 Introduction

The ideal method of specimen preparation for the electron probe microanalyser would provide: (1) good preservation of cellular ultrastructure; and (2) the retention of the naturally-occurring elements in their *in vivo* positions (at least within the limits of the resolving capabilities of the instrumentation employed). However, it is probably true to say that no single preparative technique adequately fulfils both these criteria. For example, it has been stated (Baker, 1960) that 'it is not the purpose (of fixation) to leave the chemical composition (of tissues) unchanged.' At the present time, therefore, one is too often forced to choose between the contradictory requirements of preserving ultrastructural definition and of maintaining chemical integrity. Sample preparation is the most serious impediment to achieving high-resolution localization of elements within biological tissues.

This chapter does not seek as its primary aim the provision of an extensive literature survey of this vast subject. The most that we can hope to achieve is to present an appraisal based on personal experience of the relative merits of available procedures. This is intended to serve as a guide to assist the potential user in his choice of techniques. In general, the choice depends on several factors, such as (1) the type of microprobe available; (2) the tissue; (3) the biochemical state (of binding, for example) of the element(s) of interest within that tissue; and (4) the level of resolution required of the analytical study.

Coleman and Terepka (1974), in an excellent review chapter on preparative procedures for microprobe analysis, were also unable to delineate precisely the areas of applicability of the multiplicity of available techniques. However, they sought to assist the potential user in making his decision by establishing a number of criteria that characterize a chemically well-preserved specimen, and also by providing a summary of the results of many workers on diverse biological problems with special emphasis on how these results were achieved.

It is essential in our opinion that the user asks specific questions pertinent to his biological problem before deciding on a preparative regime. To take a specific example, if he is simply interested in knowing whether the granules he sees in mitochondria during a routine TEM morphological survey are mineralized or not, then quite obviously he may prepare his specimens for the microprobe by methods similar to those employed for the earlier morphological study. However, if subsequent to finding that the granules are mineralized he begins to ask further questions of a more physiological nature (e.g. are the minerals localized in discrete granules in the *in vivo* state? Or, what is the concentration of individual elements in the mitochondrion?), then he must consider using techniques that minimize chemical disturbance. Very often it may be advisable to perform preliminary analyses to assess the degree of chemical flux during a particular preparative procedure.

The only justification for choosing a given preparative technique is that the analytical data thus obtained can be used to pursue and understand biological phenomena. But herein lies yet another major difficulty. Since the microprobe yields data unobtainable by any other analytical method, it is often impossible to judge the true validity and significance of the collected data (Coleman & Terepka, 1974). Because of this inability to extrapolate from observed elemental concentrations to *in vivo* elemental concentrations after specimen manipulation, and because no single preparative technique adequately ensures the simultaneous exploitation of the dual microprobe facilities of structural and chemical characterization, we wish to pursue the philosophy during this review that a given biologically-orientated microprobe problem be approached, wherever possible, with a multiplicity of different and complimentary preparative procedures. Thus we may accumulate a bank of interpretable artefacts from which biological processes can be reconstructed.

In terms of the instrument to be used for microprobe analysis, it is not always possible to choose the most appropriate machine for a particular problem. More choice may be exercised in the preparation of material for the microprobe. For this reason, the principles, as opposed to the details, of the available preparative procedures are discussed; thus the reader may appreciate that each procedure can often be modified in order that specimens may be visualized and analysed in EPMA, TEM, SEM, and STEM instruments. Each procedure is considered within a separate section, although these sections do not necessarily form discrete units. For example, 'resin embedding' and 'freezing' are discussed in more than one section, so the reader is advised to read the chapter as a whole to obtain a more complete picture of the problems.

4.2 Tissue preparative techniques

4.2.1 BULK ANALYSIS

Microprobe systems facilitate the study of the elemental composition of microsamples of materials. A necessary pre-requisite to a better understanding of local element compartments in biological material is a knowledge of the bulk chemistry of the tissue of interest. Unfortunately, such data are seldom available, especially in the case of invertebrate tissues and most experimentally induced situations. Even when available, the data have been derived by techniques which differ considerably in principle from X-ray analysis and, therefore, make comparisons rather difficult.

Biological microanalysts have in general neglected this aspect. Davies and Morgan (1976) described a method of preparing biological tissue samples of the order of 10^{-4}–10^{-5} g dry weight for quantitative analysis in TEM. The method was reported to be far less tedious, since it required no fluid dilutions, and it had a favourable accuracy (error $\simeq \pm 5\%$ for most of the naturally occurring biological elements) compared to more conventional analytical procedures, such as flame spectrophotometry. This technique could readily be modified for the preparation of tissue samples for probe analysis in EPMA and SEM systems.

4.2.2 FLUID ANALYSIS

Techniques for the electron microprobe analysis of pico- and nanolitre fluid volumes, sampled directly from a biological source via individually calibrated pipettes, have been established by a number of workers (for a detailed review of these techniques see Chapter 7). Essentially the procedure requires that: (1) a known volume of fluid is deposited and dried on solid specimen blocks under controlled conditions to produce minimal crystal size, which in turn reduces topographic shielding effects and assists beam penetration; (2) X-ray microanalysis of the resultant salt–crystal mixture is performed with an electron microprobe; (3) the intensity of the X-ray signals from each element of interest is compared with the X-ray signal emitted from crystals obtained by similar means from standard solutions of known composition (Lechene, 1974). Analysis of such specimens has invariably been undertaken in EPMA or SEM systems.

A technique was recently described (Davies & Morgan, 1976) for spraying microlitre fluid volumes to produce droplets of unspecified volume on the relatively fragile surface of coated TEM grids, so that the relative concentrations of each element within individual droplets could be determined by X-ray analysis.

4.2.3 EXTRACTION–REPLICATION

This technique (described and discussed in detail by Henderson & Griffiths, 1972) is often used in materials science, but has also been used for preparing certain biological, especially pathological, tissues for microprobe analysis. The technique, initially developed for the study of surface features, has been modified (Henderson, 1969; Henderson, Harse & Griffiths, 1969; Henderson, Gough & Harse, 1970; Henderson, 1971; Henderson, Joslin, Turnbull & Griffiths, 1971) to facilitate the retrieval and subsequent chemical characterization of foreign particles incidentally extracted during the replication of tissues. Mineral-rich foreign particles have been analysed and identified in cervical carcinoma, lung tissue, osteoarthritic cartilage and synovial membrane, and in mesothelioma tissue (Henderson & Griffiths, 1972).

In the Henderson modification of the extraction–replication technique the tissue samples, either fresh unfixed tissue or (usually) de-waxed histological sections, are partially embedded in an acetone-softened cellulose acetate film on a glass microscope slide. The cellulose acetate is then allowed to harden. The protruding specimens are then outlined with strips of Scotch tape to form a shallow well and to aid the removal of the replica. A 5% solution of the water-soluble plastic polyvinyl alcohol (PVA) is then poured into the well to cover the tissue samples, and slow solidification is allowed to take place overnight. The hardened PVA film replica of the tissue surface is stripped off. Small pieces of tissue and any embedded foreign particles may also be extracted with the PVA film. The location of the particles can be determined by staining and visualizing adjacent sections.

Ultrathin sectioning of both naturally occurring and pathological mineral fibres and particles in biological tissues often proves difficult. Frequently the particles are fragmented and/or displaced during sectioning. In addition their chemistry may be modified during preliminary preparation procedures, and during section flotation (Termine, 1972). The extraction–replication method, especially when applied to fresh unfixed samples, has undoubtedly proved valuable not only in preserving such mineralized structures for microprobe analysis, but also, albeit to a lesser extent, for providing evidence regarding their *in situ* location. A more precise localization of highly mineralized structures may be achieved from a high resolution study of thin sections of tissue infiltrated with Spurr's low viscosity resin (Spurr, 1969).

4.2.4 WHOLE MOUNTING AND AIR DRYING

The technique essentially involves the direct mounting of fresh whole cells on to appropriate EM specimen supports. Mounting may be achieved variously by manual placement, smearing, and centrifugation or spraying from fluid suspension. The simplicity of the technique is a great advantage where routine sampling of cell populations is required since preparation time is minimal.

From a strictly morphological viewpoint this technique has severe limitations. Most soft tissues are grossly deformed after preparation (Boyde & Wood, 1969), which results from the uncontrolled evaporation of water from cytoplasmic components having differing water content. From an analytical viewpoint, however, these shrinkage artefacts may be considered advantageous. Unfixed, air-dried cells shrink to reveal cell boundaries (Boyde & Wood, 1969), sub-surface bodies such as nuclei or mitochondria (Boyde, James, Tresman & Willis, 1968; Boyde, Grainger & James, 1969), and the A, I, and Z bands in striated muscle (Boyde & Williams, 1968).

Whole mounting, air-drying procedures have been used extensively for preparing a variety of fluid-suspended cells for microprobe analysis, e.g. erythrocytes (Duprez & Vignes, 1967; Carroll & Tullis, 1968; Beaman, Nishiyama & Penner, 1969; Kimzey & Burns, 1973; Gullasch & Kaufmann, 1974; Kirk, Crenshaw & Tosteson, 1974; Roinel, Passow & Malorey, 1974; Roinel & Passow, 1975; Colvin, Sowden & Male, 1975; Lechene, Bronner & Kirk, 1976), mast cells in peritoneal fluid (Padawer, 1974), sperm cells (Chandler, 1973; Werner & Gullasch, 1974; Chandler, 1975*a*; Maynard, Elstein & Chandler, 1975; Chandler & Battersby, 1976), the protozoan *Tetrahymena pyriformis* (Gullasch & Kaufmann, 1974), and vorticellid spasmonemes (Routledge, Amos, Gupta, Hall & Weis-Fogh, 1975). In addition, the technique has been adopted for the preparation of fresh tissue spreads so that dense cytoplasmic bodies could be chemically defined by electron probe microanalysis, e.g. zymogen granules in mouse pancreas (Takaya, 1975*a*), dense bodies in human platelets (Takaya, 1975*b*), neurosecretory granules (Takaya, 1975*c*), and the nuclei of subcutaneous connective tissue cells (Takaya, 1975*d*) and Ehrlich tumour cells (Mizuhira, 1976). A simple modification of tissue smearing is the air-drying of unfixed thick cryostat sections. This procedure has been used (Morgan & Davies, 1976) for the localization and microprobe analysis in an SEM of the calcium phosphate inclusions of earthworm chloragogenous tissue.

Three major impediments may arise during the course of preparing and microprobe analysis of whole-mounted, air-dried specimens:

(1) *Chemical contamination*

In the specific case of fluid-suspended cells, it is highly likely that cell surfaces will be contaminated by elements in the surrounding fluid environment. Early workers using this method of preparation disregarded this problem (Duprez & Vignes, 1967; Carroll & Tullis, 1968; Beaman, Nishiyama & Penner, 1969). More recently, workers studying the human erythrocyte model have attempted to wash the cells free of ionic contamination with various isotonic media (Kimzey & Burns, 1973; Kirk, Crenshaw & Tosteson, 1974; Roinel, Passow & Malorey, 1974; Roinel & Passow, 1975; Lechene,

Bronner & Kirk, 1976). Roinel's group, for example, successfully washed red cell populations (whose intracellular electrolyte composition had previously been experimentally adjusted *in vitro*) with lithium chloride/sucrose solutions. The rationale for using this particular washing medium is that the sodium contaminating the cell surfaces is replaced by lithium ions (Li^+ ions having been used as 'markers' for Na^+ ions in several physiological situations), which in turn would not interfere with the collected X-ray spectra for intracellular sodium. Such washing procedures may not, however, be generally applicable for freeing cells derived directly from *in vivo* sources of their surface contaminants, since the use of 'balanced' washing media requires that the internal composition of the cells is known beforehand, and that the cell population is chemically homogeneous.

Contamination is an equally serious problem in the analysis of cytoplasmic structures in air-dried fresh tissue smears, or in thick air-dried frozen sections where section thickness exceeds the diameter of the structure of interest. Here the contaminants are those elements in the general cytoplasm that deposit on the structure of interest during drying. Microprobe analysis of cytoplasmic structures prepared by air-drying must be accompanied by rigorous analysis of the adjacent cytoplasm before even general qualitative conclusions regarding the chemical nature of such structures may be reached. Alternatively, the viability of the preparative procedure as applied to a given cell type may be assessed by comparison with cells prepared by the infinitely more demanding technique of cryo-ultramicrotomy (Chandler & Battersby, 1976).

(2) *Chemical redistribution*

Two groups of workers (Gullasch & Kaufmann, 1974; Werner & Gullasch, 1974; Chandler & Battersby, 1976) have systematically sought to investigate the possibility of elemental redistribution during the air-drying of fluid-suspended cells. Gullasch and Kaufmann (1974) concluded that air-dried erythrocytes had an electrolyte distribution pattern which failed to correspond with the pattern present in similar cells prepared by cryoprocedures. Moreover, they found that the intracellular electrolyte compartments in cryo-prepared *T. pyriformis* were absent from air-dried cells. Werner and Gullasch (1974) reached similar conclusions regarding rat spermatozoa. In contrast, Chandler and Battersby (1976), in a more extensive quantitative study of human spermatozoa, observed that the subcellular elemental distribution in air-dried cells was similar to that of ultrathin frozen and freeze-dried sections of the same material.

Until these diametrically opposed views are resolved we must tentatively conclude that the technique of tissue whole mounting followed by air-drying is adequate for (1) maintaining the total chemistry of cells; and (2) for preserving the integrity of certain discrete, often highly mineralized, cytoplasmic constituents; but (3) is probably unsuitable for preserving local intracellular electrolyte compartments.

(3) *Specimen thickness considerations*
Since it is seldom possible to control (within certain broad limits) the thickness of whole-mounted, air-dried specimens, and since the thickness of a specimen can exert a profound effect on the generation and collection of X-rays from irradiated tissue microvolumes, a serious consideration of this aspect of quantitative analysis is warranted (see Section 4.4).

4.2.5 HEAT FIXATION

Heat fixation represents a modification of the whole mounting–air drying procedure. Cells are smeared thinly on to silicon discs, are then blotted free of surface fluid, and finally are passed through a propane flame. The technique was originally developed (Coleman, Nilsson, Warner & Batt, 1972, 1973*a*, 1973*b*; Coleman, Nilsson & Warner, 1974) so that the refractile cytoplasmic granules of *T. pyriformis* and *Amoeba proteus* could be resolved and analysed. The authors found that techniques such as whole mounting/air-drying and freeze-drying yielded preparations which were either too thick or morphologically distorted. Heat fixation has been severely criticized because it is chemically destructive and may, therefore, induce chemical redistribution (see audience discussion of paper presented by Coleman, Nilsson & Warner, 1974). Nevertheless the technique enabled its protagonists to identify chemical differences in the refractive granules of organisms grown in culture media of various molar ratios. So, whatever the theoretical objections, the continued use of the technique appears to be justified by virtue of its practical performance. Specimens prepared by heat fixation would still be subject to contamination by endogenous and exogenous elements (see Section 4.2.4 above).

4.2.6 CONVENTIONAL PROCEDURES

It is being increasingly recognized that the aqueous and organic media commonly used in most conventional (wet chemistry) procedures for preparing biological tissues for ultrastructural studies in electron microscopes tend to redistribute and extract many of the naturally occurring elements. For example, it was shown (Elbers, 1966) by a microconductometric technique that on fixation in osmium tetroxide the cell membrane of *Limnaea* eggs became completely permeable to ions within a few seconds. Membrane permeability changed much more slowly during glutaraldehyde fixation. These and similar observations do not necessarily, however, preclude the use of so-called conventional preparative procedures in conjunction with electron microprobe analysis. A literature survey (Saubermann, 1975) revealed several examples where very valuable physiological information has been obtained from biological samples prepared thus. Indeed, where a high standard of structural preservation is essential it could be argued that certain high-resolution analytical studies could not have been performed were it not for the availability of such techniques. It is, therefore, important to appreciate the potential and serious chemical limitations imposed by preparative procedures that involve exposure to histological fluids, etc. so that the biological significance of data thus derived may be critically assessed.

An extensive discussion of the influence of fixatives, buffers, dehydrating agents, embedding media, etc. on the morphological preservation of biological tissues is already available (Hayat, 1970). The present account will be restricted to consideration of those papers that deal with the effect of processing fluids on the chemical composition and integrity of tissues.

Loss and dislocation of tissue components

Processing fluids may either extract elements from, or redistribute elements within, biological tissues. Both artefacts are extremely serious in the context of microprobe analysis where the ultimate aim is to procure a simultaneous structural and valid chemical profile of the tissue. The translocation of elements within tissues exposed to processing fluids is particularly difficult to assess because it requires that the original distribution pattern be known. The absolute loss of elements is, however, relatively easy to monitor and may be taken as an approximate indicator of the degree of translocation of given elements. Such an assumption may be justified in many cases. This does not preclude the possibility that where no significant loss of a cation is detected the ion has not been displaced from its *in vivo* site to a strongly anionic site, such as the cell nucleus. It is recommended, therefore, that wherever possible the elemental distribution patterns obtained in tissues prepared by fixing, dehydrating, embedding, etc. are checked against distribution patterns in similar tissues prepared by methods that impose fewer chemical restraints, e.g. cryo-ultramicrotomy.

Several authors have supplied data on the effect of particular preparative regimes on the elemental composition of given tissues (see Table 4.1). From such studies a preparative regime may be deemed 'good' or 'bad', but such a judgement only holds for those conditions under which the procedure was tested. The conclusion should not be extended to include that same element in a different tissue, or even in the same tissue under different physiological conditions, and certainly not to include other elements whether in the same tissue or not.

Macromolecules

The influence of processing fluids on macromolecular tissue components is of direct interest since it is often assumed that 'tightly bound elements', as

Table 4.1 Summary of element losses recorded during various conventional EM preparative procedures

Element	*Tissue*	*Technique*	*% Loss*	*Reference*
^{45}Ca (I)*	Chorioallantoic membranes	6% Acrolein in 0.1 M cacodylate–HCl buffer, pH 7.2, buffer wash, 2% OsO_4, graded alcohols	75	Coleman & Terepka (1972)
		6% Acrolein in 0.1 M cacodylate–HCl+1% sodium oxalate, buffer+1% oxalate wash, 2% OsO_4+1% oxalate, graded alcohols	12	
Ca (N)*	Uterus	1% OsO_4 in water	0	Garfield, Henderson & Daniel (1972)
		1% OsO_4+2% pyroantimonate	≃ 85	
		3% GA in water	≃ 30	
		3% GA+pyroantimonate	≃ 55	
Ca (N)	Smooth muscle	5% GA in physiological saline, pH 7.2, 36 °C or 4 °C, 4 h	(increased, by ≃ 35% over controls	Schoenberg, Goodford, Wolowyk & Wootton (1973)
Ca (N)	Aorta	3% GA in PO_4 buffer, pH 7.4, 1% OsO_4, PO_4 buffer wash, alcohols, propylene oxide	51	Morgan, Davies & Erasmus (1975)
	Tendon	3% GA in PO_4 buffer, pH 7.4, 1% OsO_4, PO_4 buffer wash, alcohols, propylene oxide	57	
	Aorta	3% GA in PO_4 buffer	47	
	Tendon	3% GA in PO_4 buffer	64	
^{45}Ca (I)	Pancreas	0.5%, 1%, or 3% GA	42–58 (depending on conc. of fix)	Howell & Tyhurst (1976)
^{86}Sr (I)	Isolated mitochondria	6.25% GA in 0.1 M PO_4 buffer, pH 7.4	25–35	Greenawalt & Carafoli (1966)
		12.5% GA in 0.1 M PO_4 buffer, pH 7.4	25	
		12.5% GA+1 mM succinate	13	
		12.5% GA+3 mM ATP	26	
		12.5% GA+10 mM succinate+3 mM ATP	28	
		12.5% GA+1% OsO_4 in 0.1 M PO_4 buffer pH 7.4	63	
		1% OsO_4 in veronal–acetate buffer, pH 7.4	50	
		10% formaldehyde±succinate & ATP	40	
Mg (N)*	Uterus	1% OsO_4 in water	≃ 10	Garfield, Henderson & Daniel (1972)
		1% OsO_4 in 2% pyroantimonate	≃ 35	
		3% GA in water	≃ 20	
		3% GA+pyroantimonate	≃ 30	

Mg (N)	Smooth muscle	5% GA in physiological saline pH 7.2, 36 °C or 4 °C, 4 h	0	Schoenberg, Goodford, Wolowyk & Wootton (1973)
Mg (N)	Ehrlich ascites	5% GA+1% OsO_4 in Krebs–Ringer PO_4 buffer, pH 7.45	≏ 55	Penttila, Kalimo & Trump (1974)
K (N)*	Uterus	1% OsO_4 in water 3% GA in water	≏ 75 ≏ 95	Garfield, Henderson & Daniel (1972)
K (N)	Erythrocytes	2% formaldehyde in PO_4 buffer, pH 7.2 4% formaldehyde in PO_4 buffer, pH 7.2 1.65% GA in PO_4 buffer, pH 7.2 3.3% GA in PO_4 buffer, pH 7.2 2% acetaldehyde in PO_4 buffer, pH 7.2	} ≏ 70	Vassar, Hards, Brooks, Hagenberger & Seaman (1972)
K (N)	Smooth muscle	5% GA in physiological saline, pH 7.2, 36 °C or 4 °C, 2 h	87.5	Schoenberg, Goodford, Wolowyk & Wootton (1973)
Rb (I)	Leaf tissue	1% OsO_4 in cacodylate–acetate	90	Hall, Yeo & Flowers (1974)
K (N)	Ehrlich ascites	5% GA+1% OsO_4 in Krebs–Ringer PO_4 buffer, pH 7.45	≏ 90	Penttila, Kalimo & Trump (1974)
Na (N)*	Uterus	1% OsO_4 in water 1% OsO_4+2% pyroantimonate 3% GA in water 3% GA+2% pyroantimonate	≏ 75 ≏ 75 ≏ 95 ≏ 90	Garfield, Henderson & Daniel (1972)
Na (N)	Smooth muscle	5% GA in physiological saline, pH 7.2, 36 °C or 4 °C, 4 h	(increased by ≏ 200% over controls)	Schoenberg, Goodford, Wolowyk & Wootton (1973)
^{36}Cl (I)	Root tissue	1% OsO_4 in 0.1 M cacodylate–acetate buffer, pH 6.5+either 0.5% Ag acetate or 1% Ag lactate, buffer wash, acetone (+brief wash in 0.05 N HNO_3), propylene oxide, Spurr resin	< 4	Läuchli, Stelzer, Guggenheim & Henning (1974)
^{36}Cl (I)*	Leaf tissue	1% OsO_4 in cacodylate–acetate containing 30 mM Ag acetate, buffer wash, acetone, Pallaghy resin	≏ 70–95 (depending on fix time and sample size)	Harvey, Flowers & Hall (1976)

Table 4.1 Summary of element losses recorded during various conventional EM preparative procedures (cont.)

Element	*Tissue*	*Technique*	*% Loss*	*Reference*
^{65}Zn (I)	Corn, leaf, barley root & Chlorella	0.3% or 1% GA in PO_4 buffer	62–70	DeFilippis & Pallaghy (1975)
		0.3% or 1% GA in PO_4 buffer + 30% sodium sulphide	7.6–9.5	
^{203}Hg (I)	Corn, leaf, barley root & Chlorella	0.3% or 1% GA in PO_4 buffer	34–41	DeFilippis & Pallaghy (1975)
		0.3% or 1% GA in PO_4 buffer + 30% sodium sulphide	72–81	
Cd (I)	Bivalve tissues	GA	28	George, Nott, Pirie & Mason (1976)
		GA + H_2S	14	
		GA + H_2S, GA-urea embedding	20	
		GA + H_2S, alcohol	33	
		GA + H_2S, alcohol, Spurr resin	41	
^{31}Si (I)*	Diatoms	2% GA, wash, 1% OsO_4, alcohol	25	Mehard & Volcani (1975)
	Spleen		35	
	Lung		55	
	Liver		70	
	Kidney		90	
	Liver & diatom		40–60	
	Mitochondria			
^{68}Ge*	Diatoms	2% GA, wash, 1% OsO_4, alcohol	60	Mehard & Volcani (1975)
	Spleen		35	
	Liver mitochondria		60	
	Diatom nuclei		80	
	Diatom vesicles		65	
	Diatom microsomes		60	

GA, glutaraldehyde; I, element experimentally introduced into tissue; N, naturally occurring element; * authors have recorded the loss of element during the various stages of their preparative regime(s).

distinct from those in the ionic or free state, may be successfully monitored by microprobe analysis of specimens prepared by wet chemical procedures. But, this assumes that the macromolecules themselves are not extracted and translocated. There exists, however, much data which indicate that a significant proportion of proteins, lipids, and carbohydrates are lost from biological tissues during fixation, dehydration, and embedding. Such losses would obviously interfere both with the precise localization of macromolecules by the identification of histochemical reaction products, and with the localization of structurally and electrostatically bound elements associated with the macromolecules. The extent of these losses is not only dependent on the type, the time, and the temperature of fixation, but also on the specific dehydration and embedding media subsequently used. Hayat (1970) presents an excellent literature review on this subject, but some examples are presented here to illustrate the magnitude of the problem.

The degree of extraction of lipids from tissues appears to vary with their degree of saturation and binding, and also on the type of tissue involved. For example, a lipid loss of 21% during preparation of osmium tetroxide-fixed small intestine was reported by Dermer (1968). Virtually no lipids are lost from amoebae during glutaraldehyde fixation, but up to 90% are lost during subsequent dehydration (Korn & Weisman, 1966).

Proteins are also extracted during EM preparation. Rat liver mitochondria lose approximately 22% of their total protein content during osmium tetroxide fixation, and a further 12% during alcohol dehydration (Dallam, 1957). Fixation in glutaraldehyde for 4 hours at 22°C reduced the insulin content of rat pancreas to 60.3%; dehydration in ethanol further reduced the insulin level to 27.5% (Grillo, Ogunnaike & Faoye, 1971).

Most carbohydrates in osmium tetroxide-fixed tissues are extracted during subsequent washing and dehydration (Hayat, 1970).

It is clear, therefore, that even the less readily exchangeable, relatively large molecular constituents of biological tissues are greatly affected by exposure to EM processing fluids.

Calcium, strontium, magnesium

Brierley and Slautterback (1964) found that the fixation of ^{32}P-labelled, isolated mitochondria with unbuffered osmium tetroxide removed over 80% of the P_i associated with magnesium, 60% of that associated with manganese, and 40–50% of that associated with calcium. Fixation in 'buffered osmium tetroxide', 'buffered osmium tetroxide containing calcium', and brief fixation in unbuffered osmium tetroxide were found to yield similar results. A very extensive systematic study of the effect of various fixative/buffer regimes on the ^{85}Sr-content of pre-loaded, isolated liver mitochondria was described by Greenawalt & Carafoli (1966). They concluded that large amounts (13–63%

of the total ^{85}Sr taken up, depending on the fixation method) of radioactivity was lost from the mitochondria with all of the methods of fixation tested (see Table 4.1). The striking feature of the data is the rapidity (from 3 to 10 minutes depending on method) with which these losses occurred. They also quoted unpublished observations that mitochondria loaded with calcium retained 95% of their label after fixation in 12.5% glutaraldehyde. Coleman and Terepka (1972) in a ^{45}Ca isotope study of the chick chorioallantoic membranes found that nearly 75% of the calcium content was removed during fixation (in acrolein and O_sO_4), and dehydration. Hohman and Schraer (1972) noticed in preliminary, non-quantitative experiments on the effects of fixing and embedding ^{45}Ca-labelled avian shell gland and rat liver mitochondria, that a significant quantity of the radioisotope was lost during glutaraldehyde and osmium tetroxide fixation and alcohol dehydration. It has also been shown (Howell & Tyhurst, 1976) that fixation in 0.5%, 1%, and 3% glutaraldehyde results in an overall loss of 42–58% of the ^{45}Ca originally present in labelled rat islets of Langerhans before fixation, while 2% osmium tetroxide, either as a primary fixative or as a post-fixative after glutaraldehyde, caused significant additional losses.

The above observations on the magnesium–calcium–strontium electrolyte series were undertaken on tissues pre-loaded with radioisotopes. Similar observations have been made on endogenous elements. Morgan and Bellamy (1973) and Morgan, Davies and Erasmus (1975) found that 50–60% of the calcium of both rat thoracic aortas and tail tendons was removed during double fixation in 3% glutaraldehyde and 1% osmium tetroxide in 0.1 M phosphate buffer, followed by alcohol dehydration and propylene oxide infiltration. The bulk of the calcium loss occurred during initial fixation in glutaraldehyde. In general, these observations substantiated the calcium distribution studies of Agostini and Hasselbach (1971) on fragmented sarcoplasmic reticulum; Yarom, Peters, Scripps and Rogel (1974), Yarom, Peters and Hall (1974), and Yarom, Hall and Peters (1975) on myocardial tissues; and Oschman, Hall, Peters and Wall (1974) working on the squid axon. Moreover, alcohol dehydration produces calcium efflux even after the cation has been immobilized with pyroantimonate (Yarom, Hall & Peters, 1975). About 52 to 55% of the magnesium content of tumour cells was lost during glutaraldehyde and/or osmium tetroxide fixation (Penttila, Kalimo & Trump, 1974). An isolated observation (Garfield, Henderson & Daniel, 1972) contrary to those made by many other authors on various types of tissue, claimed that little calcium or magnesium is lost from rat uterine tissues during glutaraldehyde or osmium tetroxide fixation followed by alcohol dehydration.

The work of Krames and Page (1968) is extremely interesting in that it shows a difference between the effects of osmium tetroxide and aldehyde fixatives on the permeability of cardiac tissues to divalent cations (cf. Elbers, 1966).

When osmium tetroxide-fixed tissues were perfused with a physiological salt solution, there resulted a striking net uptake of calcium compared with unfixed controls. At a concentration of 32.6 mM OsO_4 the ventricles accumulated an average of 25×their normal calcium content; at osmic acid concentrations of 3.27 mM and 32.7 mM the ventricles also lose a small but significant amount of magnesium. In comparison, formaldehyde-fixation brought about a small net uptake of calcium and no significant loss of magnesium. Similarly, Schoenberg, Goodford, Wolowyk and Wootton (1973) found that the fixation of saline-equilibrated guinea-pig smooth muscle in 5% glutaraldehyde in physiological saline resulted in a steady increase in tissue calcium concentration to a level 350% higher than controls after 4 h. There was little or no change in the magnesium content under similar conditions.

Sodium, potassium, chlorine, iodine

Because of their obvious physiological significance the monovalent electrolytes have been the subject of several microprobe studies. These elements are often highly exchangeable and easily displaced by exposure to fluids. It is not surprising, therefore, that many attempts have been made to monitor and eliminate the flux of these elements during EM preparation.

Early studies (Boothroyd, 1964, 1968; Thomas, 1964, 1969) implied that large quantities of sodium and potassium were lost from tissues during exposure to processing fluids. Fixation in glutaraldehyde with post-fixation in osmium tetroxide removed the bulk of the sodium from the cytoplasm of *Amphiuma* red cells (Andersen, 1967), and rat uterine cells (Garfield, Henderson & Daniel, 1972). Formaldehyde – and osmium tetroxide – fixed cardiac tissues exhibit a marked exchange of cellular K^+ for extracellular Na^+ when exposed to physiological saline, the effect increasing with the concentration of the fixative (Krames & Page, 1968). Under similar experimental conditions, *ca.* 90% of the initial potassium content of saline-equilibrated taenia coli muscles was lost after 4 h fixation in 5% glutaraldehyde in physiological saline at 36 °C (≏ 40% loss during the first 10 min of fixation), whilst a sodium uptake of > 200% was recorded, after 4 h (≏ 150% increase during first 10 min), by Schoenberg, Goodford, Wolowyk and Wootton (1973). A fixative consisting of 20% glutaraldehyde, 10% formalin, 10% acetic acid, and 60% methanol removed all the sodium from the follicles of the rat thyroid gland (Robison & Davis, 1969; Robison, 1973).

Excellent kinetic studies on the loss of potassium (and magnesium) from Ehrlich ascites tumour cells were described by Penttila, Kalimo and Trump (1974). They found that 85–90% of intracellular potassium was released during the first 2 min when the cells were fixed with 5% glutaraldehyde and/or 1% OsO_4, or 3% glutaraldehyde/1% OsO_4 at 0 °C or 37 °C, or with 3% glutaraldehyde alone at 37 °C. After this initial loss the potassium concentration remained relatively constant over the next 24-hour observation period. With glutaraldehyde as the only fixative, the rate of potassium efflux was increased as either the concentration of the fixative or the temperature of fixation rose (cf. Schoenberg, Goodford, Wolowyk & Wootton, 1973).

Vassar, Hards, Brooks, Hagenberger and Seaman (1972) described the effect of three different aldehyde fixatives on the potassium content of human erythrocytes. Acetaldehyde, formaldehyde, and glutaraldehyde caused substantial potassium leakage during fixation. The final concentration of potassium in all the fixatives reached the same level, but whereas the leakage with glutaraldehyde was immediate, that with formaldehyde was more gradual, and that with acetaldehyde reached a steady state only after 2 h. Thus the rate of potassium efflux was directly proportional to the rate of fixative penetration.

Chloride ions are extremely mobile in aqueous media (Harvey, Flowers & Hall, 1976; Harvey, Hall & Flowers, 1976). Chlorine presents severe difficulties, therefore, both in terms of maintaining it at its *in vivo* loci during specimen preparation (Van Steveninck, Van Steveninck, Hall & Peters, 1974), and in terms of limiting its volatilization in the beam of the electron microprobe (see Section 4.3.3).

The lability of halides during specimen processing depends largely on their state of binding within tissues. Iodine, which is tightly bound within the follicles of the rat thyroid gland, was unaffected by exposure to several preparative regimes (Robison & Davis, 1969; Robison, 1973). On the other hand, it was interesting that sodium content varied according to the specific preparative method used.

Phosphorus and sulphur

Phosphorus and sulphur are primarily structural elements associated with lipids, proteins, and polysaccharides, and are, therefore, considered to be relatively non-exchangeable. Nevertheless phosphorus- and sulphur-containing constituents are lost from tissues during EM preparation.

Phospholipids, for example, are extracted even during the recommended glutaraldehyde/osmium tetroxide fixation. Morgan and Huber (1967) prepared samples of lung containing tritiated choline-labelled phospholipids for EM by four different fixation methods, followed in each case by alcohol/propylene oxide dehydration and Epon 812 embedding. Losses of radioactivity were 66.9% of total radioactivity when fixed in *s*-collidine-buffered OsO_4, 46.5% in 4% buffered formalin followed by OsO_4, 39% in 2% glutaraldehyde followed by OsO_4, and 51.3% by tricomplex fixation in palladium chloride–osmium tetroxide–sodium phosphotungstate. Chemical analysis (Dermer, 1968) showed that 18% of the phospholipid in intestinal slices was lost during preparation, whilst 25% of the phospholipid of amoebae was extracted with various fixative and dehydration procedures (Korn & Weisman, 1966). In addition, many common embedding media such as Epon, Araldite and methacrylates are powerful lipid solvents [see Hayat (1970) for references]. The rapidity with which phosphorus-containing molecules can be lost was demonstrated by Penttila, Kalimo and Trump (1974). All the ATP disappeared from tumour cells during the first 2 min of fixation in either 3% glutaraldehyde/1% OsO_4 or 1% OsO_4 at 37 °C.

Available microprobe data suggest that nuclear phosphorus (distributed primarily in nucleoprotein) is not as labile as cytoplasmic phosphorus (distributed primarily in membrane lipids). Fixation in glutaraldehyde with post-fixation in osmium tetroxide removed practically all the phosphorus from the cytoplasm of *Amphiuma* red cells, whilst nuclear phosphorus was relatively unaffected (Andersen, 1967). A similar differential effect of processing fluids on nuclear and cytoplasmic phosphorus occurs in arterial smooth muscle cells (Morgan, Davies & Erasmus, 1975). Höhling and Nicholson (1975) demonstrated that all the phosphorus is washed out of predentine sections during contact with either water or alcohol. These authors concluded that in order to monitor phosphorus in their specimens it was necessary to use freezing techniques.

That some sulphur-containing macromolecules are also lost during EM processing was shown by Luft and Wood (1963), who labelled tissue proteins with ^{35}S-methionine. At least 8% of the label was lost during exposure to 8 different fixative and buffer systems, and an additional 4% was lost during dehydration. Osmium would seem to have a more damaging affect on the protein content of tissues than aldehyde fixatives, since there is good biochemical evidence that oxidizing agents can cleave disulphide bridges [see Hayat (1970) for references]. Glutaraldehyde, on the other hand, insolubilizes many proteins through the formation of irreversible cross-links.

Heavy metals

Certain heavy metals are occasionally found in normal and pathological tissues at concentrations detectable by the electron microprobe. Heavy metals are often considered to be relatively non-labile, but this is not necessarily so in all cases. Tapp (1975) indicated that copper was lost from the copper-rich granules of *Drosophila* mid-gut epithelial cells during glutaraldehyde fixation. Thomas (1964) implied a similar loss based on observations of microincinerated sections. George, Nott, Pirie and Mason (1976) measured the loss of introduced radioactive cadmium from the tissues of the marine bivalve, *Mytilus edulis*. Glutaraldehyde fixation removed approximately 28% of the cadmium (whilst the addition of H_2S as a precipitant to the fixative reduced the loss of 14%). Large losses of both ^{65}Zn (62–70%) and ^{203}Hg (34–43%) from plant tissues were recorded during 2% glutaraldehyde fixation

(DeFilippis & Pallaghy (1975). Subsequent treatment (i.e. buffer washing, 1% OsO_4 post-fixation, alcohol dehydration, and resin embedding) contributed additional losses. The loss during glutaraldehyde fixation was reduced in the case of zinc (to 7.6–9.5%) but increased in the case of mercury (72–81%) by saturating the fixative with sulphide precipitating ions. In comparison, freeze-substitution resulted in a total retention of zinc and a loss of < 5% mercury.

Silicon and germanium

Silicon is being increasingly implicated as a metabolically significant element in biological systems. Mehard and Volcani (1975) presented an excellent paper on the effect of an EM preparative procedure (involving double fixation in 2% glutaraldehyde and 1% OsO_4, dehydration in alcohols, washing in propylene oxide, and embedding in Epon) on the retention of ^{31}Si, and its chemical analogue ^{68}Ge, by various rat tissues and diatoms.

Although both ^{31}Si and ^{68}Ge were incorporated into the wall structure of the diatoms, a total of approximately 25–30% of the ^{31}Si and 55–60% of the ^{68}Ge were extracted during the various processing steps, including alcohol dehydration and Epon 812 embedding. Liver samples retained 30–35% of the ^{31}Si and ^{68}Ge, the spleen about 65% of the ^{31}Si, lung tissue 45%, and the kidney only 10%. Silicon and germanium were more easily mobilized during Epon embedding than during Spurr-resin embedding. These workers proceeded to show that the losses could be almost completely eliminated by adopting a freeze-substitution procedure (see Section 4.2.8).

Mineralized deposits

This is a somewhat artificial textual category since it discusses different forms of some of the elements already mentioned.

The problem of elemental loss from biological minerals has long since been recognized by histologists. Renaud (1959) showed that neutral formalin fixation was entirely unsuitable for the histochemical visualization of calcified deposits in cardiac tissues, whilst 80% alcoholic formalin reduced the crystal dissolution.

In the vertebrate skeleton calcium phosphate exists in two forms: (1) the apatite phase which occupies an increasingly large volume as the skeleton matures; and (2) an amorphous phase which is always present but decreases in volume with maturation. As much as 5–100% (depending on the degree of hydration) of the total calcium phosphate in the amorphous precipitate is lost during a routine EM preparative procedure (Termine, 1972). Quantitatively similar findings were presented by Posner (1972). A further artefact is the transformation of the amorphous into a crystalline phase during processing (Termine, 1972). In general apatite crystals are relatively insoluble in processing fluids. Nevertheless, apatite material may be lost from thin bone sections during water flotation and staining (Boothroyd, 1964; cf. Harvey, Hall & Flowers, 1976).

Introduction of extraneous elements

Besides disturbing the chemical integrity of biological tissues, the processing fluids involved in tissue preparation for the EM often introduce elements extraneous and/or foreign to the cells. These elements are undesirable because: (1) they invalidate the microprobe observations if the element concerned is also naturally occurring; (2) if foreign to the tissue, their characteristic emission lines may overlap those of certain elements of interest, a problem frequently encountered in energy-dispersive analysis; or (3) they may contribute substantially to the background signal if they emit high-energy X-rays.

It is possible to distinguish two categories of introduced elements: (1) most obviously, those that are chemical constituents or contaminants of given processing fluids; and (2) elements within the extracellular fluid of the tissue which are translocated to the intracellular compartment due either to a change in membrane permeability, or to an increased affinity of cellular macromolecules for an element as a direct consequence of the process of fixation.

Examples of the latter form of introduction are furnished by the data of Krames and Page (1968), and Schoenberg, Goodford, Wolowyk and Wootton (1973). Aldehyde and osmium fixed cells lose their ability to maintain K/Na gradients across their plasma membranes. The fixed cells passively accumulate sodium concomitant with a loss of potassium when exposed to physiological saline. Furthermore, osmium-fixed cells have an increased affinity for calcium due, it was implied (Krames & Page, 1968), to an interaction between calcium, osmic acid, and the osmic acid-reactive groups within the cells. In our own experience, osmium-fixed nuclei invariably contain more calcium than nuclei prepared by cryo-procedures without exposure to processing fluids. Significantly, cytoplasmic calcium was detected in myocardial tissue only after osmium tetroxide/pyroantimonate fixation; calcium was not detected in the cytoplasm of cells fixed in glutaraldehyde or pyroantimonate alone (Yarom, Hall & Peters, 1975).

Truly extraneous elements may be introduced into biological specimens at a number of stages during processing.

(1) *Fixation*

Many samples of commercial glutaraldehyde contain calcium (Oschman & Wall, 1972). Tissues can take up calcium during fixation (Schoenberg, Goodford, Wolowyk & Wootton, 1973). This has led some workers to use redistilled glutaraldehyde (Oschman & Wall, 1972; Skaer, Peters & Emmines, 1974).

Fixative solutions are usually prepared in a buffer vehicle. The buffers themselves often contain elements that may also be found in the tissue. Phosphate buffers may contribute sodium and phosphorus to tissues. Other buffers (e.g. cacodylate containing arsenic) contribute elements whose emission lines overlap those of given elements being investigated. For example the Mα line of osmium at 1.91 keV may overlap the Kα line of phosphorus at 2.01 keV. The problem of overlapping peaks is a much more serious impediment when using energy-dispersive (solid state) spectrometers (with a resolving power of $\simeq$ 150 eV at 5.9 keV) than with the infinitely higher resolution ($\simeq$ 10 eV depending on the specific crystal chosen) crystal spectrometer systems (Chandler, 1975*b*).

(2) *Resins*

Resin embedding is very useful in the context of microprobe analysis because: (1) it simplifies the manipulation and sectioning of specimens; (2) sections may be stained either before, or, preferably after, chemical analysis; (3) the embedded specimen, being about 80% plastic, results in a more homogeneous sample which simplifies quantitation and the interpretation of data (Ingram & Ingram, 1975).

The disadvantages of plastic embedding are similarly manifold: (1) morphological changes may result from the surface tension forces at the advancing plastic front; (2) the resin effectively dilutes cellular constituents by increasing specimen mass, thus adversely affecting the limits of detection (Ingram & Ingram, 1975); (3) the plastic extracts and redistributes certain plastic-soluble constituents (Posner, 1972; Morgan, Davies & Erasmus, 1975; Yarom, Hall & Peters, 1975), and may physically mobilize certain other constituents; and (4) many otherwise satisfactory embedding resins such as Araldite (Andersen, 1967; Davies & Erasmus, 1973), and Epon 812 and Spurr resin (Van Steveninck, Van Steveninck, Hall & Peters, 1974) contain chlorine and, in the case of Araldite, sulphur (Davies & Erasmus, 1973). These various problems have been overcome in some instances by certain specific developments. For example, Yarom, Hall and Peters (1975) embedded their tissue in a polymerizable mixture of 50% glutaraldehyde and urea in which dehydration may be omitted (Pease & Peterson, 1972), so that calcium efflux occurring during dehydration and Epon 812 embedding could be avoided (cf. Mehard & Volcani, 1975). Other workers interested in chlorine distribution have developed low-chlorine resins: Epon 826 (Ingram & Hogben, 1968), glutaraldehyde/urea and glutaraldehyde/carbohydrazide (Van Steveninck, Van Steveninck, Hall & Peters, 1974), and Pallaghy's (1973) modification of Spurr's resin.

Precipitation reactions

Attempts have been made to counter the leaching influences of processing fluids by employing various precipitation reactions. Most of these reactions involve the *in situ* reaction of an endogenous tissue ion with an introduced heavy metal salt, thus producing a visible electron dense deposit. For example, Cl^- ions have been precipitated with gold (Läuchli, Stelzer, Guggenheim & Henning, 1974; Läuchli, 1975), and silver salts (Van Steveninck, Chenoweth & Van Steveninck, 1973; Harvey, Flowers & Hall, 1976), Na^+ with pyroantimonate (Tandler, Libanati & Sauchis, 1970), PO_4^{3-} with lead salts (Tandler & Solari, 1969), and Ca^{2+} with oxalate (Diculescu, Popescu, Ionescu & Butucescu, 1971).

The precipitating agent is usually included in the pre-fixative and often also in some or all of the subsequent preparative fluids. Ideally, both the fixative molecules and the precipitating agent should have similar penetration kinetics, so that the capture of endogenous ions and the fixation of their micro-environment are simultaneous events. Whether this in fact occurs in any of the reactions is not known.

The effectiveness of the inclusion of certain precipitating agents in the prevention of ion loss during specimen preparation has been quantitatively assessed by a few workers. For example, rat uterine tissues fixed in the presence of potassium pyroantimonate retained much of their calcium and magnesium (Garfield, Henderson & Daniel, 1972). It should be noted that calcium-rich precipitates in pyroantimonate-fixed tissues can be solubilized during alcohol dehydration (Yarom, Hall & Peters, 1975), and uranyl acetate and lead citrate staining (Yarom, Peters, Scripps & Rogel, 1974). Surprisingly, the pyroantimonate-fixed uterus was found to retain little ($\simeq$ 10%) of its original sodium content (Garfield, Henderson & Daniel, 1972). Less than 4% of the total ^{36}Cl absorbed by the roots of barley seedlings was lost during fixation in 1% OsO_4 containing either 0.5% Ag lactate in 0.1 M cacodylate acetate buffer, and subsequent dehydration and embedding (Läuchli, Stelzer, Guggenheim & Henning, 1974). However, other workers (Harvey, Flowers & Hall, 1976) have cast doubt on the silver precipitation procedure for chlorine [cf. the ineffectiveness of mercury precipitation with sulphide (DeFilippis & Pallaghy, 1975)]. The inclusion of 30 mM silver acetate in the osmium fixative still resulted in a loss of $\simeq$ 90% of the ^{36}Cl-label from leaf tissue during conventional EM processing (Harvey, Flowers & Hall, 1976). The loss was particularly rapid during fixation and buffer washing stages, and was only marginally reduced by shorter fixation times and increasing specimen size.

The rationale for some of these precipitating reactions is presented in depth elsewhere in this volume (Chapters 3 and 6). Although it is our intention to avoid the controversy surrounding the specificity and, therefore, the applicability of these reactions, it is pertinent to observe: (1) that it is often an advantage to be able to analyse simultaneously (temporally and/or spatially) the distribution of more than one element in many biological investigations, this would by definition be precluded where a tissue had been exposed to a specific precipitation procedure; and (2) the formation of discrete precipitates implies that elements are drawn to focal points from their *in vivo* locations in the surrounding tissue, but whether this ion migration occurs across distances within the resolving power of existing microprobe instruments is unestablished.

4.2.7 CRITICAL POINT DRYING

Critical point drying involves the sequential replacement of tissue water by exposure to: (1) graded alcohols or acetone; and (2) transitional fluids such as amyl acetate or Freon 113. Finally, the transitional fluid is replaced by a liquid gas (amyl acetate by liquid CO_2, and Freon 113 by Freon 13) which is then vapourized by raising its temperature to about 10 °C above its critical point (31.1 °C for CO_2, 28.9 °C for Freon 13) in an enclosed pressure chamber. The advantage of the technique over freeze-drying, for example, is that artefacts caused by ice-crystal formation and surface tension forces are eliminated. Critical point drying is, therefore, a valuable technique for the preparation of a variety of plant and animal tissues for SEM morphological study. However, most specimens usually need to be fixed prior to dehydration so that metabolic processes are arrested and structural stability improved. Critical point drying is, therefore, an essentially 'wet-chemistry' procedure, and from a microprobe analysis viewpoint suffers many of the chemical limitations described in Section 4.2.5 (above) on conventional procedures.

A good review of the critical point drying technique was presented by Cohen (1974).

4.2.8 FREEZE SUBSTITUTION

In freeze substitution the sample is rapidly frozen and the ice is slowly replaced by an anhydrous solvent at low temperature (Fernandez-Moran, 1960, 1961; Rebhun, 1961; Bullivant, 1965; Pearse, 1968). The dehydrated specimen is then usually infiltrated with an embedding medium. Freeze substitution seems, therefore, to offer a possibility of a compromise between an adequate preservation of cellular ultrastructure and minimal loss and translocation of endogenous elements. A review of morphological aspects of freeze substitution together with a comprehensive bibliography was provided by Pease (1973).

Freezing

A detailed discussion of the various methods of freezing and of the chemical properties of freezing and cryo-protective agents is beyond the scope of this chapter. Reviews on this subject are available elsewhere (Sjöstrand, 1967; Rebhun, 1972).

There is no single recommended freezing procedure, but the chosen procedure must be both: (1) simple, so that the time involved between the retrieval of the specimen and its quench freezing is minimized; and (2) rapid, so that ice crystal formation and the resultant tissue damage and element displacement is also minimized. Unfortunately, the low thermal conductivity of biological tissue seriously limits the rate of freezing. Nevertheless, the cooling rate can be improved by: (1) increasing the temperature gradient between the specimen and the cooling agent, e.g. by using melting nitrogen slush (Umrath, 1974) or a liquid nitrogen cooled metal surface (Eränkö, 1954; Van Harreveld & Crowell, 1964; Van Harreveld, Crowell & Malhotra, 1965; Van Harreveld, Trubatch & Steiner, 1974; Dempsey & Bullivant, 1976*a*, 1976*b*); (2) the use of cryo-protectives such as aqueous solutions of glycerol or DMSO; and (3) limiting specimen size.

Specimen size is not, however, an important practical consideration because the well-preserved region of animal tissue specimens frozen in the absence of a cryo-protective or an artificial nucleating agent, such as chloroform (Boyde & Wood, 1969), is confined to a narrow layer of the order of 10 to 12 μm wide near the tissue/coolant boundary (Eränkö, 1954; Van Harreveld & Crowell, 1964; Van Harreveld, Crowell & Malhotra, 1965; Christensen, 1971; Dempsey & Bullivant, 1976*a*). The depth of this may be greater in certain plant tissues (Hereward & Northcote, 1972).

The criteria for considering the superficial layer to be well preserved have been based on morphological appearance and not on the validity of the distribution of its chemical components. It is generally assumed that those areas possessing the least mechanical disruption due to ice crystal growth during freezing also possess element distribution patterns that most closely resemble the *in vivo* state. However, it should be borne in mind that this surface layer also represents that tissue volume which is most likely to be disturbed during dissection and accidental drying prior to freezing. The observations of Rebhun and co-workers are interesting in this context (see Rebhun, 1972). They concluded that ice crystal formation in marine eggs becomes less probable as water is removed from cells either osmotically or by partial air drying. Thus, under these circumstances, those areas having the best structural integrity would have the least meaningful chemistry. However, Dempsey and Bullivant (1976*a*, 1976*b*) were careful to limit surface drying before freezing their specimens against a liquid nitrogen-cooled polished copper block by ensuring that exposed tissue surfaces were kept immersed in water, phosphate buffer, or blood, and yet were still able to observe an undamaged superficial layer 12 μm wide.

The possibility of elemental loss from tissues during quench freezing has been investigated by three groups of workers. No loss of sodium, potassium, and chlorine from botanical tissue was detected during freezing in 8% (v/v) methylcyclohexane/2-methylbutane cooled by liquid nitrogen (Harvey, Hall & Flowers, 1976), or of calcium from rodent aorta during liquid nitrogen freezing (Morgan, Davies & Erasmus, 1975), or of ^{68}Ge from diatoms and rat tissues and isolated organelles during isopentane freezing (Mehard & Volcani, 1975).

Dehydration

A variety of organic solvents such as acetone (Malhotra & Van Harreveld, 1965; Rebhun, 1965*a*; Fisher, 1972; Fisher & Housley, 1972; Hereward & Northcote, 1972; Kuhn, 1972; DeFilippis & Pallaghy, 1973; Pallaghy, 1973; Steinbiß & Schmitz, 1973), diethyl ether (Lüttge & Weigl, 1965; Läuchli, Spurr & Wittkopp, 1970; Läuchli, Spurr & Epstein, 1971; Spurr, 1972; Steinbiß & Schmitz, 1973; Mehard & Volcani, 1975, 1976), ethanol (Neeracher, 1966; Van Harreveld & Steiner, 1970), ethylene glycol (Pease, 1967), methanol (Fisher & Housley, 1972), propylene oxide (Fisher & Housley, 1972), n-hexane (Neumann, 1973, 1974), tetrahydrofuran (Rebhun & Sander, 1971), and acetone/acrolein mixtures (Steinbiß & Schmitz, 1973; DeFilippis & Pallaghy, 1975; Van Zyl, Forrest, Hocking & Pallaghy, 1976) have been employed to remove water from frozen specimens at low temperatures. In general, small pieces of tissue need to be substituted for periods of several days at temperatures of about −70 °C to −80 °C (Rebhun, 1965*b*; Van Harreveld, Crowell & Malhotra, 1965), the period required depending to some extent on sample size and the substituting fluid employed. For example, extraction studies involving Sudan Black B in agar blocks as a crude tissue model have indicated that acetone probably substitutes tissue much faster than propylene oxide (Fisher & Housley, 1972) and diethyl ether (Harvey, Hall & Flowers, 1976).

There have been many claims in the literature regarding the improved morphological qualities of tissues exposed to particular substituting fluids. However, Rebhun (1972) reviewing his own experiments on sea urchin eggs wherein several solvents were assessed, and Van Harreveld and Steiner (1970) who compared acetone- and ethanol-substituted central nervous tissues, concluded that no significant morphological improvements could be detected attributable to a given fluid. The inclusion of osmium tetroxide in the substituting fluids does not significantly improve the ultrastructural result, but it does assist in the location of samples after resin embedding (Rebhun, 1972). In fact, osmium does not interact with tissues during substitution, but osmium oxidation occurs post-substitution at temperatures above about −25 °C (Van Harreveld, Crowell & Malhotra, 1965; Hereward & Northcote, 1972). Furthermore, the addition of osmium never improves, but occasionally diminishes, the ion-retention properties of a range of common substitution fluids (Van Zyl, Forrest, Hocking & Pallaghy, 1976). Shrinkage artefacts that may arise during freeze substitution can be almost completely eliminated (in plant tissues) if all solvents, including embedding media, are kept anhydrous over activated molecular sieves (Fisher, 1972). Aluminium oxide has also been used to provide anhydrous fluids (Läuchli, Spurr & Wittkopp, 1970). The maintenance of anhydrous conditions during processing helps to preserve the electrolyte composition of freeze-substituted specimens (see below).

From the viewpoint of X-ray microprobe analysis, the choice of substitution solvent is primarily governed by its properties of element retention. Other considerations such as its freezing point and substitution time are of secondary importance. Several workers have attempted to measure the loss of elements from botanical tissues during freeze substitution. Gielink, Sauer and Ringoet (1966) showed autoradiographically that the leaching of water-soluble and exchangeable calcium from oat plant tissues could be prevented by acetone substitution at −80 °C. Läuchli, Spurr and Wittkopp (1970) found little loss of introduced ^{86}Rb from corn roots during ether substitution and infiltration with Spurr's low viscosity resin. In comparison, Spurr (1972) found that most of the sodium and calcium content of the cell walls and cytoplasm of tomato leaves (measured in EMMA-4) was extracted during substitution in diethyl ether at −80 °C for 8 days, followed by Spurr resin infiltration. However, these losses were averted by the addition of benzamide to the substitution fluid. Other additives tested by Spurr (1972) were picrolonic acid, 2,6-dinitrophenol, alizarin, oxalic acid, and cyanuric acid, but each was found to be ineffective.

The above studies did not monitor the over-all loss of tissue elements during substitution, and were restricted to the study of botanical tissue. A further limitation of Spurr's (1972) microprobe study is that it considered a heterogeneous element distribution after processing to be a strong validation of the employed processing regime. It is conceivable, however, that a heterogeneous distribution of elements could exist on tissues whose total element content had been substantially depleted during processing. Fortunately, four groups of authors have quantitatively and systematically studied the loss of elements from both animal (Ostrowski, Komender, Koscianek & Kwarecki, 1962*a*, 1962*b*; Mehard & Volcani, 1975; Van Zyl, Forrest, Hocking & Pallaghy, 1976), and plant tissues (DeFilippis & Pallaghy, 1975; Harvey, Hall & Flowers, 1976; Van Zyl, Forrest, Hocking & Pallaghy, 1976) during exposure to various substitution procedures. The findings of these authors are summarized in Table 4.2. Before these important papers are considered individually, it should be pointed out that the most valuable contribution of the more recent reports (DeFilippis & Pallaghy, 1975; Mehard & Volcani, 1975; Harvey, Hall & Flowers, 1976; Van Zyl, Forrest, Hocking & Pallaghy, 1976) is that they convincingly show that the intervention of complexing additives (Spurr, 1972)

Table 4.2 Summary of published data on the retention of elements during various freeze-substitution procedures

Element	*Tissue*	*Procedure*	*% Retention*	*Reference*
P	Rat liver	Samples frozen in isopentane: Substitution in		Ostrowski, Komender, Koscianek & Kwarecki (1962*a*)
		methanol, −79 °C, 6–9 days	61.3	
		ethanol, −79 °C, 6–9 days	79.7	
		butanol, −79 °C, 6–9 days	83.0	
		propanol, −79 °C, 6–9 days	83.7	
		acetone, −79 °C, 6–9 days	98.4	
N	Rat liver	Samples frozen in isopentane: Substitution in		Ostrowski, Komender, Koscianek & Kwarecki (1962*a*)
		methanol, −79 °C, 6–9 days	95.5	
		ethanol, −79 °C, 6–9 days	97.2	
		butanol, −79 °C, 6–9 days	97.6	
		propanol, −79 °C, 6–9 days	99.4	
		acetone, −79 °C, 6–9 days	100.0	
P	Rat liver	Samples frozen in isopentane: Substitution in		Ostrowski, Komender, Koscianek & Kwarecki (1962*b*)
		acetone, −79 °C, 3 days	97.5	
		acetone, −20 °C, 3 days	98.5	
		acetone, +4 °C, 3 days	94.1	
		methanol, −79 °C, 3 days	60.6	
		methanol, −20 °C, 3 days	46.1	
		methanol, +4 °C, 3 days	44.8	
N	Rat liver	Samples frozen in isopentane: Substitution in		Ostrowski, Komender, Koscianek & Kwarecki (1962*b*)
		acetone, −79 °C, 3 days	100.9	
		acetone, −20 °C, 3 days	101.2	
		acetone, +4 °C, 3 days	100.3	
		methanol, −79 °C, 3 days	96.7	
		methanol, −20 °C, 3 days	96.5	
		methanol, +4 °C, 3 days	95.7	

^{68}Ge	Diatom	Freezing in isopentane/liquid N_2	100	Mehard & Volcani (1975)
		Substitution in *dry ether*, −80 °C, 3 days	100	
		vinyl cyclohexene dioxide	85	
		Spurr embedding medium	80	
	*Liver, spleen, kidney, lung	Freezing in isopentane/liquid N_2	110	
		Substitution in *dry ether*, −80 °C, 3 days	104	
		vinyl cyclohexene dioxide	110	
		Spurr embedding medium	106	
	*Isolated liver nuclei, mitochondria, microsomes	Freezing in isopentane/liquid N_2	98	
		Substitution in *dry ether*, −80 °C, 3 days	100	
		vinyl cyclohexene dioxide	105	
		Spurr resin embedding	97	
^{65}Zn		Samples frozen in liquid propane:		DeFilippis & Pallaghy (1975)
	Agar block	*dry ether*, −78 °C, 3–5 days	100	
	Carrot root	*dry ether*, −78 °C, 3–5 days	100	
	Carrot root	*dry 20% acrolein in ether*, −78 °C, 3–5 days	100	
	Barley root	*dry 20% acrolein in ether*, −78 °C, 3–5 days	100	
^{203}Hg		Samples frozen in liquid propane:		DeFilippis & Pallaghy (1975)
	Agar block	*dry ether*, −78 °C, 3–5 days	98.4	
	Carrot root	*dry ether*, −78 °C, 3–5 days	96.1	
	Carrot root	*dry 20% acrolein in ether*, −78 °C, 3–5 days	95.1	
	Barley root	*dry 20% acrolein in ether*, −78 °C, 3–5 days	95.9	
^{22}Na		Samples frozen in liquid propane:		Van Zyl, Forrest, Hocking & Pallaghy (1976)
	Barley root	*dry methanol*, −78 °C, 6 days	1	
	Cockroach nerve	*dry methanol*, −78 °C, 6 days	7	
	Agar block	*dry methanol*, −78 °C, 6 days	0	
	Barley root	*dry acetone*, −78 °C, 6 days	75	
	Cockroach nerve	*dry acetone*, −78 °C, 6 days	61	
	Agar block	*dry acetone*, −78 °C, 6 days	12	
	Barley root	*dry ethanol*, −78 °C, 6 days	10	
	Cockroach nerve	*dry ethanol*, −78 °C, 6 days	23	
	Agar block	*dry ethanol*, −78 °C, 6 days	2	

Table 4.2 Summary of published data on the retention of elements during various freeze-substitution procedures (cont.)

Element	*Tissue*	*Procedure*	*% Retention*	*Reference*
	Barley root	*dry acrolein*, −78 °C, 6 days	70	
	Cockroach nerve	*dry acrolein*, −78 °C, 6 days	36	
	Agar block	*dry acrolein*, −78 °C, 6 days	2	
	Barley root	*dry ether*, −78 °C, 6 days	99	
	Cockroach nerve	*dry ether*, −78 °C, 6 days	99	
	Agar block	*dry ether*, −78 °C, 6 days	100	
	Barley root	*dry 20% acrolein in ether*, −78 °C, 6 days	99.5	
	Cockroach nerve	*dry 20% acrolein in ether*, −78 °C, 6 days	97.6	
^{36}Cl		Samples frozen in liquid propane:		Van Zyl, Forrest, Hocking & Pallaghy (1976)
	Barley root	*dry 20% acrolein in ether*, −78 °C, 6 days	96	
	Cockroach nerve	*dry 20% acrolein in ether*, −78 °C, 6 days	94	
^{86}Rb		Samples frozen in liquid propane:		Van Zyl, Forrest, Hocking & Pallaghy (1976)
	Barley root	*dry 20% acrolein in ether*, −78 °C, 6 days	101	
	Cockroach nerve	*dry 20% acrolein in ether*, −78 °C, 6 days	104	
Na	Leaf tissue	Substitution in *dry acetone*, −72 °C (1 day), −40 °C (2 days)	> 98	Harvey, Hall & Flowers (1976)
		Substitution in *dry ether*, −72 °C (1 day) −40 °C (20 days)	> 99	
^{22}Na†	Leaf tissue	Substitution in *dry acetone*+Spurr-resin embedding regime	96.85	
		Substitution in *dry ether*+Spurr-resin embedding regime	99.43	
K+Rb	Leaf tissue	Substitution in *dry acetone*, −72 °C (1 day), −40 °C (2 days)	> 98.5	Harvey, Hall & Flowers (1976)
		Substitution in *dry ether*, −72 °C (1 day), −40 °C (20 days)	> 98.5	
^{36}Cl†	Leaf tissue	Substitution in *dry acetone*+Spurr-resin embedding regime	95.18	Harvey, Hall & Flowers (1976)
		Substitution in *dry ether*+Spurr-resin embedding regime	99.25	

* The retention figures presented are mean values derived from the data on individual tissues and organelle fractions recorded by Mehard and Volcani (1975); † the loss of ^{22}Na and ^{36}Cl isotopes during the various stages of freezing, substitution, resin infiltration and embedding (and during water flotation of sections) was also monitored by Harvey, Hall and Flowers (1976).

is unnecessary since the leaching of electrolytes can be restricted by the rigorous maintenance of anhydrous conditions during all stages of substitution and embedding (cf. DeFilippis & Pallaghy, 1973; Pallaghy, 1973).

Ostrowski, Komender, Koscianek and Kwarecki (1962*a*) showed that the loss of phosphorus from rat liver over a period of 6–9 days at −79 °C was ≏ 40% in methanol, ≏ 20% in ethanol, ≏ 17% in both butanol and propanol, and negligible in acetone. In addition the loss of liver nitrogen was less than 5% in all the media studied. In a subsequent paper (Ostrowski, Komender, Koscianek & Kwarecki, 1962*b*) the superiority of acetone over methanol was confirmed, but, surprisingly, it was found that substitution at −79 °C, −20 °C, and +4 °C did not significantly affect the phosphorus and nitrogen content of liver samples, especially those exposed to acetone.

Rat liver, spleen, kidney, and lung samples retained all of their ^{86}Ge-label during a freeze-substitution regime involving quenching in isopentane, dehydration in anhydrous diethyl ether at −80 °C, and embedding in Spurr's resin, medium E (Mehard & Volcani, 1975). The loss of label from isolated liver nuclei mitochondria and microsomes prepared similarly was almost negligible, but a loss of approximately 20% (much of which may have been due to the loss of cells during transfer through processing fluids) was recorded for a pelleted sample of diatoms.

Harvey, Hall and Flowers (1976) monitored the loss of sodium, potassium, rubidium, and chlorine from *Suaeda maritima* plants during the various stages of two freeze-substitution regimes, which included substitution in anhydrous acetone and diethyl ether, respectively. The plants chosen for study accumulated high concentrations of electrolytes, thus small percentage losses during processing could be accurately measured. It was found that in all cases the retention of cations was better than 96%; and the chlorine retention in acetone was ≏ 95% and in ether ≏ 99%. Thus, it was concluded that either solvent might be used for the localization of sodium or potassium, but that ether was more reliable than acetone for chloride localization. In addition, these authors compared the solubilities of pure chloride and nitrate salts of sodium, potassium, rubidium, magnesium, and calcium in acetone and diethyl ether at −70 °C and −40 °C. In general, the solubility of the various salts was greater at the higher temperature. No major differences in the solubilities of most of the salts in the two solvents were recorded, except for magnesium and calcium nitrates, where the solubility in acetone was much greater than in ether.

Van Zyl, Forrest, Hocking and Pallaghy (1976) compared the loss of ^{22}Na and ^{86}Rb from agar blocks, barley roots, and cockroach nerve tissue during freeze substitution at −79 °C in anhydrous acetone, acrolein, diethyl ether, n-hexane, ethanol, and methanol. The results presented by these authors are detailed in Table 4.2. Their general conclusion was that diethyl ether was the only solvent that did not induce a significant loss of ^{22}Na and ^{86}Rb from both tissues and model system during substitution (cf. Harvey, Hall & Flowers, 1976). In order to improve the ultrastructural appearance of the tissues the authors chose to substitute their specimens in a mixture of 20% acrolein in diethyl ether. The loss of ^{86}Rb, ^{22}Na and ^{36}Cl from tissues substituted by an acrolein/ether mixture was usually less than 6%, the loss slightly greater from nerve tissue than barley roots – these findings corresponded with those of DeFilippis and Pallaghy (1975) on zinc and mercury retention in plant tissues during a similar substitution regime. In addition, the preservation of cellular ultrastructure compared favourably with that seen in conventionally fixed and stained material, and represented an improvement over that obtained by substitution in pure ether, acetone, or n-hexane. Preliminary microprobe analyses (Van Zyl, Forrest, Hocking & Pallaghy, 1976) showed that the electrolytes were heterogeneously distributed within cellular compartments in both the plant and animal tissues. The authors concluded, therefore, that their technique preserved cellular ultrastructure and did not cause a significant loss of diffusable electrolytes. The authors wisely advised that 'it is important to strive for 100% retention of ions since methods which allow even small losses of ions from tissues raise uncertainties because the origin of the ion loss from within the cell cannot be established.' In other words, although the absolute loss of electrolytes may be minimal this does not preclude major shifts in the inter- and intracellular electrolyte compartments. In a recent communication, Forrest and Marshall (1976) reported similar element distribution patterns in specimens prepared by freeze substitution and in specimens sectioned and maintained during analysis in the frozen–hydrated state. Clearly, the translocation of labile components does not necessarily occur during freeze substitution.

It is possible to draw certain firm conclusions regarding the practical requirements for successful freeze substitution:

(1) Tissue frozen rapidly by direct contact with a quenching agent (or metal surface cooled by a quenching agent), or, as an inferior alternative, by direct contact with frozen substitution medium, is less damaged than fresh tissue immersed in substitution medium at the chosen substitution temperature.

(2) Unquestionably the best substitution fluid for the retention of labile elements is diethyl ether, with acetone being a second choice. The ultrastructural preservation of ether substituted tissues can be significantly improved with a minimal disruption of chemical integrity by including 20% or $< 20\%$ acrolein in the ether.

(3) All processing fluids must be kept absolutely anhydrous.

(4) Freeze-substituted tissues may be advantageously infiltrated with low viscosity resins. A potential source of element flux is the exposure of the dehydrated tissues to substitution solvents at or near room temperature

necessitated by the need to infiltrate and embed prior to sectioning. Artefacts of this nature could be avoided by sectioning the substituted tissues at or below the substitution temperature in a cryo-ultramicrotome.

4.2.9 FREEZE DRYING

Freeze drying as a technique for histological examination dates back to the turn of the present century (Ingram, Ingram & Hogben, 1974). It may be defined as a method for dehydration of frozen tissue and consists of the sublimation of water vapour from the specimen at low temperature (Rebhun, 1972). Although the morphological appearance of freeze-dried material is often distorted, the technique has recently received a great deal of attention in the context of preparing biological material for electron microprobe analysis. The attraction of the technique in this regard is that aqueous and/or organic solvents are not generally involved, and hence the possibility of retaining meaningful element distribution profiles is enhanced.

Several types of specimen may be prepared for EM visualization by freeze drying (for definitions and discussion of specimen thickness the reader is referred to Section 4.2.4):

(1) *Bulk specimens.* These are usually mounted on a specimen holder after quenching, freeze-drying, and (optionally) plastic embedding; and, since they are by definition too thick to allow the transmission of either photons or electrons, are visualized directly in SEM.

(2) *Thick or thin sections of plastic impregnated material.* Here, bulk specimens are freeze dried, plastic embedded and sectioned for visualization in TEM, SEM, or STEM.

(3) *Thick or thin sections of frozen material.* Sections are cut of the frozen bulk specimen, and the sections themselves are freeze dried prior to their visualization in TEM, SEM, or STEM (see Chapter 5 for a detailed discussion of this procedure).

Excellent reviews of the use of freeze drying for the preparation of biological tissues specifically for morphological examination have been presented by Sjöstrand (1967), Pearse (1968), Boyde and Wood (1969), and Burstone (1969).

Freezing

The problems of freezing have been defined in some detail in the section on freeze substitution (4.2.8). It may be pertinent to reiterate that whilst a number of freezing methods have been described in the literature, it is unwise to recommend a single regime for all specimens and experimental situations. Given a certain tissue and a given biological problem it may be necessary to develop a new freezing procedure, such as, for example, the novel methods used by Bullivant (1970), Monroe, Gamble, LaFarge, Gamboa, Morgan, Rosenthal and Bullivant (1968), Sjöstrom, Johansson and Thornell (1974), and

Bachmann and Schmitt (1971) to arrest their particular specimens in a given physiological 'state'. It is possible to generalize in the case of bulk specimens which are to be observed directly (without embedding), and point out that if the specimen is quench frozen by contact with a cooled metal surface, then it is that region of the specimen contiguous with the coolant that is most likely to possess the least dislocated element distribution pattern, albeit the most distorted structure. Therefore, such specimens may most advantageously be frozen by immersion.

The use of cryo-protectives cannot be recommended in conjunction with freeze drying since these agents will tend to concentrate on the surface of the specimen as the drying process proceeds (Boyde & Wood, 1969; Rebhun, 1972). This may result in poor structural definition, and introduce the possibility of contamination of the specimen chamber of the microscope (Boyde & Wood, 1969) and of ionic translocation by osmotic forces across the specimen surface towards the increasingly more concentrated cryo-protective solution.

Dehydration and resin embedding

The rate of sublimation of ice is a function of the vapour pressure of ice at the specimen temperature (this decreases sharply with decrease in temperature), and of the partial pressure of water vapour above the ice (Boyde & Wood, 1969). The combination of specimen temperature and the necessary drying time are, therefore, very important practical considerations.

Although a drying temperature of around −120 °C should theoretically provide for minimal tissue damage, in practice drying temperatures of around −60 °C to −80 °C are more commonly employed (Ingram, Ingram & Hogben, 1974; Ingram & Ingram, 1975). The period of drying will not only depend on the drying temperature [e.g. 1 mm thick liver samples are completely dried after 45 min at −10 °C, 2 h at −40 °C, and 5.5 h at −60 °C (Pearse, 1968)], but also on the size of the specimen [at −40 °C the drying time for liver samples 1 mm thick is 2 h, 2 mm is 3.75 h, 3 mm is 4.5 h, and 4 mm is 6.5 h (Pearse, 1968)]. The quality of the tissue matrix does not exert a great effect on the rate of drying. For example, the rate of drying of spleen, kidney, and liver samples was very similar over a fairly wide range of drying temperatures (Pearse, 1968).

There appears to be little advantage in using drying pressures in excess of 10^{-2} to 10^{-3} mmHg, since the rate of water removal from the greater mass of a specimen below its superficial layers is dependent on the rate of water transport through the specimen (Sjöstrand, 1967; Malhotra, 1968; Boyde & Wood, 1969; Ingram & Ingram, 1975).

Although the process of freeze drying is often considered to be non-destructive (Meryman, 1966), it has been found by a number of workers to introduce morphological damage (Bell, 1956; Mackenzie, 1965; Pearse, 1968; Burstone, 1969). Volume shrinkage is a frequently encountered artefact. Miyamoto and Moll (1971) showed that freeze-dried lung tissue embedded in paraffin wax and de-waxed had shrunk to 74% of its frozen volume. Ingram, Ingram and Hogben (1974) found that drops of 20% serum albumin shrank 32% after freeze drying and plastic embedding. Quantitatively similar observations were recorded for *Amphiuma* whole blood, mouse liver, and skeletal muscle (Ingram, Ingram & Hogben, 1974). Volume shrinkage can be reduced by fixing the dried tissues with osmium before embedding (Hanzon & Hermodsson, 1960; Ingram, Ingram & Hogben, 1974). Exposing the freeze-dried tissue to aqueous fixative solutions would, however, detract from the prime advantage of the technique. Fixing in a vapour phase would be more appropriate for microprobe analysis.

It has been said (Ingram, Ingram & Hogben, 1974) that specimen shrinkage during preparation must be measured should the biologist wish to convert his microprobe data obtained from freeze-dried tissue to the infinitely more meaningful concentration values in the original wet specimen. The conversion to wet weight values is, however, fraught with difficulties which arise primarily from our inability to extrapolate from a bulk specimen to the microvolume within that specimen which is being sampled by the microprobe. In other words, the percentage water content of the bulk tissue sample represents a mean of the percentage water content of the constituent cells and organelles, which individually may have widely differing values.

The leaching effects of embedding media have not received much attention (see also Section 4.2.6, above). Morgan, Davies and Erasmus (1975) showed in a quantitative microprobe study that Araldite extracted most of the electrolyte from freeze-dried rat aortas. In contrast, Ingram, Ingram and Hogben (1974) and Ingram and Ingram (1975) found no evidence of electrolyte leaching and redistribution in mouse kidney cells, frog red cells, frog skin epithelial cells, adenopodia, and mud puppy retinal cells, all of which were embedded with Epon 826 after being freeze dried. It may be, therefore, that certain embedding media are more potent electrolyte solvents than others (cf. Yarom, Peters & Hall, 1974; Yarom, Hall & Peters, 1975).

Applications

The freeze drying of thin frozen sections is becoming a routine procedure in many laboratories (see Chapter 5 for references). Apart from Ingram and co-workers (1974, 1975), whose work was referred to briefly in the preceding sections, surprisingly few workers have published X-ray microprobe data obtained from freeze-dried tissue blocks or whole mounts [but see Ryder and Bowen's work (1977) on copper distribution across the egg shell of a slug; and Trösch and Lindemann's work (1975) on the electrolyte composition of condensed and de-condensed insect salivary gland chromosomes], or obtained from freeze-dried, resin-embedded tissues sectioned for TEM observations [but see Skaer, Peters and Emmines's work (1974) on the calcium and phosphorus composition of human platelets]. This is surprising in view of the relative simplicity of the technique, the widespread use of the technique for preparing biological specimens for SEM morphological examination, and the limited involvement of processing fluids. The following points may be offered as some of the reasons: although the technique lends itself especially to the preparation of bulk specimens for the SEM, the analysis of such specimens is complicated by such considerations as (1) relatively poor structural and analytical resolution; (2) indeterminable beam penetration and, thus, a contribution to the analysis by uncharacterized sub-surface features; (3) X-ray absorption within the specimen; and (4) topographic shadowing effects. Even so, Ingram, Ingram and Hogben (1974) concluded that the freeze drying technique [in conjunction with the use of appropriately characterized standards in freeze-dried albumin block (Ingram, Ingram & Hogben, 1972; see also Dörge, Rick, Gehring, Mason & Thurau, 1975; Lehrer & Berkley, 1972; Spurr, 1975)] can produce specimens that are sufficiently stable under the electron beam to allow for analytical studies with an error of the order of 5 to 10%. However, these impressive figures refer to the accuracy and/or reproducibility of the analytical system employed, but not, unfortunately, to the accuracy with which the obtained data approach the *in vivo* concentration values in the vertebrate tissues studied. This, again, is a situation where the microprobe analyst is unable to assess the true accuracy and biological validity of his obtained data, since they are not verifiable by any other method. However, the fact that Ingram, Ingram and Hogben (1974) were (at least) able to monitor substantial levels of heterogeneously distributed sodium, potassium, and chlorine points to the potential efficacy of the procedure they describe. Since these workers always analysed either resin-embedded blocks of tissue or electron-opaque thick sections cut from the blocks, the method could easily be adopted for preparing thinner sections for observation and analysis in TEM or STEM. A recent paper (Zs.-Nagy, Pieri, Giuli, Bertoni-Freddari & Zs.-Nagy, 1977) describes a relatively simple method of freeze drying freeze-fractured tissue blocks. Major tissue components such as nuclei, cytoplasm, and red cells could be recognized in specimens prepared by this method. The authors also attempted to measure experimentally the depth of penetration of an electron beam into tissue specimens, and modified the Hall quantitative procedure (see Section 4.4) to allow for the measurement of absolute concentrations of sodium, potassium, and chlorine.

A modified freeze drying procedure has recently been described (Frederik & Klepper, 1976) that facilitates the autoradiographic localization of steroids in tests under the light microscope and TEM. Briefly, the freeze-dried tissues were osmium-vapour fixed, embedded in Epon, and thin sections were cut at a specimen temperature of −70 °C in a cryo-ultramicrotome. The ultrastructural preservation of the tissue thus prepared was impressive. The technique avoids any direct contact of the tissue with aqueous media (organic flotation media may or may not be used to aid section retrieval and flattening). It offers good prospects for the autoradiographic localization of steroids and other water-soluble components (Frederik & Klepper, 1976), and warrants attention as a method suitable for preparing biological soft tissues for microprobe analysis.

4.3 Specimen changes during microprobe analysis

Having taken great care to avoid disturbing the chemistry of a specimen during its preparation for microprobe analysis, it is equally important to reduce the likelihood of specimen damage and changes during its visualization and analysis in the microscope. The problem of specimen/electron beam interaction is of such profound importance, albeit a relatively neglected one, that it is useful to reiterate much of what has already been admirably presented by Chandler (1975*b*).

An electron beam can produce three major changes in the integrity of an organic specimen: (1) radiation damage; (2) contamination; (3) loss of materials.

4.3.1 RADIATION DAMAGE

During the first milliseconds after the exposure of an organic matrix to an electron beam of an EM, chemical bonds within the matrix are ruptured and radiation products are formed (Bahr, Johnson & Zeitler, 1965; Reimer, 1965; Stenn & Bahr, 1970; Glaeser, Cosslett & Valdre, 1971; Thach & Thach, 1971; Glaeser, 1977). The most obvious consequences of this interaction are a loss of mass due to the escape of such radiation products as CO_2, H_2, and NO_2, and a significant change in the electron diffraction pattern.

Radiation damage cannot be reduced by lowering the specimen temperature. For example, no observable difference was found in the susceptibility of crystalline *l*-valine and crystalline nucleoside adenosine to radiation damage at room temperature and when cooled on a liquid helium stage in a high voltage EM operated at 190 kV and at a magnification of ×5400 (Glaeser, Cosslett & Valdre, 1971). Essentially, therefore, it is impossible to prevent the formation of radiation products by lowering the specimen temperature, but it is feasible to reduce the rate of escape of the products from the irradiated microvolumes. In other words, 'at low temperature you don't *reduce* the damage, but you only *see* the damage when you bring the specimen up to a higher temperature' (comment in response to paper presented by Hall and Gupta, 1974*a*). Radiation damage thus becomes a practical consideration to the microprobe analyst only inasmuch as it predisposes the organic matrix to subsequent mass and element losses.

4.3.2 CONTAMINATION

Contamination in this context refers to the deposition of hydrocarbons on the surface of the specimen and near to the point of incidence of the electron beam. These hydrocarbons are derived from the oil and grease vapours that circulate even in the best vacuum systems. They are usually deposited as a halo – the so-called 'contamination ring' – at the cooler periphery of the irradiated area.

The rate of contamination with currently used instruments, in the absence of anti-contaminating devices, has been estimated to be of the order of 10 and 100 molecular layers a second (Echlin & Moreton, 1974). This highly significant local increase in specimen mass presents the microprobe analyst with three major problems (Chandler, 1975*b*):

(1) In TEM it affects the quality of the image. The problem is not as serious in SEM and EPMA systems.

(2) It can absorb a proportion of the low energy X-rays emerging from the specimen.

(3) It makes a significant contribution to the background X-ray signal. This affects qualitative studies by reducing the peak/background ratio, and quantitative studies via the resultant overestimation (especially in thin specimens) of mass thickness as measured from the continuum signal.

A reduced contamination rate can be achieved by: firstly, improving the quality of the EM vacuum system, and secondly, by trapping the vacuum vapours on the surfaces of liquid nitrogen cooled anti-contamination devices. It is particularly important to eliminate contamination during the analysis of specimens at low temperature, e.g. biological sections in the frozen-hydrated state (Saubermann & Echlin, 1975), because the cooled specimen itself will act as a cold trap.

4.3.3 LOSS OF MATERIALS

Electrons lose some of their energy to an irradiated specimen in the form of heat. Biological material has a relatively poor thermal conductivity and is, therefore, very susceptible to being heated. Specimen size does, however, modulate the degree of heating. For example, the bulk of a thick specimen is far removed from the conductive surfaces of the coating film and support material, and thus it heats up to a greater extent than thin specimens.

Local temperature increases in the specimen can be reduced by various procedures:

(1) It is usual to coat all specimens with a conductive coat of materials such as carbon, aluminium, gold, etc. For most microprobe applications,

especially if the full advantages of energy-dispersive spectrometers are to be exploited, it is best to choose coating and support materials of low atomic number in order to minimize the background signal and possible interference with certain characteristic radiations (Hodges & Muir, 1974).

(2) The specimen may be cooled to well below room temperature and transferred via special transfer devices to a microscope specimen stage, usually maintained at or near liquid nitrogen temperature. The first cooling stages, some sophisticated (e.g. Echlin & Moreton, 1974; Hutchinson, Bacaner, Broadhurst & Lilley, 1974), and others simpler (e.g. Turner & Smith, 1974), were designed for the large specimen chambers of various SEMs. Technically more demanding cold stage designs have fairly recently been described for the more confined specimen chambers of TEMs (e.g. Valdre & Horne, 1975).

(3) Microscope operating conditions may be set to minimize the heating effect. This may be achieved in the analysis of thin specimens, for example, by operating at high kV (> 40 kV), low beam current densities, and with a short analysis time. Soft biological tissues contain elements at relatively low concentrations, however, so in order to generate statistically significant spectra, microscope operating conditions are necessarily compromised.

When materials are quantified by procedures that express element concentrations as ratios of the local mass density (see Section 4.4), it is obviously important to avoid the removal of materials (light organic constituents and heavier analysable elements) from the specimen during its visualization and analysis.

Loss of organic constituents

A focused electron beam can remove a substantial proportion (10–90%) of the mass from an irradiated volume of an organic specimen (Bahr, Johnson & Zeitler, 1965; Hall & Gupta, 1974*b*), thus potentially introducing an intolerable error into any quantitative determination involving mass fractions. Organic material is lost from an irradiated specimen both as an immediate and rapid consequence of the formation of radiation products (Stenn & Bahr, 1970; Hall & Gupta, 1974*a*, 1974*b*), and, to a lesser and slower extent, due to a local rise in temperature (Höhling, Hall, Kriz, von Rosenstiel, Schnermann & Zessack, 1973).

Mass loss curves for uncooled, frozen air-dried sections mounted on celloidin films show a rapid loss of mass, with a stable residual mass reached after a dose of approximately 4×10^{-10} C/μm^2 (Hall & Gupta, 1974*b*). Pure Araldite sections and (occasionally) sections of Araldite-embedded material displayed a much slower loss of mass. It was also found that a celloidin film was completely removed in the electron beam.

The rapid loss of radiation products is not inhibited by coating the specimens with a thick layer of aluminium (Hall & Gupta, 1974*b*), whilst the slow thermal removal of organic mass under high beam currents is reduced by such coatings (Höhling, Hall, Kriz, von Rosenstiel, Schnermann & Zessack, 1973). Interestingly, it has been shown (Fuchs & Lindemann, 1975) that the local heating effect in ice crystals is reduced by a factor of 10 if their surface is coated with a 60 μm carbon layer. The thermal loss of mass is predictably reduced by lowering the specimen temperature (Höhling, Hall, Kriz, von Rosenstiel, Schnermann & Zessack, 1973), and so, surprisingly, is the loss of radiation products (Hall & Gupta, 1974*a*). The latter effect is puzzling but it has been suggested that the inhibition may occur in the step between bond rupture and the formation of volatile radiation products.

From a practical viewpoint, therefore, beam induced mass loss can be significantly reduced by:

(1) Cooling the specimen. An interesting point raised in the discussion of the paper presented by Hall and Gupta (1974*a*) is that the lowering of specimen temperature in order to preserve mass can lead to an increased rate of local contamination; thus leading to a net increase in local mass. One cannot overemphasize, therefore, the need to minimize the rate of specimen contamination within the microprobe.

(2) Employing a moving beam as opposed to a stationary probe (Hall & Gupta, 1974*a*; Fuchs & Lindemann, 1975). This facility, even more so perhaps than its unique imaging capacity, may yet prove to be the major advantage of the STEM mode over TEM.

Loss of analysable elements

The local heating of a specimen by electron beam interaction may cause the internal migration and/or partial or complete volatilization of certain elements. An example of the former phenomenon was described by Hodson and Marshall (1971), who showed that bound Na^+ and K^+ ions diffused out of an irradiated microvolume of a thin film, and returned to their original locations, but in a more loosely bound state, when the electron beam was removed. This confirmed an earlier observation by Borom and Hanneman (1966) on bulk glass samples.

The important instrumental parameter governing the degree of heating within a given specimen is the beam current density within the electron probe. Too often the beam current and beam diameter values are not provided, so that a comparison of the observations by various authors on the volatilization of different elements from a variety of different samples is seldom possible. Indeed, there are many microprobe instruments commercially available but not furnished with a means of measuring beam current [see Nicholson and Schreiber (1975) for suggestions for the incorporation of a Faraday cage into the Philips EM 301].

Elements may be volatilized even from pure mineral or salt crystal mixtures.

Aluminium was volatilized after about 100 sec from kaolin particles at a beam current density of about 11×10^5 pA/μm^2 in EMMA 4 (Chandler, 1973). Potassium and sodium can be volatilized from the minerals of sanidine and albite when analysed with a small spot and low current in EMMA; although the same minerals were stable at much higher currents with a large beam in EPMA (discussion of paper by Hall and Gupta, 1974*a*). Roinel (1975) found that with a 80 μm beam and a 200 μA current ($\simeq$ 40 pA/μm^2) in a CAMECA MS-46 microprobe the chlorine count rate obtained from a pure, dried mineral salt solution decreased rapidly. A similar but slower rate of loss was recorded for bromine and iodine. Under these same analytical conditions the count rates for sodium, magnesium, phosphorus, potassium, and calcium in pure standards (and of chlorine when in plasma, urine, or tubular fluids) were perfectly stable for more than 1000 sec irradiation. Morgan and Davies (unpublished) have assessed the degree of loss of elements from dried fluid standards after 200 sec irradiation in TEM. They found that given elements were beginning to volatilize at specific beam current density values: sulphur at 1.8×10^3 pA/μm^2, and phosphorus at 5.3×10^3 pA/μm^2. In addition, chlorine was volatilized down to a threshold of about 45% of its original concentration even at the lowest current density values employed (0.35×10^3 pA/μm^2), whilst calcium and magnesium were stable at the highest current density value (7.1×10^3 pA/μm^2).

Mercury, even when bound to macromolecules, is highly volatile under vacuum and in an electron beam (Hodges & Muir, 1974). Potassium was thermally volatilized from freeze-dried tobacco leaf stomata within 50 sec in a 1–2 μm beam with a current of 50 μA (maximum current density $\simeq 63.3 \times 10^3$ pA/μm^2) in an Acton EPMS (Sawhney & Zelitch, 1969). The data obtained by Chandler and Battersby (1976) represent the only systematic study on the effect of beam current (and accelerating voltage) on the stability of endogenous elements in biological soft tissues. Above certain threshold levels of beam current, the volatilization of chlorine (0.70×10^6 pA/μm^2), potassium (1.41×10^6 pA/μm^2), sodium, sulphur, and zinc (all at 2.82×10^6 pA/μm^2) was recorded after 100 sec irradiation of freeze-dried sections of sperm heads; calcium and phosphorus were relatively stable even at 7×10^6 pA/μm^2. These results are qualitatively similar to those of Roinel (1975), and Morgan and Davies (unpublished), but note that volatilization occurs at higher current densities when the elements are surrounded by a solid matrix.

The thermal removal of elements is also a function of specimen density (i.e. a function of the ability of a specimen to absorb heat). Elements are lost more readily from the denser sperm head than from the midpiece (Chandler, 1975*b*) and the surrounding seminal fluid (Chandler & Battersby, 1976). Similarly, the rate of chlorine loss from standard droplets is directly proportional to the concentration of salts within the droplets (Roinel, 1975). Elements are also volatilized much more readily from large (i.e. also thick) standard droplets in TEM than from smaller droplets (Morgan & Davies, unpublished).

Element volatilization rates can be significantly reduced by coating the specimen with a *continuous* film of carbon (Morgan & Davies, unpublished). Roinel (1975) achieved stable chloride counting rates by routinely adding 0.5 g/l urea to each of her standard mineral solutions. The reason for this stabilization is unclear, but Roinel suggested that the urea may interact with the electron beam effectively to produce a contamination surface layer of carbon. Whether a similar effect occurs in other matrices, including tissue samples, has yet to be established.

4.4 Specimen thickness considerations

Biological samples may be analysed as infinitely thick specimens, thick specimens on bulk substrates, or thin specimens on very thin supports (Hall, 1975). The method adopted for converting measured X-ray intensities to actual elemental concentrations depends largely on specimen thickness, and for this reason it is necessary to define what is meant by a 'thin' or a 'thick' specimen. Detailed definitions and discussions may be found in other texts (Hall, 1971, 1975; Russ, 1974; Russ, in this volume, Chapter 2; Warner & Coleman, 1974).

Infinitely thick specimens are those whose thickness exceeds the depth of penetration of the electron beam. Quantitative analysis of such specimens is usually undertaken by the classical ZAF correction procedure; where the Z correction accounts for effects due to the mean atomic number of the specimen matrix, A accounts for internal absorption of the generated X-rays within the specimen, and F accounts for the possibility of X-rays generated from the atoms of one element themselves being able to excite X-ray emission from the atoms of other elements. Cobet's simplification of the ZAF procedure represents an alternative formulation specifically applicable to biological specimens (see Hall, 1975).

Moderately thick specimens are defined as those whose thickness is only slightly less than the depth of penetration of the electron beam. Thus, the electron beam penetrates into the supporting substrate material. Several methods representing modifications of the basic ZAF procedure are available for the quantitative analysis of such specimens (Hall, 1975), some of which are readily applicable to biological material (e.g. Warner & Coleman, 1974).

Thin specimens are defined as those whose thickness is much less than the penetrative range of an incident electron beam. In such specimens, the characteristic X-ray intensity for a given element is linearly proportional to specimen thickness (Russ, 1972, 1974; Hall, 1975). The theory of quantitative analysis of thin specimens is, consequently, much simpler than for bulk specimens, because the atomic number effect (Z) is irrelevant when the method is based on ratios of simultaneously generated X-ray intensities, internal absorption of X-rays (A) is minimal (although it may be contributory in the case of low energy X-rays such as sodium Kα rays), and the fluorescence effect (F) is negligible since primary X-rays effectively escape from a thin specimen with a high probability of avoiding secondary X-ray generation (Hall, 1975). Thin specimens may be quantitatively analysed by two methods: (1) the determination of the relative amounts of two or more elements, i.e. the direct element ratio model (Russ, 1974); and (2) the determination of the mass fraction of a single element (Hall, 1975).

Before either of these quantitative procedures can be applied it must be established whether a given specimen conforms to the 'thinness' assumption as defined by the linear relationship of X-ray emission and thickness. In practice the Bremsstrahlung radiation may be conveniently taken as an indirect measure of local thickness (for applications, see Russ, 1972, 1974). Several practical considerations emerge from the thinness assumption:

(1) The Bremsstrahlung signal provides a measure of specimen density or of total mass within an irradiated volume. One cannot assume, therefore, that all biological specimens (and constituent parts of such specimens) of less than a given absolute thickness are 'thin' in the analytical sense. It is conceivable, for example, that certain dense bodies in an otherwise thin section may be analytically thick (the linearity relationship having failed due to X-ray absorption).

(2) For the purpose of comparison, specimens and standards should be analysed under identical instrumental conditions. For example, lowering the accelerating voltage may result in a non-linear change in the characteristic radiation: Bremsstrahlung ratio due to improved X-ray yields.

(3) Specimen thickness may also be considered in terms of the specific energies of characteristic radiations. Calcium Kα rays escape much more readily (with lower internal absorption) than sodium Kα rays (Russ, 1974). Thus, a specimen that is still thin in the context of quantifying calcium Kα radiation may be too thick with respect to sodium Kα radiation.

(4) A further practical consideration is the problem of imaging the specimen, especially in TEM. Because of the relatively low concentration of elements in most biological tissues it is often necessary to analyse sections of maximal thickness but which still conform to the analytical definitions of thinness. However, if this means that morphological features cannot be identified the thickness must be reduced. A useful rule-of-thumb is that a specimen thin enough to visualize in TEM is thin enough to analyse. Approximations of the upper limit of analytically useful section thicknesses have been published. The upper limit for sections of epoxy resin analysed at 40 kV has been estimated to be about 0.5 μm, but this may be extended to at least 1 μm if the elemental ratio model is applied (Russ, 1974). Hall (1975)

suggested an upper limit of at least 5 μm for soft biological tissues analysed at 30 kV or higher. Specimens near these upper limits of thinness may most appropriately be imaged and analysed in SEM or STEM modes.

4.5 Summary

It is impossible to draw firm conclusions on the efficacy of the manifold available preparative procedures in the context of preserving chemical integrity of biological tissues, because our understanding of the processes involved is incomplete. It is evident, however, that specimen preparation does not produce a straightforward introduction or extraction of elements, but a complex series of sequential fluxes in all directions, accompanied by local changes in ionic affinity, and a production of a profile of artificial loci. In addition, the extent of these disruptive influences depends on a number of parameters, not least of which is the exact nature and physiological state of the tissue examined.

In the light of the literature surveyed the present authors feel drawn to the opinion that the technique that offers the best long-term possibility of studying physiological processes is that of cryo-ultramicrotomy. More established techniques do not necessarily, however, have no contribution to make in the field of biological X-ray microanalysis in the foreseeable future. Given the present deficiencies of cryo-sectioning, especially the limited structural resolution due to the inherent lack of contrast in biological soft tissues, it seems that the more established techniques have certain outstanding advantages and, therefore, warrant the investment of time and effort in certain areas to yield biologically viable improvements. A list of the major advantages and disadvantages of given procedures and suggestions for their improvement is presented below:

(1) *Bulk analysis.* Advantage is that it yields base-line data on tissue microsamples and thus facilitates the understanding of subsequent high resolution microprobe investigations on 'intact' tissue specimens.

(2) *Fluid analysis.* Techniques for accurate quantitative analysis of nano- and picolitre fluid samples of biological origin in SEM and EPMA are well established. Sampling techniques of comparable sophistication are needed to facilitate such analysis in TEM.

(3) *Whole mounting.* Outstanding advantage is the rapidity with which large numbers of specimens can be prepared. Lends itself particularly to the study of individual cells in fluid suspension. Disadvantages are the poor morphological definition that it affords, and the inevitable surface contamination of cells and organelles by extracellular and cytoplasmic fluids, respectively. Need to investigate (a) methods of quantitatively compensating for surface contamination, and (b) methods of washing cell surfaces free of adhering contaminants.

(4) *Conventional* (*wet chemistry*) *method.* Advantages are the excellent morphological definition it offers, and the relative physical stability of the specimen in the electron beam. In addition, high resolution qualitative studies are rendered possible where it is acceptable to load the processing fluids with specific ions thus producing transcellular ionic equilibrium [e.g. the study of Ca^{2+}, Sr^{2+}, and Ba^{2+} uptake by mitochondria (Somlyo, Somlyo, Devine, Peters & Hall, 1974)]. Uncompromising disadvantage is the involvement of fluids at all stages during preparation which leads to a leaching and displacement of (at least) the more mobile tissue constituents. The efficiency of the various specific precipitation reactions in reversing these element fluxes has not been quantitatively assessed in most cases.

(5) *Freeze substitution.* An impressive wealth of evidence, especially from botanical sources, shows that certain freeze substitution regimes preserve subcellular ionic compartments. Freeze substitution may offer a viable compromise between the retention of structural and chemical integrity. Disadvantages of the technique are that it is time consuming, and that the analytically and morphologically well-preserved region of the specimen is limited to a narrow band approximately one or two cell diameters deep near the specimen/freezing agent interface. The technique needs to be (a) more widely assessed on animal tissues, and (b) modified by sectioning the dehydrated specimen at or near the substituting temperature in a cryo-ultramicrotome, so that element distribution in the sample more closely approaches the *in vivo* pattern, so that processing time is reduced.

(6) *Freeze drying.* Advantages are its relative simplicity, the fact that preparative fluids (except, optionally, embedding media) are not required, and that it lends itself to the preparation of several different types of specimen for eventual analysis in TEM, SEM, or STEM. Freeze drying and sectioning of embedded tissue blocks has been superceded by the freeze drying of unembedded frozen sections.

The present authors are of the opinion, therefore, that there exists at the present time no single preparative procedure which may be unreservedly recommended for all X-ray microprobe applications in biology. Given this premise it appears that the most sensible approach to a given analytical problem is to adopt, where feasible, a series of complimentary preparative regimes. For example, a typical sequence of experiments might consist of:

(1) *Bulk analysis* – either by a conventional technique such as flame spectrophotometry, or preferably via a microprobe modification.

(2) *Morphological study* – in TEM and/or SEM. This would greatly assist the identification of certain interesting structural features in frozen sections.

(3) *Preliminary microprobe survey* – either in single fixed, unstained, resin-embedded tissue specimens, or on tissues prepared by a 'compromise procedure' such as 'wet' cryo-microtomy or freeze substitution.

(4) *Analysis of unfixed, frozen tissues.* This is by far the most chemically rigorous and technologically demanding procedure. It may be used, therefore, either: (a) in a qualitative study to confirm observations recorded during the course of the above preliminary scheme; or (b) in a quantitative analysis of physiological processes as the routine preparative procedure, whilst the earlier preliminary studies may be used to assist in the identification of structures and in the interpretation of chemical data.

4.6 References

AGOSTINI, B. & HASSELBACH, W. (1971) Electron cytochemistry of calcium uptake in the fragmented sarcoplasmic reticulum. *Histochemie*, **28**, 55–67.

ANDERSEN, C. A. (1967) An introduction to the electron probe microanalyser and its application to biochemistry. In *Methods of Biochemical Analysis*, ed. Glick, D. Vol. 15, pp. 147–270. New York: Intersciences.

BACHMANN, L. & SCHMITT, W. W. (1971) Improved cryofixation applicable to freeze-etching. *Proceedings of the National Academy of Sciences of the United States of America*, **68**, 2149–2152.

BAHR, G. F., JOHNSON, F. B. & ZEITLER, E. (1965) The elementary composition of organic objects after electron irradiation. *Laboratory Investigation*, **14**, 1115–1133.

BAKER, J. R. (1960) *Cytological Technique.* 4th Edition, p. 18. London: Methuen & Co., Ltd.

BEAMAN, D. R., NISHIYAMA, R. H. & PENNER, J. A. (1969) The analysis of blood diseases with the electron microprobe. *Blood. The Journal of Haematology*, **34**, 401–413.

BELL, L. G. E. (1956) Freeze-drying. In *Physical Techniques in Biological Research*, ed. Oster, G. & Pollister, A. W. Vol. 3, pp. 1–27. New York: Academic Press.

BOOTHROYD, B. (1964) The problem of demineralization in thin sections of fully calcified bone. *Journal of Cell Biology*, **20**, 165–173.

BOOTHROYD, B. (1968) The adaptation of the technique of micro-incineration to electron microscopy. *Journal of The Royal Microscopical Society*, **88**, 529–544.

BOROM, M. P. & HANNEMAN, R. E. (1966) Local composition changes in alkali silicate glasses during electron microprobe analysis. *General Electric Technical Information Series*, No. 66-C-484, December.

BOYDE, A., JAMES, D. W., TRESMAN, R. L. & WILLIS, R. A. (1968) Outgrowth from chick embryo spinal cord *in vitro*, studied with the scanning electron microscope. *Zeitschrift für Zellforschung und Mikroskopische Anatomie*, **90**, 1–18.

BOYDE, A. & WILLIAMS, J. C. P. (1968) Surface morphology of frog striated muscle as prepared for and examined in the scanning electron microscope. *Journal of Physiology*, **197**, 10P.

BOYDE, A., GRAINGER, F. & JAMES, D. W. (1969) Scanning electron microscopic observations of chick embryo fibroblasts *in vitro* with particular reference to the movement of cells under others. *Zeitschrift für Zellforschung und Mikroskopische Anatomie*, **94**, 46–55.

BOYDE, A. & WOOD, C. (1969) Preparation of animal tissues for surface-scanning electron microscopy. *Journal of Microscopy*, **90**, 221–249.

BRIERLEY, G. P. & SLAUTTERBACK, D. B. (1964) Studies on ion transport. IV. An electron microscope study of the accumulation of Ca^{2+} and inorganic phosphate by heart mitochondria. *Biochimica et Biophysica Acta*, **82**, 183–186.

BULLIVANT, S. (1965) Freeze-substitution and supporting techniques. *Laboratory Investigation*, **14**, 1178–1195.

BULLIVANT, S. (1970) Present status of freezing techniques. In *Some Biological Techniques in Electron Microscopy*, ed. Parsons, D. F., pp. 101–146. New York: Academic Press.

BURSTONE, M. S. (1969) Cryobiology techniques in histochemistry, including freeze drying and cryostat techniques. In *Physical Techniques in Biological Research*, ed. Pollister, A. W. 4th Edition, Vol. 3, Part 3, *Cells and Tissues*, pp. 1–94. New York: Academic Press.

CARROLL, K. G. & TULLIS, J. L. (1968) Observations on the presence of titanium and zinc in human leucocytes. *Nature*, **217**, 1172–1173.

CHANDLER, J. A. (1973) Recent developments in analytical electron microscopy. *Journal of Microscopy*, **98**, 359–378.

CHANDLER, J. A. (1975*a*) Application of X-ray microanalysis to pathology. *Journal de Microscopie et de Biologie Cellulaire*, **22**, 425–432.

CHANDLER, J. A. (1975*b*) Electron probe microanalysis in cytochemistry. In *Techniques of Biochemical and Biophysical Morphology*, ed. Glick, D. & Rosenbaum, R. Vol. 2, pp. 308–433. New York: John Wiley & Sons, Inc.

CHANDLER, J. A. & BATTERSBY, S. (1976) X-ray microanalysis of ultrathin frozen and freeze-dried sections of human sperm cells. *Journal of Microscopy*, **107**, 55–65.

CHRISTENSEN, A. K. (1971) Frozen thin sections of fresh tissue for electron microscopy, with a description of pancreas and liver. *Journal of Cell Biology*, **51**, 772–804.

COHEN, A L. (1974) Critical point drying. In *Principles and Techniques of Scanning Electron Microscopy. Biological Applications*, ed. Hayat, M. A., Vol. 1, pp. 44–112. New York, Cincinnati, Toronto, London, Melbourne: Van Nostrand Reinhold Company.

COLEMAN, J. R., NILSSON, J. R., WARNER, R. R. & BATT, P. (1972) Qualitative and quantitative electron probe analysis of cytoplasmic granules in *Tetrahymena pyriformis. Experimental Cell Research*, **74**, 207–219.

COLEMAN, J. R., NILSSON, J. R., WARNER, R. R. & BATT, P. (1973*a*) Electron probe analysis of refractive bodies in *Amoeba proteus. Experimental Cell Research*, **76**, 31–40.

COLEMAN, J. R., NILSSON, J. R., WARNER, R. R. & BATT, P. (1973*b*) Effects of calcium and strontium on divalent ion contents of refractive granules in *Tetrahymena pyriformis. Experimental Cell Research*, **80**, 1–9.

COLEMAN, J. R., NILSSON, J. R. & WARNER, R. R. (1974) Electron probe analysis of calcium-rich lipid droplets in protozoa. In *Microprobe Analysis as Applied to Cells and Tissues*, ed. Hall, T., Echlin, P. & Kaufmann, R. Pp. 313–335. London: Academic Press.

COLEMAN, J. R. & TEREPKA, A. R. (1972) Electron probe analysis of the calcium distribution in cells of the embryonic chick chorioallantoic membrane. I. A critical evaluation of techniques. *Journal of Histochemistry and Cytochemistry*, **20**, 401–403.

COLEMAN, J. R. & TEREPKA, A. R. (1974) Preparatory methods for electron probe analysis. In *Principles and Techniques of Electron Microscopy. Biological Applications*, ed. Hayat, M. A. Vol. 4, pp. 159–207. New York, Cincinnati, Toronto, London, Melbourne: Van Nostrand Reinhold Company.

COLVIN, J. R., SOWDEN, L. C. & MALE, R. S. (1975) Variability of the iron, copper, and mercury contents of individual red blood cells. *Journal of Histochemistry and Cytochemistry*, **23**, 329–341.

DALLAM, R. D. (1957) Determination of protein and lipid lost during osmic acid fixation of tissues and cellular particulates. *Journal of Histochemistry and Cytochemistry*, **5**, 178–181.

DAVIES, T. W. & ERASMUS, D. A. (1973) Cryo-ultramicrotomy and X-ray microanalysis in the transmission electron microscope. *Science Tools*, **20**, 9–13.

DAVIES, T. W. & MORGAN, A. J. (1976) The application of X-ray analysis in the transmission electron analytical microscope (TEAM) to the quantitative bulk analysis of biological microsamples. *Journal of Microscopy*, **107**, 47–54.

DEFILIPPIS, L. F. & PALLAGHY, C. K. (1973) Effect of light on the volume and ion relations of chloroplasts in detached leaves of *Elodea densa*. *Australian Journal of Biological Sciences*, **26**, 1251–1265.

DEFILIPPIS, L. F. & PALLAGHY, C. K. (1975) Localization of zinc and mercury in plant cells. *Micron*, **6**, 111–120.

DEMPSEY, G. P. & BULLIVANT, S. (1976*a*) A copper block method for freezing non-cryoprotected tissue to produce ice crystal-free regions for electron microscopy. I. Evaluation using freeze-substitution. *Journal of Microscopy*, **106**, 251–260.

DEMPSEY, G. P. & BULLIVANT, S. (1976*b*) A copper block method for freezing non-cryoprotected tissue to produce ice crystal-free regions for electron microscopy. II. Evaluation using freeze fracturing with a cryo-ultramicrotome. *Journal of Microscopy*, **106**, 261–271.

DERMER, G. B. (1968) An autoradiographic and biochemical study of oleic acid absorption by intestinal slices including determinations of lipid loss during preparation for electron microscopy. *Journal of Ultrastructure Research*, **22**, 312–325.

DICULESCU, I., POPESCU, L. M., IONESCU, N. & BUTUCESCU, N. (1971) Ultrastructural study of calcium distribution in cardiac muscle cells. *Zeitschrift für Zellforschung und Mikroskopische Anatomie*, **121**, 181–198.

DÖRGE, A., RICK, R., GEHRING, K., MASON, J. & THURAU, K. (1975) Preparation and applicability of freeze-dried sections in the microprobe analysis of biological soft tissue. *Journal de Microscopie et de Biologie Cellulaire*, **22**, 205–214.

DUPREZ, A. & VIGNES, A. (1967) Determination ponctuelle du potassium intracellulaire d'hematies humaines. Analyze a la microsonde electronique de Castaing. *Comptes Rendus des Seances de la Societe de Biologie et de ses Filiales*, **161**, 1358–1360.

ECHLIN, P. & MORETON, R. (1974) The preparation of biological materials for X-ray microanalysis. In *Microprobe Analysis as Applied to Cells and Tissues*, ed. Hall, T., Echlin, P. & Kaufmann, R. Pp. 159–174. London & New York: Academic Press.

ELBERS, P. F. (1966) Ion permeability of the egg of *Limnaea stagnalis* L. in fixation for electron microscopy. *Biochimica et Biophysica Acta*, **112**, 318–329.

ERÄNKÖ, O. (1954) Quenching of tissues for freeze-drying. *Acta Anatomica*, **22**, 331–336.

FERNANDEZ-MORAN, H. (1960) Low-temperatures preparation techniques for electron microscopy of biological specimens based on rapid freezing with liquid helium. II. *Annals of the New York Academy of Sciences*, **85**, 689–713.

FERNANDEZ-MORAN, H. (1961) Lamellar systems in myelin and photoreceptors as revealed by high resolution electron microscopy. In *Macromolecular Complexes*, ed. Edds, M. V. Pp. 113–159. New York: Ronald Press.

FISHER, D. B. (1972) Artefacts in the embedment of water-soluble compounds for light microscopy. *Plant Physiology*, **49**, 161–165.

FISHER, D. B. & HOUSLEY, T. L. (1972) The retention of water-soluble compounds during freeze substitution and micro-autoradiography. *Plant Physiology*, **49**, 166–171.

FORREST, Q. G. & MARSHALL, A. T. (1976) Comparative X-ray microanalysis of frozen-hydrated and freeze-substituted specimens. *Proceedings of the Sixth European Congress on Electron Microscopy*, Jerusalem, 1976.

FREDERIK, P. M. & KLEPPER, D. (1976) The possibility of electron microscopic autoradiography of steroids after freeze drying of unfixed testes. *Journal of Microscopy*, **106**, 209–219.

FUCHS, W. & LINDEMANN, B. (1975) Electron beam X-ray microanalysis of frozen biological bulk specimen below 130K. I. Instrumentation and specimen preparation. *Journal de Microscopie et de Biologie Cellulaire*, **22**, 227–232.

GARFIELD, R. E., HENDERSON, R. M. & DANIEL, E. E. (1972) Evaluation of the pyroantimonate technique for localization of tissue sodium. *Tissue and Cell*, **4**, 575–589.

GEORGE, S. G., NOTT, J. A., PIRIE, B. J. S. & MASON, A. Z. (1976) A comparative quantitative study of cadmium retention in tissues of a marine bivalve during different fixation and embedding procedures. *Proceedings of the Royal Microscopical Society, Micro 76 Supplement*, **11**, Part 5, p. 42.

GIELINK, A. J., SAUER, G. & RINGOET, A. (1966) Histoautoradiographic localization of calcium in oat plant tissues. *Stain Technology*, **41**, 281–286.

GLAESER, R. M. (1977) Radiation damage at low temperatures. *Proceedings of the Royal Microscopical Society, Abstracts of the First International Meeting on Low Temperature Biological Microscopy, Cambridge 1977*, **12**, Part 2.

GLAESER, R. M., COSSLETT, V. E. & VÄLDRE, U. (1971) Low temperature electron microscopy: radiation damage in crystalline biological materials. *Journal de Microscopie et de Biologie Cellulaire*, **12**, 133–138.

GREENAWALT, J. W. & CARAFOLI, E. (1966) Electron microscope studies on the active accumulation of Sr^{++} by rat-liver mitochondria. *Journal of Cell Biology*, **29**, 37–61.

GRILLO, T. A. I., OGUNNAIKE, P. O. & FAOYE, S. (1971) Effects of histological and electron microscopical fixatives on the insulin content of the rat pancreas. *Journal of Endocrinology*, **51**, 645–649.

GULLASCH, J. & KAUFMANN, R. (1974) Energy-dispersive X-ray microanalysis in soft biological tissues: relevance and reproducibility of the results as depending on specimen preparation (air drying, cryo-fixation, cool-stage techniques). In *Microprobe Analysis as Applied to Cells and Tissues*, ed. Hall, T., Echlin, P. & Kaufmann, R. Pp. 175–190. London & New York: Academic Press.

HALL, J. L., YEO, A. R. & FLOWERS, T. J. (1974) Uptake and localization of rubidium in the halophyte *Suaeda maritima. Zeitschrift für Pflanzenphysiologie*, **71**, 200–206.

HALL, T. A. (1971) The microprobe assay of chemical elements. In *Physical Techniques in Biological Research*, ed. Oster, G. 2nd Edition, Vol. 1A, pp. 157–275. New York & London: Academic Press.

HALL, T. A. (1975) Methods of quantitative analysis. *Journal de Microscopie et de Biologie Cellulaire*, **22**, 271–282.

HALL, T. A. & GUPTA, B. L. (1974*a*) Measurement of mass loss in biological specimens under an electron microbeam. In *Microprobe Analysis as Applied to Cells and Tissues*, ed. Hall, T., Echlin, P. & Kaufmann, R. Pp. 147–158. London & New York: Academic Press.

HALL, T. A. & GUPTA, B. L. (1974*b*) Beam-induced loss of organic mass under electron-microprobe conditions. *Journal of Microscopy*, **100**, 177–188.

HANZON, V. & HERMODSSON, L. H. (1960) Freeze drying of tissues for light and electron microscopy. *Journal of Ultrastructure Research*, **4**, 332–348.

HARVEY, D. M. R., FLOWERS, T. J. & HALL, J. L. (1976) Localization of chloride in leaf cells of the halophyte *Suaeda maritima* by silver precipitation. *The New Phytologist*, **77**, 319–323.

HARVEY, D. M. R., HALL, J. L. & FLOWERS, T. J. (1976) The use of freeze substitution in the preparation of plant tissue for ion localization studies. *Journal of Microscopy*, **107**, 189–198.

HAYAT, M. A. (1970) *Principles and Techniques of Electron Microscopy. Biological Applications.* Vol. 1, pp. 5–107. New York, Cincinnati, Toronto, London & Melbourne: Van Nostrand Reinhold Company.

HENDERSON, W. J. (1969) A simple replication technique for the study of biological tissues by electron microscopy. *Journal of Microscopy*, **89**, 369–372.

HENDERSON, W. J. (1971) Some application of an extraction replication technique for the study of biological materials. *Micron*, **2**, 250–266.

HENDERSON, W. J., GOUGH, J. & HARSE, J. (1970) Identification of mineral particles in pneumoconiotic lungs. *Journal of Clinical Pathology*, **23**, 104–109.

HENDERSON, W. J., HARSE, J. & GRIFFITHS, K. (1969) A replication technique for the identification of asbestos fibres in mesothelioma. *European Journal of Cancer*, **5**, 621–624.

HENDERSON, W. J., JOSLIN, C. A. F., TURNBULL, A. C. & GRIFFITHS, K. (1971) Talc and carcinoma of the ovary and cervix. *Journal of Obstetrics and Gynaecology of the British Commonwealth*, **78**, 266–272.

HENDERSON, W. J. & GRIFFITHS, K. (1972) Shadow casting and replication. In *Principles and Techniques of Electron Microscopy. Biological Applications*, ed. Hayat, M. A. Vol. 2, pp. 149–193. New York, Cincinnati, Toronto, London & Melbourne: Van Nostrand Reinhold Company.

HEREWARD, F. V. & NORTHCOTE, D. H. (1972) A simple freeze substitution method for the study of ultrastructure of plant tissue. *Experimental Cell Research*, **70**, 73–80.

HODGES, G. M. & MUIR, M. D. (1974) X-ray spectroscopy in the scanning electron microscope study of cell and tissue culture material. In *Microprobe Analysis as Applied to Cells and Tissues*, ed. Hall, T., Echlin, P. & Kaufmann, R. Pp. 277–291. London & New York: Academic Press.

HODSON, S. & MARSHALL, J. (1971) Migration of potassium out of electron microscope specimens. *Journal of the Royal Microscopical Society*, **93**, 49–53.

HÖHLING, H. J., HALL, T. A., KRIZ, W., VON ROSENSTIEL, A. P., SCHNERMANN, J. & ZESSACK, U. (1973) Loss of mass in biological specimens during electron-probe X-ray microanalysis. In *Modern Techniques in Physiological Sciences*, ed. Gross, J.

F., Kaufmann, R. & Wetterer, E. Pp. 335–344. London & New York: Academic Press.

HÖHLING, H. J. & NICHOLSON, W. A. P. (1975) Electron microprobe analysis in hard tissue research: specimen. *Journal de Microscopie et de Biologie Cellulaire*, **22**, 185–192.

HOHMAN, W. & SCHRAER, H. (1972) Low temperature ultramicroincineration of thin sectioned tissue. *Journal of Cell Biology*, **55**, 328–354.

HOWELL, S. L. & TYHURST, M. (1976) 45-Calcium localization in islets of Langerhans, a study by electron-microscope autoradiography. *Journal of Cell Science*, **21**, 415–422.

HUTCHINSON, T. E., BACANER, M., BROADHURST, J. & LILLEY, J. (1974) Elemental microanalysis of frozen biological thin sections by scanning electron microscopy and energy selective X-ray analysis. In *Microprobe Analysis as Applied to Cells and Tissues*, ed. Hall, T., Echlin, P. & Kaufmann, R. Pp. 191–200. London & New York: Academic Press.

INGRAM, M. J. & HOGBEN, C. A. M. (1968) Procedures for the study of biological soft tissue with the electron microprobe. In *Developments in Applied Spectroscopy*, ed. Baer, W. K., Perkins, A. J. & Grove, E. L. Pp. 43–64. New York: Plenum Press.

INGRAM, F. D., INGRAM, M. J. & HOGBEN, C. A. M. (1972) Quantitative electron probe analysis of soft biological tissue for electrolytes. *Journal of Histochemistry and Cytochemistry*, **20**, 716–722.

INGRAM, F. D., INGRAM, M. J. & HOGBEN, C. A. M. (1974) An analysis of the freeze-dried, plastic embedded electron probe specimen preparation. In *Microprobe Analysis as Applied to Cells and Tissues*, ed. Hall, T., Echlin, P. & Kaufmann, R. Pp. 119–146. London & New York: Academic Press.

INGRAM, F. D. & INGRAM, M. J. (1975) Quantitative analysis with the freeze-dried plastic embedded tissue specimen. *Journal de Microscopie et de Biologie Cellulaire*, **22**, 193–204.

KIMZEY, S. L. & BURNS, L. C. (1973) Electron probe microanalysis of cellular potassium distribution. *Annals of the New York Academy of Sciences*, **204**, 486–501.

KIRK, R. G., CRENSHAW, M. A. & TOSTESON, D. C. (1974) Potassium content of single human red cells measured with an electron probe. *Journal of Cellular and Comparative Physiology*, **84**, 29–36.

KORN, E. D. & WEISMAN, R. A. (1966) I. Loss of lipids during preparation of amoebae for electron microscopy. *Biochimica et Biophysica Acta*, **116**, 325–335.

KRAMES, B. & PAGE, E. (1968) Effects of electron microscopic fixatives on cell membranes of the perfused rat heart. *Biochimica et Biophysica Acta*, **150**, 23–31.

KUHN, C. (1972) A comparison of freeze substitution with other methods for preservation of the pulmonary alveolar lining layer. *American Journal of Anatomy*, **133**, 495–507.

LÄUCHLI, A. (1975) Precipitation technique for diffusable substances. *Journal de Microscopie et de Biologie Cellulaire*, **22**, 239–246.

LÄUCHLI, A., SPURR, A. R. & EPSTEIN, E. (1971) Lateral transport of ions into the xylem of corn roots. II. Evaluation of a stelar pump. *Plant Physiology*, **48**, 118–124.

LÄUCHLI, A., SPURR, A. R. & WITTKOPP, R. W. (1970) Electron probe analysis of freeze-substituted, epoxy resin embedded tissue for ion transport studies in plants. *Planta*, **95**, 341–350.

LÄUCHLI, A., STELZER, R., GUGGENHEIM, R. & HENNING, L. (1974) Precipitation techniques as a means for intracellular ion localization by use of electron probe analysis. In *Microprobe Analysis as Applied to Cells and Tissues*, ed. Hall, T., Echlin, P. & Kaufmann, R. Pp. 107–118. London & New York: Academic Press.

LECHENE, C. (1974) Electron probe microanalysis of picolitre liquid samples. In *Microprobe Analysis as Applied to Cells and Tissues*, ed. Hall, T., Echlin, P. & Kaufmann, R. Pp. 351–367. London & New York: Academic Press.

LECHENE, C. P., BRONNER, C. & KIRK, R. G. (1976) Electron probe microanalysis of chemical elemental content of single human red cells. *Journal of Cellular and Comparative Physiology*, **90**, 117–126.

LEHRER, G. M. & BERKLEY, C. (1972) Standards for electron probe microanalysis of biologic specimens. *Journal of Histochemistry and Cytochemistry*, **20**, 710–715.

LUFT, J. H. & WOOD, R. L. (1963) The extraction of tissue protein during and after fixation with osmium tetroxide in various buffer systems. *Journal of Cell Biology*, **19**, 46A.

LÜTTGE, U. & WEIGL, J. (1965) Zur Mikroautoradiographic wasserlöslicher Substanzen. *Planta*, **64**, 28–36.

MACKENZIE, A. P. (1965) Factors affecting the mechanism of transformation of ice into water vapour in the freeze-drying process. *Annals of the New York Academy of Sciences*, **125**, 522–547.

MALHOTRA, S. K. (1968) Freeze-substitution and freeze-drying in electron microscopy. In *The Interpretation of Cell Structure*, ed. McGee-Russell, S. M. & Ross, K. F. A. Pp. 11–21. London: Edward Arnold, Ltd.

MALHOTRA, S. K. & VAN HARREVELD, A. (1965) Some structural features of mitochondria in tissues prepared by freeze-substitution. *Journal of Ultrastructure Research*, **12**, 473–487.

MAYNARD, P. V., ELSTEIN, M. & CHANDLER, J. A. (1975) The effect of copper on the distribution of elements in human spermatozoa. *Journal of Reproduction and Fertility*, **43**, 41–48.

MEHARD, C. W. & VOLCANI, B. E. (1975) Evaluation of silicon and germanium retention in rat tissues and diatoms during cell and organelle preparation for electron probe microanalysis. *Journal of Histochemistry and Cytochemistry*, **23**, 348–358.

MEHARD, C. W. & VOLCANI, B. E. (1976) Silicon in rat liver organelles: electron probe microanalysis. *Cell and Tissue Research*, **166**, 255–263.

MERYMAN, H. T. (1966) Freeze-drying. In *Cryobiology*, ed. Meryman, H. T. Pp. 609–663. London & New York: Academic Press.

MIYAMOTO, Y. & MOLL, W. (1971) Measurements of dimensions and pathway of red cells in rapidly frozen lungs *in situ*. *Respiration Physiology*, **12**, 141–156.

MIZUHIRA, V. (1976) Elemental analysis of biological specimens by electron probe microanalysis. *Acta Histochemica et Cytochemica*, **9**, 69–87.

MONROE, R. G., GAMBLE, W. J., LAFARGE, C. G., GAMBOA, R., MORGAN, C. L., ROSENTHAL, A. & BULLIVANT, S. (1968) Myocardial ultrastructure in systole and diastole using ballistic cryofixation. *Journal of Ultrastructure Research*, **22**, 22–36.

MORGAN, A. J. & BELLAMY, D. (1973) Microanalysis of the elastic fibres of rat aorta. *Age and Ageing*, **2**, 61.

MORGAN, A. J., DAVIES, T. W. & ERASMUS, D. A. (1975) Changes in the concentration and distribution of elements during electron microscope preparative procedures. *Micron*, **6**, 11–23.

MORGAN, A. J. & DAVIES, T. W. (1976) X-ray analysis in TEM and SEM of calcium metabolism in the earthworm *Lumbricus terrestris*, using strontium as a marker. *Proceedings of the Royal Microscopical Society, Micro 76 Supplement*, **11**, Part 5, p. 33.

MORGAN, T. E. & HUBER, G. L. (1967) Loss of lipid during fixation for electron microscopy. *Journal of Cell Biology*, **32**, 757–760.

NEERACHER, H. (1966) Transportuntersuchungen an *Zea mays* mit Hilfe von THO and Mikroautoradiographie. *Berichte der Schweitzerischen Botanischen Gesellschaft*, **75**, 303–346.

NEUMANN, D. (1973) Zur Darstellung pflanzlicher Gewebe nach Gefriersubstitution unter besonderer B'rücksichtigung der Strukturerhaltung der Plastiden. *Acta Histochemica*, **47**, 278–288.

NEUMANN, D. (1974) Ionenbestimung in Zellkompartimenten durch energiedispersive Röntgenanalyse. *Naturwissenschaften*, **5**, 166–171.

NICHOLSON, W. A. P. & SCHREIBER, J. (1975) Electron microprobe microanalysis in the electron microscope. *Journal de Microscopie et de Biologie Cellulaire*, **22**, 169–176.

OSCHMAN, J. L., HALL, T. A., PETERS, P. D. & WALL, B. J. (1974) Microprobe analysis of membrane-associated calcium deposits in squid giant axon. *Journal of Cell Biology*, **61**, 156–165.

OSCHMAN, J. L. & WALL, B. J. (1972) Calcium binding to intestinal membranes. *Journal of Cell Biology*, **55**, 58–73.

OSTROWSKI, K., KOMENDER, J., KOSCIANEK, H. & KWARECKI, K. (1962*a*) Quantitative investigation of the P and N loss in the rat liver when using various media in the freeze-substitution technique. *Experientia*, **18**, 142–144.

OSTROWSKI, K., KOMENDER, J., KOSCIANEK, H. & KWARECKI, K. (1962*b*) Quantitative studies on the influence of the temperature applied in freeze-substitution on P, N, and dry mass losses in fixed tissues. *Experientia*, **18**, 227–228.

PADAWER, J. (1974) Identification of mast cells in the scanning electron microscope by means of X-ray spectrometry. *Journal of Cell Biology*, **61**, 641–648.

PALLAGHY, C. K. (1973) Electron probe microanalysis of potassium and chloride in freeze-substituted leaf sections of *Zea mays*. *Australian Journal of Biological Sciences*, **25**, 1015–1034.

PEARSE, A. G. E. (1968) *Histochemistry. Theoretical and Applied.* 3rd Edition, Vol. 1, Chap. 3, Freeze-drying of biological tissues, pp. 27–58. Chap. 4, Freeze-substitution of tissues and sections, pp. 59–69. London: J. & A. Churchill Ltd.

PEASE, D. C. (1967) Eutectic ethylene glycol and pure propylene glycol as substituting media for the dehydration of frozen tissue. *Journal of Ultrastructure Research*, **21**, 75–97.

PEASE, D. C. (1973) Substitution techniques. In *Advanced Techniques in Biological Electron Microscopy*, ed. Koehler, J. K. Pp. 35–66. Berlin, Heidelberg, New York; Springer-Verlag.

PEASE, D. C. & PETERSON, R. G. (1972) Polymerizable glutaraldehyde-urea mixtures as polar, water-containing embedding media. *Journal of Ultrastructure Research*, **41**, 133–159.

PENTTILA, A., KALIMO, H. & TRUMP, B. F. (1974) Influence of glutaraldehyde and/or osmium tetroxide on cell volume, ion content, mechanical stability, and membrane permeability of Ehrlich ascites tumor cells. *Journal of Cell Biology*, **63**, 197–214.

POSNER, A. S. (1972) In *Comparative Molecular Biology of Extracellular Matrices*, ed. Slavkin, H. C., p. 437. New York & London: Academic Press.

REBHUN, L. I. (1961) Applications of freeze-substitution to electron microscope studies of invertebrate oocytes. *Journal of Biophysics and Biochemical Cytology*, **9**, 785–798.

REBHUN, L. I. (1965*a*) Freeze substitution as a function of water concentration in cells. *Federation Proceedings*, **24** (Suppl. 15), S217–S232.

REBHUN, L. I. (1965*b*) (In a general discussion on freeze-substitution) *Federation Proceedings*, **24** (Suppl. 15), S235.

REBHUN, L. I. (1972) Freeze-substitution and freeze-drying. In *Principles and Techniques of Electron Microscopy. Biological Applications*, ed. Hayat, M. A. Vol. 2, pp. 1–49. New York, Cincinnati, Toronto, London & Melbourne: Van Nostrand Reinhold Company.

REBHUN, L. I. & SANDER, G. (1971) Electron microscope studies of frozen-substituted marine eggs. 1. Conditions for avoidance of ultracellular ice crystallization. *American Journal of Anatomy*, **130**, 1–16.

REIMER, L. (1965) Irradiation changes in organic and inorganic objects. *Laboratory Investigation*, **14**, 1082–1096.

RENAUD, S. (1959) Superiority of alcoholic over aqueous fixation in the histochemical detection of calcium. *Stain Technology*, **34**, 267–271.

ROBISON, W. L. (1973) Applications of the electron microprobe to the analysis of biological materials. In *Microprobe Analysis*, ed. Andersen, C. E. Pp. 271–321. New York, London, Sydney & Toronto: John Wiley & Sons.

ROBISON, W. L. & DAVIS, D. (1969) Determination of iodine concentration and distribution in rat thyroid follicles by electron probe microanalysis. *Journal of Cell Biology*, **43**, 115–121.

ROINEL, N. (1975) Electron microprobe quantitative analysis of lyophilised 10^{-10} volume samples. *Journal de Microscopie et de Biologie Cellulaire*, **22**, 261–268.

ROINEL, N., PASSOW, H. & MALOREY, P. (1974) The study of the applicability of the electron microprobe to a quantitative analysis of K and Na in single human red blood cells. *FEBS* (*Federation of European Biochemical Societies*) *Letters*, **41**, 81–84.

ROINEL, N. & PASSOW, H. (1975) Analyse quantitative au moyen de la microsonde électronique du contenu en sodium et potassium d'hématies humaines isolées. *Journal de Microscopie et de Biologie Cellulaire*, **22**, 475–478.

ROUTLEDGE, L. M., AMOS, W. B., GUPTA, B. L., HALL, T. A. & WEIS-FOGH, T. (1975) Microprobe measurements of calcium binding in the contractile spasmoneme of a vorticellid. *Journal of Cell Science*, **19**, 195–201.

RUSS, J. C. (1972) Obtaining quantitative X-ray analytical results from thin sections in the electron microscope. In *Thin-Section Microanalysis*, ed. Russ, J. C. & Panessa, B. J. Pp. 115–132. Raleigh, North Carolina: EDAX Laboratories.

RUSS, J. C. (1974) The direct element ratio model for quantitative analysis of thin sections. In *Microprobe Analysis as Applied to Cells and Tissues*, ed. Hall, T., Echlin, P. & Kaufmann, R. Pp. 269–276. London & New York: Academic Press.

RYDER, T. A. & BOWEN, I. D. (1977) The use of X-ray microanalysis to demonstrate the uptake of the molluscicide copper sulphate by slug eggs. *Histochemistry*, **52**, 55–60.

SAUBERMANN, A. J. (1975) The application of X-ray microanalysis in physiology. *Journal de Microscopie et de Biologie Cellulaire*, **22**, 401–414.

SAUBERMANN, A. J. & ECHLIN, P. (1975) The preparation, examination, and analysis of frozen hydrated tissue sections by scanning transmission electron microscopy and X-ray microanalysis. *Journal of Microscopy*, **105**, 155–191.

SAWHNEY, B. L. & ZELITCH, I. (1969) Direct determination of potassium ion accumulation in guard cells in relation to stomatal opening in light. *Plant Physiology*, **44**, 1350–1354.

SCHOENBERG, C. F., GOODFORD, P. J., WOLOWYK, M. W. & WOOTTON, G. S. (1973) Ionic changes during smooth muscle fixation for electron microscopy. *Journal of Mechanochemistry and Cell Motility*, **2**, 69–82.

SJÖSTRAND, F. S. (1967) Freeze-drying preservation. In *Electron Microscopy of Cells and Tissues. Instrumentation and Techniques.* Vol. 1, pp. 188–221. New York: Academic Press.

SJÖSTROM, M., JOHANSSON, R. & THORNELL, L. E. (1974) Cryo-ultramicrotomy of muscles in defined state. Methodological aspects. In *Electron Microscopy and Cytochemistry*, ed. Wiesse, E., Daems, W. Th., Molenaar, J. & van Duijn, P. Pp. 387–391. Amsterdam: North-Holland Publishing Co.

SKAER, R. J., PETERS, P. D. & EMMINES, J. P. (1974) The localization of calcium and phosphorus in human platelets. *Journal of Cell Science*, **15**, 679–692.

SOMLYO, A. P., SOMLYO, A. V., DEVINE, C. E., PETERS, P. D. & HALL, T. A. (1974) Electron microscopy and electron probe analysis of mitochondrial cation accumulation in smooth muscle. *Journal of Cell Biology*, **61**, 723–742.

SPURR, A. R. (1969) A low-viscosity epoxy resin embedding medium for electron microscopy. *Journal of Ultrastructure Research*, **26**, 31–43.

SPURR, A. R. (1972) Freeze-substitution additives for sodium and calcium retention in cells studied by X-ray analytical electron microscopy. *Botanical Gazette*, **133**, 263–270.

SPURR, A. R. (1975) Choice and preparation of standards for X-ray microanalysis of biological materials with special reference to macrocyclic polyether complexes. *Journal de Microscopie et de Biologie Cellulaire*, **22**, 287–302.

STEINBIß H. & SCHMITZ, K. (1973) CO_2-Fixierung and Stofftransport in Benthischen marinen Algen. *Planta*, **112**, 253–263.

STENN, K. S. & BAHR, G. F. (1970) A study of mass loss and product formation after irradiation of some dry amino acids, peptides, polypeptides, and proteins with an electron beam of low current density. *Journal of Histochemistry and Cytochemistry*, **18**, 574–580.

TAKAYA, K. (1975*a*) Energy-dispersive X-ray microanalysis of zymogen granules of mouse pancreas using fresh air dried tissue spread. *Archivum Histologicum Japonicum*, **37**, 387–393.

TAKAYA, K. (1975*b*) Electron probe microanalysis of the dense bodies of human blood platelets. *Archivum Histologicum Japonicum*, **37**, 335–341.

TAKAYA, K. (1975*c*) Energy-dispersive X-ray microanalysis of neurosecretory granules of mouse pituitary using fresh air-dried tissue spreads. *Cell and Tissue Research*, **159**, 227–231.

TAKAYA, K. (1975*d*) Intranuclear silica detection in a subcutaneous connective tissue cell by energy-dispersive X-ray microanalysis using fresh air-dried spread. *Journal of Histochemistry and Cytochemistry*, **23**, 681–685.

TANDLER, C. J. & SOLARI, A. J. (1969) Nucleolar orthophosphate ions. Electron microscope and diffraction studies. *Journal of Cell Biology*, **41**, 91–108.

TANDLER, C. J., LIBANATI, C. M. & SAUCHIS, C. A. (1970) The intracellular localization of inorganic cations with potassium pyroantimonate. *Journal of Cell Biology*, **45**, 355–366.

TAPP, R. L. (1975) X-ray microanalysis of the mid-gut epithelium of the fruitfly *Drosophila melanogaster*. *Journal of Cell Science*, **17**, 449–459.

TERMINE, J. D. (1972) In *The Comparative Molecular Biology of Extracellular Matrices*, ed. Slavkin, H. C. Pp. 444–446. New York & London: Academic Press.

THACH, R. E. & THACH, S. S. (1971) Damage to biological samples caused by the electron beam during electron microscopy. *Biophysical Journal*, **11**, 204–210.

THOMAS, R. S. (1964) Ultrastructural localization of mineral matter in bacterial spores by microincineration. *Journal of Cell Biology*, **23**, 113–133.

THOMAS, R. S. (1969) Microincineration for electron-microscopic localization of biological minerals. In *Advances in Optical and Electron Microscopy*, ed. Barer, R. & Cosslett, V. E. Vol. 3, pp. 99–150. New York: Academic Press Inc.

TRÖSCH, W. & LINDEMANN, B. (1975) Electron-beam X-ray microanalysis of isolated, freeze-dried giant chromosomes. *Journal de Microscopie et de Biologie Cellulaire*, **22**, 487–492.

TURNER, R. H. & SMITH, C. B. (1974) A simple technique for examining fresh, frozen, biological specimens in the scanning electron microscope. *Journal of Microscopy*, **102**, 209–214.

UMRATH, W. (1974) Cooling bath for rapid freezing in electron microscopy. *Journal of Microscopy*, **101**, 103–105.

VALDRE, U. & HORNE, R. W. (1975) A combined freeze chamber and low temperature stage for an electron microscope. *Journal of Microscopy*, **103**, 305–317.

VAN HARREVELD, A. & CROWELL, J. (1964) Electron microscopy after rapid freezing on a metal surface and substitution fixation. *Anatomical Record*, **149**, 381–385.

VAN HARREVELD, A., CROWELL, J. & MALHOTRA, S. K. (1965) A study of extracellular space in central nervous tissue by freeze substitution. *Journal of Cell Biology*, **25**, 117–137.

VAN HARREVELD, A. & STEINER, J. (1970) Extracellular space in frozen and ethanol substituted central nervous tissue. *Anatomical Records*, **166**, 117–129.

VAN HARREVELD, A., TRUBATCH, J. & STEINER, J. (1974) Rapid freezing and electron microscopy for the arrest of physiological processes. *Journal of Microscopy*, **100**, 189–198.

VAN STEVENINCK, R. F. M., CHENOWETH, A. R. F. & VAN STEVENINCK, M. E.

(1973) Ultrastructural localization of ions. In *Ion Transport in Plants*, ed. Anderson, W. P. Pp. 25–37. London & New York: Academic Press.

VAN STEVENINCK, R. F. M., VAN STEVENINCK, M. E., HALL, T. A. & PETERS, P. D. (1974) A chlorine-free embedding medium for use in X-ray analytical electron microscopic localization of chlorine in biological tissues. *Histochimie*, **38**, 173–180.

VAN ZYL, J., FORREST, Q. G., HOCKING, C. & PALLAGHY, C. K. (1976) Freeze substitution of plant and animal tissue for the localization of water-soluble compounds by electron probe microanalysis. *Micron*, **7**, 213–224.

VASSAR, P. S., HARDS, J. M., BROOKS, D. F., HAGENBERGER, B. & SEAMAN, G. V. F. (1972) Physicochemical effects of aldehydes on the human erythrocyte. *Journal of Cell Biology*, **53**, 809–818.

WARNER, R. R. & COLEMAN, J. R. (1974) Quantitative analysis of biological material using computer correction of X-ray intensities. In *Microprobe Analysis as Applied to Cells and Tissues*, ed. Hall, T., Echlin, P. & Kaufmann, R. Pp. 249–268. London & New York: Academic Press.

WERNER, G. & GULLASCH, J. (1974) Röntgenmikroanalyse an Rattenspermien. *Mikroskopie*, **30**, 95–101.

YAROM, R., PETERS, P. D. & HALL, T. A. (1974) Effect of glutaraldehyde and urea embedding on intracellular ionic elements. X-ray microanalysis of skeletal muscle and myocardium. *Journal of Ultrastructure Research*, **49**, 405–418.

YAROM, R., HALL, T. A. & PETERS, P. D. (1975) Calcium in myonuclei: electron microprobe X-ray analysis. *Experientia*, **31**, 154–157.

YAROM, R., PETERS, P. D., SCRIPPS, M. & ROGEL, S. (1974) Effect of specimen preparation on intracellular myocardial calcium. Electron microscopic X-ray microanalysis. *Histochemistry*, **38**, 143–153.

ZS.-NAGY, I., PIERI, C., GIULI, C., BERTONI-FREDDARI, C. & ZS.-NAGY, V. (1977) Energy-dispersive X-ray microanalysis of the electrolytes in biological bulk specimen. I. Specimen preparation, beam penetration, and quantitative analysis. *Journal of Ultrastructure Research*, **58**, 22–33.

5

T. C. APPLETON

The contribution of cryo-ultramicrotomy to X-ray microanalysis in biology

5.1 Introduction

The conventional procedures which are used to prepare biological material for examination in the electron microscope involve the use of fixatives in aqueous solutions, dehydration through graded series of alcohol or acetone, and embedding in plastics such as Araldite, Epon, and more recently Spurr's resin. These techniques are well established and are very suitable for morphological studies where clear membrane boundaries are desired. Sections of fixed and embedded material are usually stained in either aqueous or alcoholic stains to increase the electron density of structures within the cell. Tissues treated in this way, however, are often not suitable for chemical analysis unless the interest is primarily in crystalline inclusions such as hydroxy-apatite crystals (Hall, Anderson & Appleton, 1973), which remain with little or no extractions. The majority of elements of biological interest are either freely diffusible or are present as lightly bound deposits which become labile during 'active' phases of the cell's metabolism, and so loss of any material during the preparative procedures should be avoided. Loss or movement of one element may well influence the position of another.

Some attempts have been made to reduce the loss of labile materials using a variety of techniques. Höhling, Barckhaus, Krefting and Shreiber (1976) analysed mineralized collagen fibrils after freeze-dried samples had been embedded in styrol methacrylate (Kushida, 1961) or Epon (Spurr, 1969) under vacuum. Loss of material has been 'minimized' by reducing fixation and dehydration procedures, and by using glutaraldehyde–urea (Yarom, Peters & Hall, 1974), or by forming electron-dense precipitates which may partially withstand embedding procedures (Yarom & Chandler, 1974; Yarom, Maunder, Scripps, Hall & Dubowitz, 1975). These go some way towards solving the problem but overlook the fact that damage probably occurs within the few milliseconds contact with a fluid, and before either polymerization or precipitation can take place.

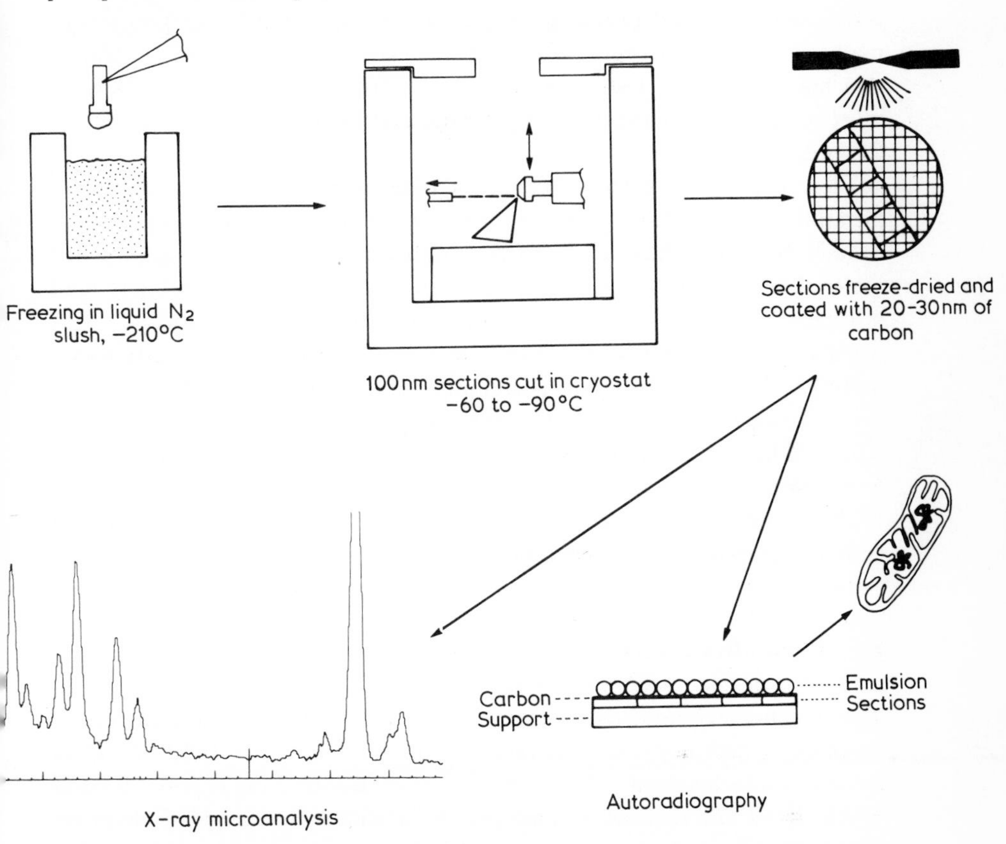

Fig. 5.1 Summary of technique using ultrathin frozen sections for X-ray microanalysis and autoradiography. (After Appleton, Shute & Baker, 1976)

Other workers have turned towards air drying as a viable alternative (Chandler & Battersby, 1976; Mizuhira, 1976) and have argued that the element of interest was not lost, but overlooked the fact that when an important physiological ion, such as potassium, is lost or shifted, the electrical balance of the cell is upset and redistribution of other ions must take place in order that the state of neutral overall charge is maintained. It may not even be enough to assume that a material is bound (Peters, Yarom, Dormann & Hall, 1976; Tsuchiya, 1976) because the really interesting biological question should be 'how did that material get there'? In trying to answer that question we should be prepared to look at any element within the context of all elements detected within the cell; it is often the most unexpected element which provides the clue to biological questions, and it is often our own arrogance and blindness which makes us miss the vital clue. Energy-dispersive analysis is sufficiently sensitive to detect physiologically significant levels of active elements, and provided we pay enough attention to the preparative methods we should accept information about all elements, hopefully confident that none have been lost or redistributed. That should be our aim, but we must remain critical about our techniques and honest in our evaluation (see Chapter 3).

This chapter will attempt to describe in critical terms the contribution of ultrathin frozen sections to X-ray microanalysis in biology, and in particular it describes one technique of cutting frozen sections of fresh frozen, unfixed material which can be examined without staining or contact with any fluid other than liquid nitrogen. The chapter will also discuss problems of interpretation (i.e. expressing the results in terms which are biologically meaningful) and some of the artefacts and pitfalls encountered. Certain aspects of the technique are described elsewhere (Appleton, 1972, 1973, 1974*a*, 1974*b*, 1977*a*). The techniques involved are not difficult but require some attention to technique and equipment.

The technique is summarized in Figure 5.1. The sections can be used for autoradiography and X-ray microanalysis.

5.2 Cryo-ultramicrotomy

5.2.1 FREEZING AND STORAGE

The freezing technique used is perhaps one of the most critical stages in any technique which hopes to localize diffusible compounds and is, of necessity, a compromise between a need for good morphological detail and a process which 'locks' the elements in their precise locations. Because physiologically active ions (e.g. sodium, potassium, and chlorine) are so mobile it is important that the time interval between excision and freezing of the material from an animal or plant be kept to the minimum. Even where tissues lie within the abdominal cavity of an animal such as a mouse it should be possible to remove and freeze small pieces of tissue within a few seconds. Cell cultures present few problems since a drop of culture can be frozen without disrupting the boundaries of any cells. 'Disruption' of a few cells in a densely packed structure such as liver may influence the elemental chemistry of neighbouring cells. Care must therefore be taken in preparing the material for freezing so that the damage is minimal and is trimmed away before sectioning (see Section 5.2.3), preparing oneself so that delays are kept to a minimum.

No freezing method available gives perfect preservation; freezing with cryo-protectives such as DMSO or glycerol provides excellent structural preservation but diffusible ions move across membrane boundaries; slow freezing methods (with or without cryo-protectives) allow cells to remain viable but probably allow redistribution of diffusible ions as the temperature is lowered; fast freezing without cryo-protectives gives good preservation with minimal ice crystal damage, but the cells may not remain in a viable state.

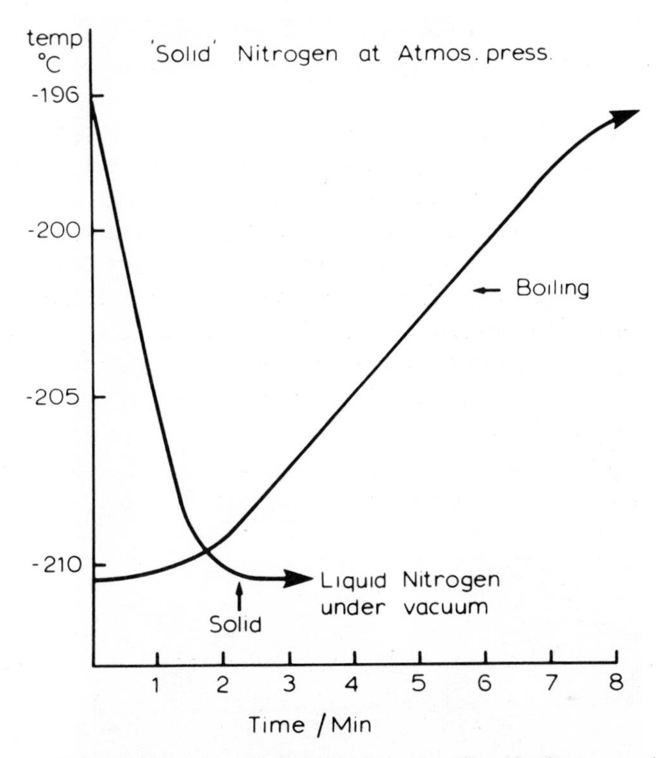

Fig. 5.2 Temperature curves for liquid nitrogen under rough vacuum (0.05 mmHg) and liquid nitrogen slush (solid). Note that there is a period of about 5–6 minutes before boiling takes place after the slush is removed from the vacuum.

There is now considerable evidence that cryo-protectives commonly used to increase morphological preservation of cell and membrane structures induce clustering of intramembraneous structures (McIntyre, Gilula & Karnovsky, 1974, Plattner, Schmitt-Fumian & Bachmann 1973), or even cell fusion (Ahkong, Fisher, Tampion & Lucy, 1975). Spray freezing of cell suspensions (Plattner, Schmitt-Fumian & Bachmann, 1973) probably gives the best preservation without cryo-protection but is not a practical proposition for most tissues. High pressure freezing (Riehle & Höchli, 1973), and propane jet-freezing (Moor, Kistler & Müller, 1976) allow vitrification with good morphological detail (Niedermeyer & Moor, 1976) and may eventually be applicable to tissues as well as cell suspensions.

Rapid freezing in liquid nitrogen slush gives preservation with minimal ice crystal damage. The preservation is sufficiently good to be able to identify cell types with sufficient confidence and to localize X-ray microanalysis to specific subcellular organelles (see Section 5.3 and 5.4). Liquid nitrogen slush

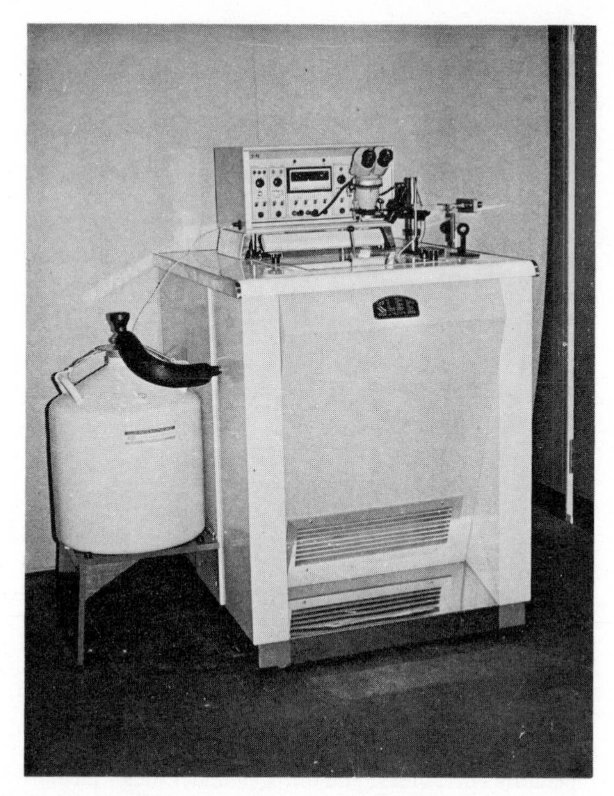

Fig. 5.3 Cryostat chamber. Note the liquid nitrogen Dewar and feed pipe to chamber. The cryostat is kept at back-up temperature of −25 to −40 °C with compressor refrigeration, and cooled to operating temperature by injection of liquid nitrogen controlled by a heater within the Dewar.

can be made in several commercially available 'slush-makers' or more simply in any evaporating unit available in most EM laboratories. Liquid nitrogen is held in a shallow cup made of expanded polystyrene – the top of an insulated baby bottle container is most suitable, 5 cm diameter and 4 cm deep – and this is placed in the evaporating chamber and evacuated with the roughing pump (0.05 mmHg). After about 1.5 minutes the liquid nitrogen becomes 'milky', and it freezes to a solid within 3 minutes. The 'slush' continues to contain small particles of solid nitrogen for 6 or 7 minutes after the container is placed at atmospheric pressure (Fig. 5.2).

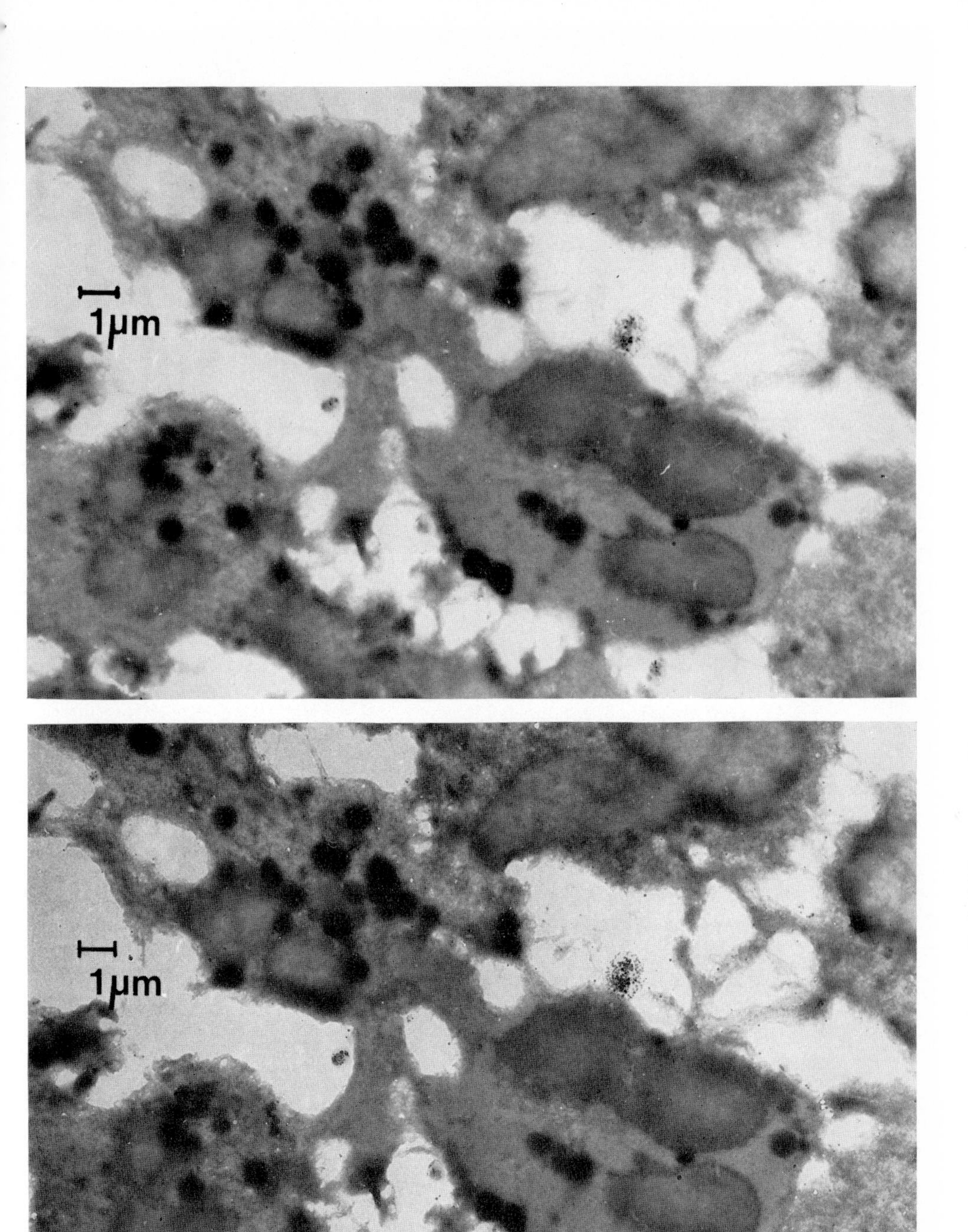

Plate 5.1 'True' focus image of ultrathin frozen section of polymorphonuclear leucocytes freeze-dried at atmospheric pressure in the cryostat for 3 hours.

Plate 5.2 Underfocused image of same area as in Plate 5.1. Note the 'phase-effect' of gross underfocusing which often makes the interpretation of unstained frozen sections easier.

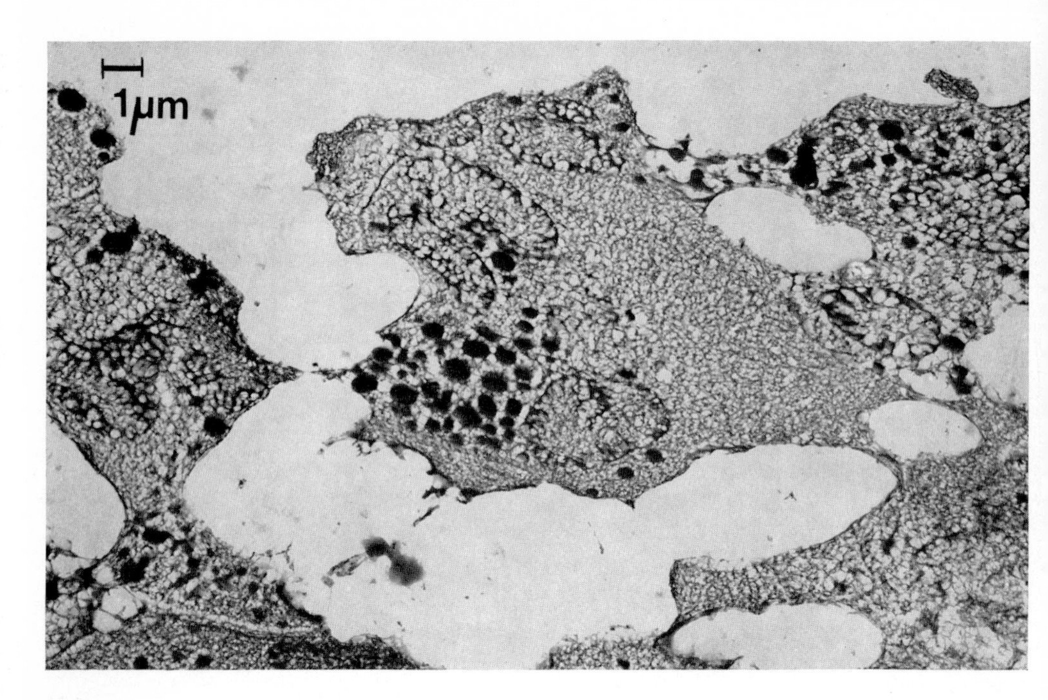

Plate 5.3 Adjacent ribbon of sections to Plates 5.1 and 5.2 but after freeze drying under vacuum at the same temperature (−75 °C). Vacuum was 10^{-4} mmHg. The lace effect is symptomatic of rapid freeze drying which causes disruption of the sections – probably as a result of the surface tension forces involved. Slow freeze drying avoids such disruption and the rate of drying must be determined for each tissue.

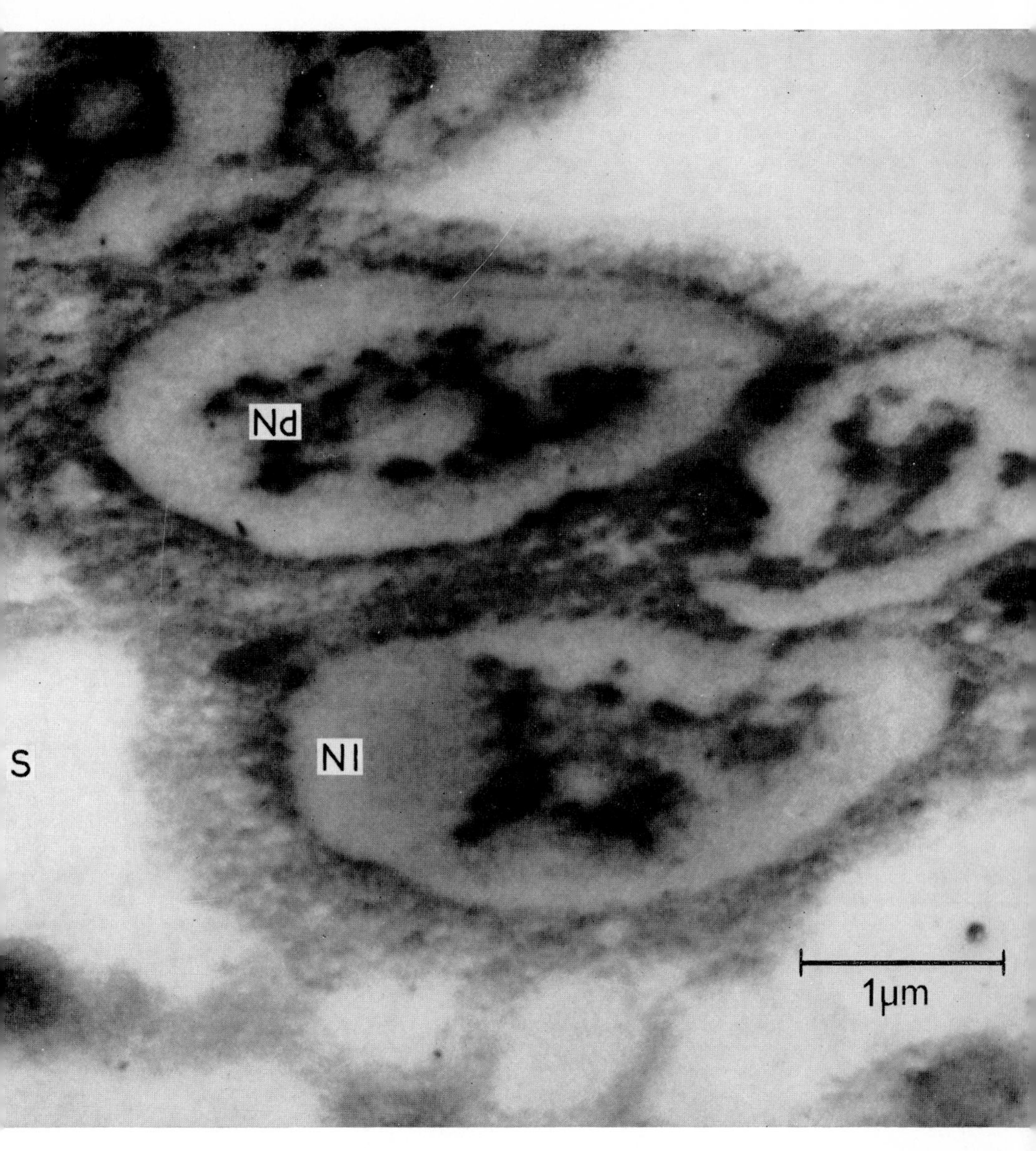

Plate 5.4 STEM (scanning transmission electron micrograph) image of frozen section of mouse spleen showing three cells with nuclei (Nl, light areas of nucleus; Nd, dark areas of nucleus), and sinusoid spaces (S). Note that it is difficult to see the boundaries between cells. Micrograph taken by courtesy of JEOL (UK) Ltd., using the 100C microscope at 100 kV and 25° tilt. Compare with Plate 5.5.

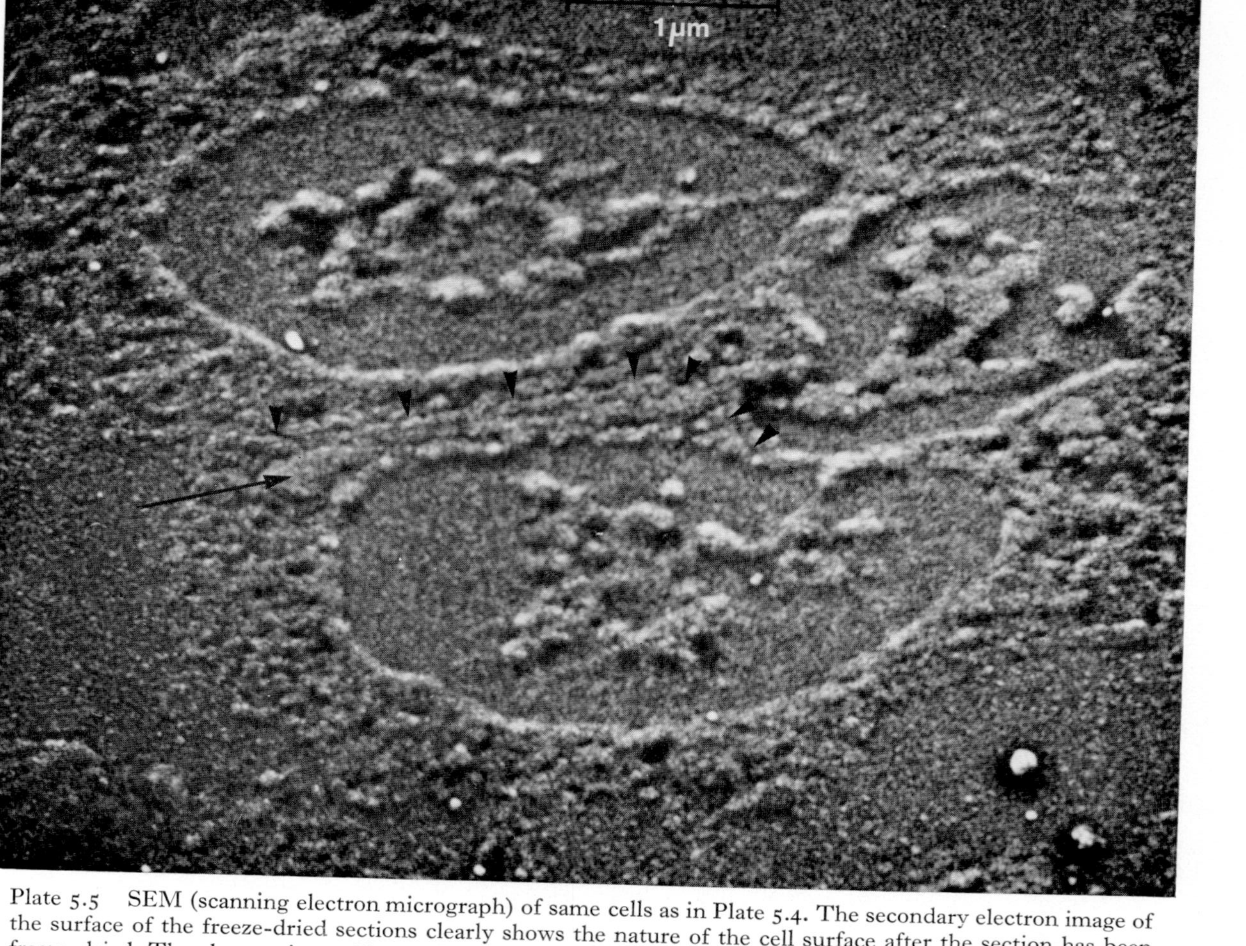

Plate 5.5 SEM (scanning electron micrograph) of same cells as in Plate 5.4. The secondary electron image of the surface of the freeze-dried sections clearly shows the nature of the cell surface after the section has been freeze-dried. The changes in profile result from the different water contents of the various components of the cell. JEOL 100C microscope (courtesy of JEOL Ltd.) 100 kV 25° tilt. Note that the junction between the cells (arrow heads) is now quite obvious. The same boundaries can be seen in Plate 5.4 once they have been seen in Plate 5.5.

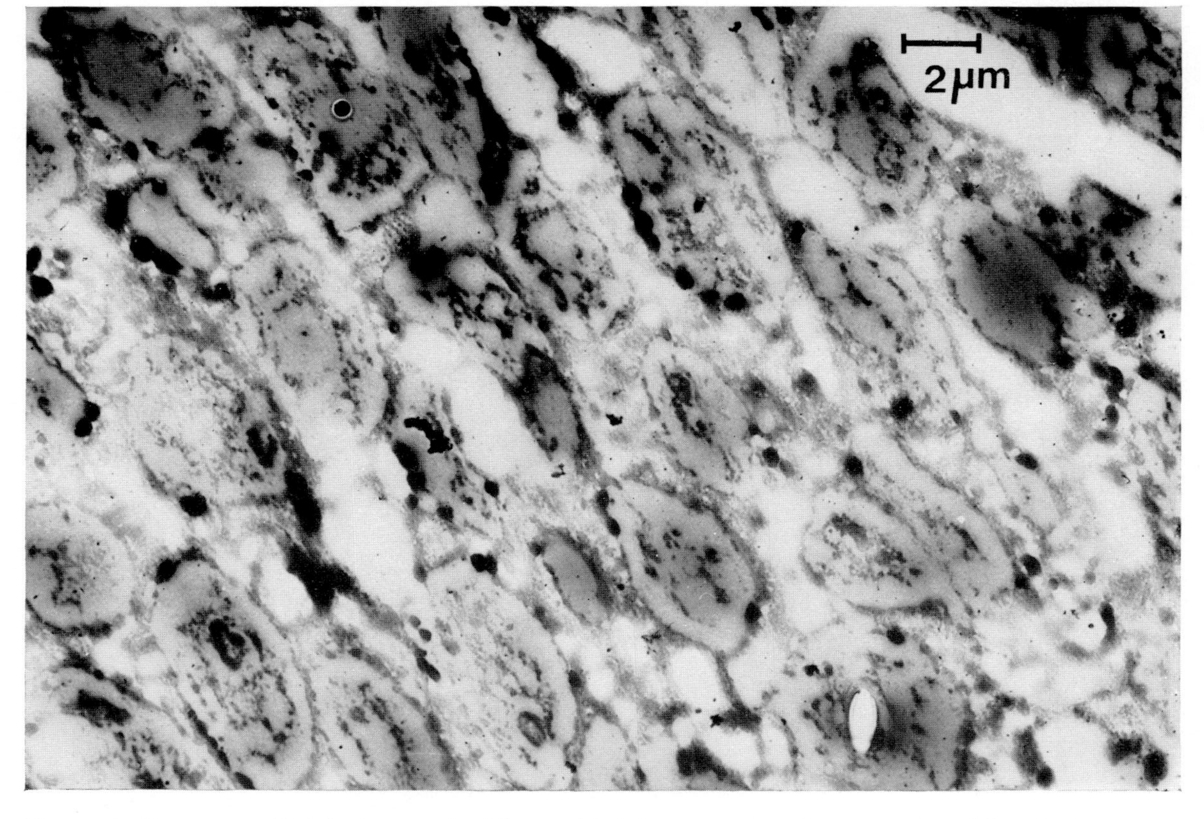

Plate 5.6 Low power TEM micrograph of freeze-dried frozen section of mouse spleen. Note the large area which can be examined. The light areas are sinusoids running through the tissue. Unstained section.

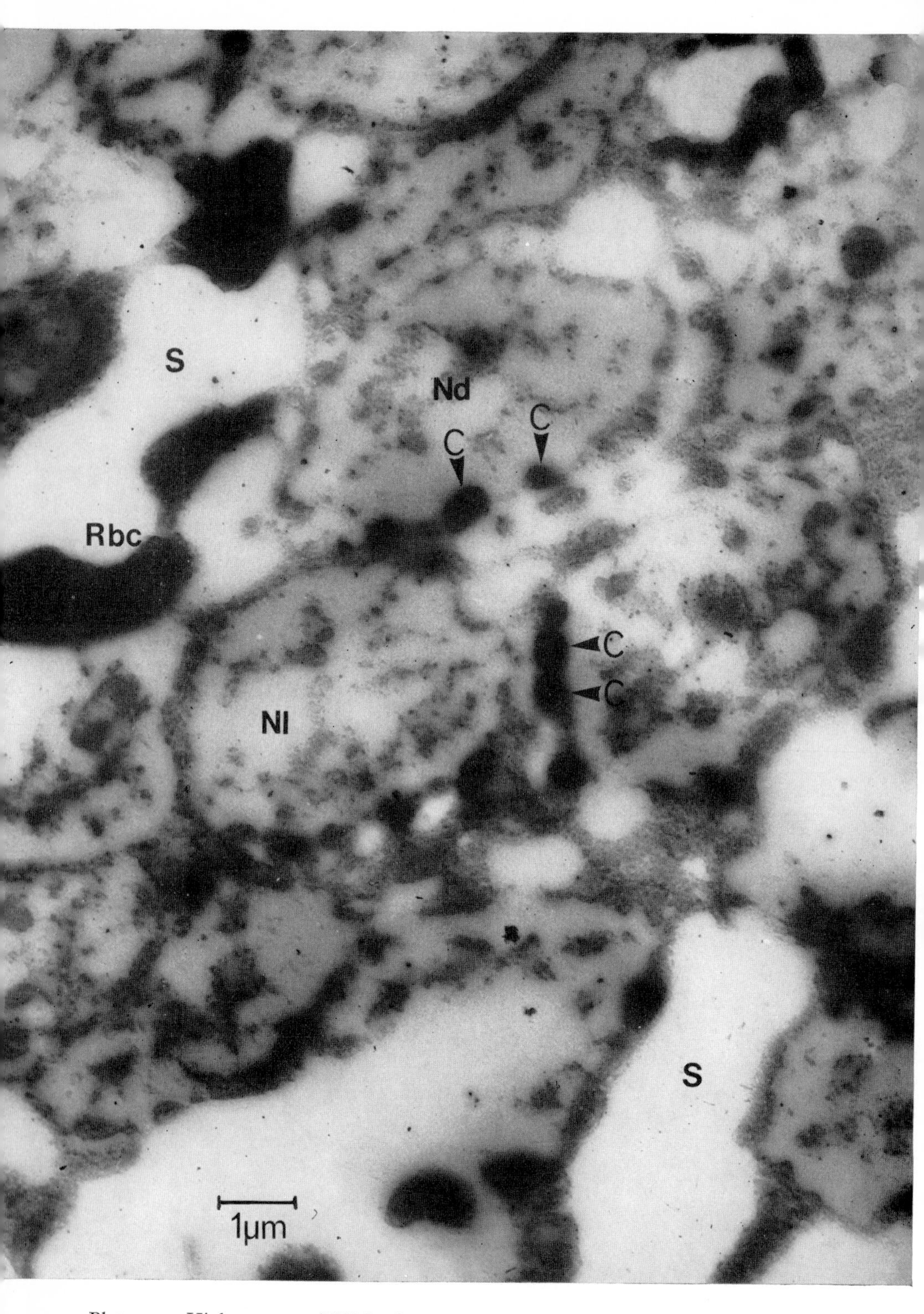

Plate 5.7 Higher power TEM micrograph of freeze-dried frozen section of mouse spleen. Nl, light areas of nucleus (A in Figs. 5.12 and 5.15); Nd, dark areas of nucleus (B in Figs. 5.12 and 5.15); arrow heads, mitochondria (C in Figs. 5.12 and 5.15); RBC, red blood cells in sinusoid (S). The section is unstained. Cells with little cytoplasm and 'indenting' mitochondria are identified as lymphocytes.

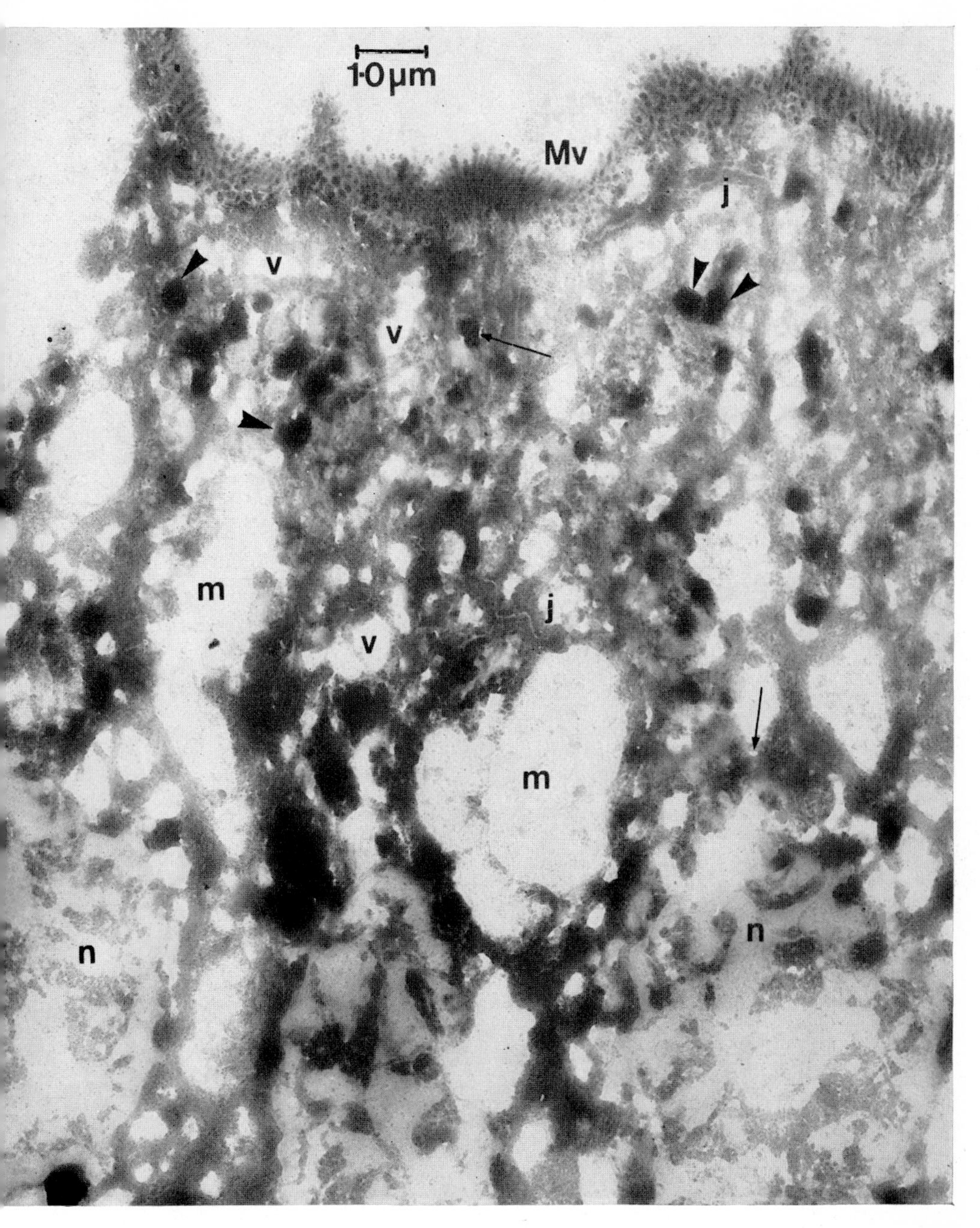

Plate 5.8 TEM micrograph of epithelial cells from the mantle epithelium of *Otala lactea* – freeze-dried frozen section of regulating animal (i.e. deprived of food or water for three weeks). Arrow heads, mitochondria; m, mucocytes; j, junctional complexes between cells; v, vesicles; n, nucleus; arrows, ice-crystal artefacts; Mv, Microvilli at apical surface. Section is unstained. Analytical results are shown in Figs. 5.13 and 5.14.

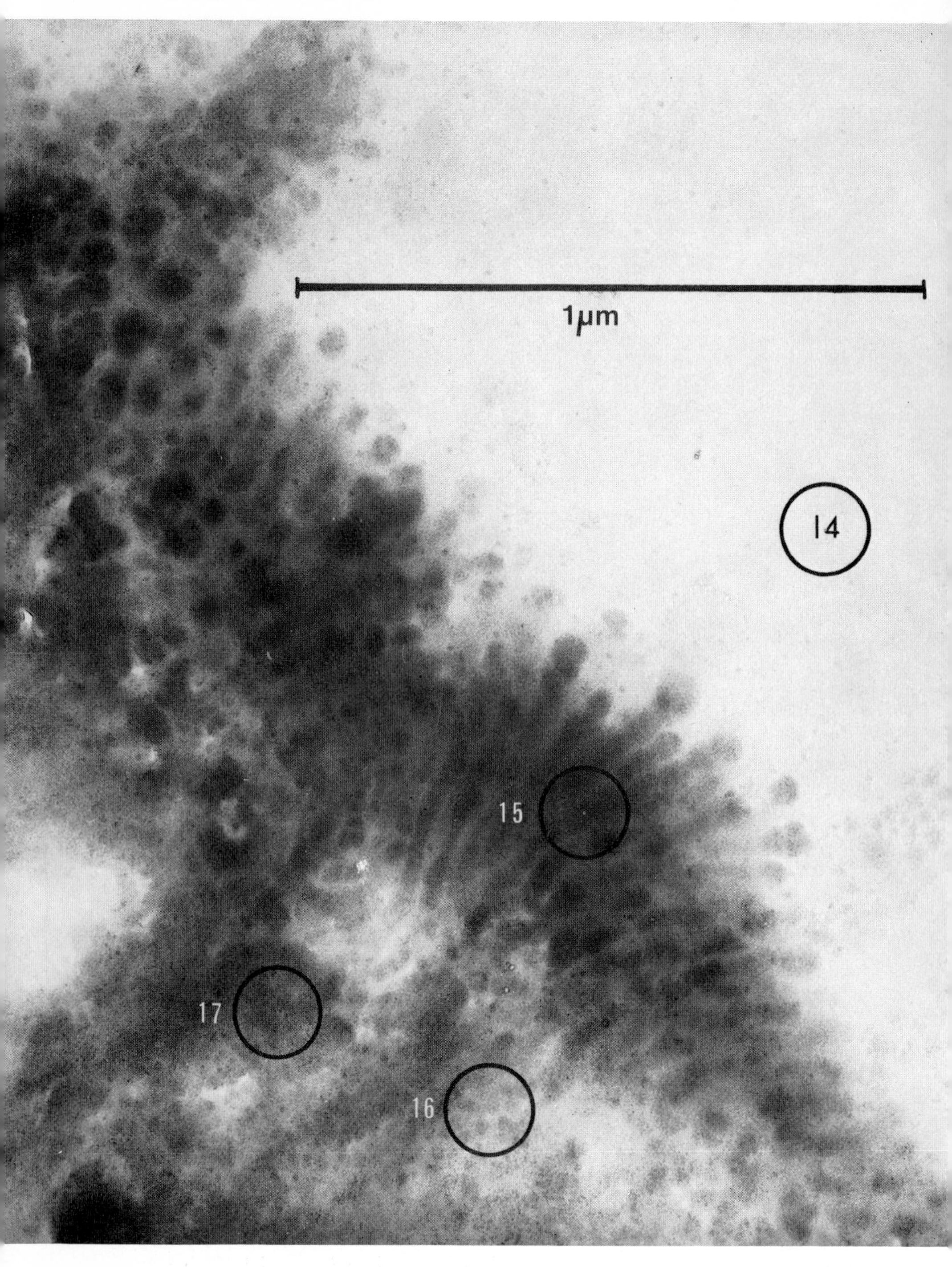

Plate 5.9 High power TEM showing detail in the microvilli of the apical region of epithelial cells from Plate 5.8. The numbered circles represent the actual areas analysed. Photograph taken on AEI CORA (courtesy of AEI Ltd.).

Small pieces of tissue (1 mm^3 or less) are placed on to the tips of brass or copper specimen holders. The tip of the specimen holder is machined to form a cup with a rough surface (Appleton, 1974*b*). The specimen holder and tissue are inverted and the tissue is plunged rapidly into the liquid nitrogen slush

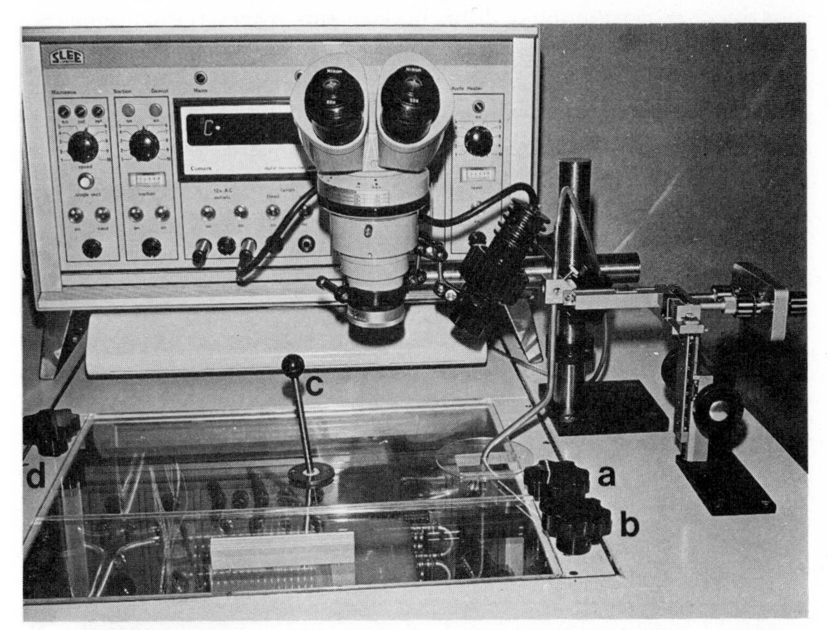

Fig. 5.4 Cryostat chamber, surface view showing external controls of knife stage (b, c), micromanipulator controls to vacuum device (see Fig. 5.5 and 5.6), temperature read-out to thermocouples, cutting speed controls, lateral knife movement (a), section thickness (d).

(−210 °C). Tissues usually adhere to the specimen holders by their inherent moisture but if necessary a small 'smear' of OCT compound (Ames) or Tissue Tec can be used to help the tissue to adhere to the surface of the holder. However any tissue which comes into contact with the mounting medium should be assumed to be unsuitable for the localization of diffusible elements.

The formation of ice crystals is at a minimum below −130 °C (Merryman, 1956; Pryde & Jones, 1957), and provided that tissues are kept below that temperature the growth of ice crystals is prevented, but it increases if the tissue is stored for any length of time above −129 °C. Tissue may therefore be stored in liquid nitrogen indefinitely without deterioration and has shown good morphological preservation with no loss or movement of diffusible elements or compounds for periods up to 2 years (Appleton, 1967). Commercially available liquid nitrogen refrigerators are available from many sources with a minimal loss-rate of liquid nitrogen.

5.2.2 TRIMMING

Trimming of the block is important for two reasons: (1) it enables ribbons of sections to be cut and handled with comparative ease; and (2) it removes 'damaged' areas of excised tissue which are likely to give misleading analytical results due to redistribution of diffusible elements around the damaged cells.

The frozen blocks are trimmed in the cryostat (Figs. 5.3–5.6) at the temperature at which they are sectioned. The block is trimmed to a pyramid (sides at 30°) with a square block face. The amount of compression can be judged by comparing the two axes of the sections as they are cut. The

Fig. 5.5 Internal view of cryostat. Note pipe for vacuum device attached to micromanipulator (Fig. 5.4), remote controls to knife stage, and thickness control of microtome, freeze drying platform in left-hand corner. The knife stage rotates to allow trimming within the cryostat. The separate freeze drying platform allows the grids to be stored at temperatures below −130 °C. The liquid nitrogen supply to the chamber passes through the insulated container which supports the freeze drying surface keeping the storage area for the grids at temperatures of −130 to −160 °C. Sections stored at these temperatures will not freeze dry. Drying is initiated by raising the temperature of the surface by means of a heater below the surface. A cold trap (−176 °C) adjacent to the surface ensures that water within the section sublimes away towards the cold trap. Freeze drying is usually carried out at the same temperature as cutting and for a period of 3 hours. The grids can then be warmed slowly to ambient temperatures by switching off the liquid nitrogen supply to the chamber, allowing the chamber to warm to 'back-up' temperature. The grids can be removed from the chamber in the presence of dry nitrogen as described in the text. Thermocouples: 1: Freeze-drying platform; 2: Air; 3: Knife stage.

conditions of sectioning (i.e. temperature, cutting speed, knife angle) are optimum when compression is at a minimal level – 10% or less. Ribbons of sections can only be cut if the leading and trailing edges of the trimmed block are parallel; the two lateral edges need not be trimmed with such care

Fig. 5.6 Detail showing the arrangement of the vacuum device for sectioning (Fig. 5.7). Note thermocouples measuring air temperature. (B) shows a vertical view of vacuum device and knife. The flattened needle is placed just behind the cutting edge of the knife, and is moved backwards via the micromanipulator as each section is cut. Thermocouples in (A) measure air and knife temperatures. See caption Fig. 5.5 for details of thermocouples.

although sectioning is frequently easier if they are trimmed to give a square block face. Trimming is facilitated by the rotating knife stage of the ultramicrotome.

It could be argued that trimming removes the tissues with the best preservation because this part of the tissue comes into contact with the liquid nitrogen slush. This part of the excised tissue is the most unreliable from the physiological point of view; trimming can be minimized with symmetrical droplets of cell cultures provided that the leading and trailing edges of the block face are parallel.

The secret of successful sectioning is in the care of trimming. Good ribbons of sections can only be cut from well trimmed blocks.

5.2.3 SECTIONING

Ultrathin frozen sections can be cut with varying degrees of difficulty on any microtome fitted with a low temperature attachment. The advantage of the attachment approach is that an existing ultramicrotome can be modified for low temperature work while retaining the ability to use the microtome for conventional use. The disadvantages are that the temperature stability, although probably adequate, is not as good as that of the cryostat and the volume available for attachments and instruments is limited. The cryostat, on the other hand, provides excellent temperature stability with a large chamber volume with minimal turbulence. Turbulence of air in the vicinity of the knife makes sectioning very difficult. The microtome in the cryostat cannot be used easily for operation at room temperature since the bearings, slides, etc. must be modified for low temperature operation. The budgets of individual laboratories will obviously govern the choice of equipment; it may not be cheaper to opt for an attachment for an existing microtome if man hours are important. Experience in this laboratory suggests that sectioning in an ultramicrotome cryostat is considerably easier than with an attachment 'kit'.

Many techniques have been described for cutting ultrathin frozen sections of a wide range of biological materials, and it is not the purpose of this chapter to provide an extensive review. However it is thought that sections cut on to fluids such as DMSO (Bernhard, 1971; Sjöström, Thornell & Cedergren, 1973; Bauer & Sigarlakie, 1973) are better suited to morphological studies and histochemistry. Diffusible materials are lost or redistributed if frozen sections of unfixed material come into contact with a fluid such as DMSO (unpublished observations). Sjöström and Thornell (1975) compared the spectra obtained over sections of frog striated muscle after different treatments. They concluded that: (1) the spectra obtained from plastic embedded material contained elements which depended on the reagents used for fixation and staining and these were absent if the reagents were omitted; (2) brief fixation in glutaraldehyde (15 min in 2.5%) caused gross ionic changes; (3) sectioning of unfixed frozen material with a trough liquid of DMSO led to extraction of elements; and (4) that the method of choice was 'dry cutting in the frozen state, and freeze-drying should be the procedure of choice if data on diffusible ions is desired.'

Opinions differ on the optimum temperature of sectioning. On the one hand Hodson and Marshall (1970, 1972), and Spriggs and Wynne-Evans (1976) suggested that there was little point in sectioning above −130 °C since redistribution due to a thawing phase would be so severe as to make analysis of diffusible ions impossible. On the other hand experience in this laboratory (Appleton, 1974*a*, 1974*b*, 1977*a*) and in America (Christensen, 1971) suggests that sections of unfixed material do not cut as true sections below about −80 °C. The fact that Hodson and Marshall (1970, 1972) were able to cut sections on to cyclohexene at temperatures below −130 °C suggests that the fluid is in some way making sectioning possible. Kirk and Dobbs (1976) showed, by examining replicas of fractured red blood corpuscles (RBC), that fracture faces are only formed at very cold temperatures (less than −70 °C) and never at temperatures above −70 °C. Replicas of DMSO-treated red blood cells prepared at knife and specimen temperatures of −65 °C showed relatively smooth surfaces. The authors concluded that 'this is strong evidence in support of the conclusion that sections are being "cut" and not fractured at warmer temperatures.'

The optimum sectioning temperature varies from tissue to tissue depending on water content and the nature of solutes, non-solutes, and macromolecules. Experience has suggested that the optimum temperature is that point at which true sections cut – i.e. ribbons of consistent thickness – and that below this temperature the 'slices' include both fractured surfaces and sections until a point where complete fractures occur. Errors of sampling are likely to be great when sectioning is erratic and includes fractured surfaces, and these errors are greater than those which one might attribute to redistribution during sectioning. X-ray microanalysis of freeze-dried frozen sections cut at their optimum sectioning temperature suggests that there is no redistribution of diffusible elements due to sectioning or freeze-drying *within the limits of resolution used* – i.e. 100 nm (Appleton 1974*a*, 1974*b*, 1977*a*). Marked gradients of highly mobile ions such as potassium and chlorine have been observed in the apical region of the regulating mantle epithelium of the snail *Otala lactea*, and the level of these ions shows a 50% fall over a distance which is less than 3 μm (Appleton & Newell, 1977*a*, 1977*b*). Analysis of freeze-dried sections of red blood cells (frozen in liquid nitrogen slush) gave a potassium to sodium ratio of 7.33:1 where the expected ratio was 7.35:1 (Appleton & Steward, 1977), whereas the ratio was 6.4:1 when the blood was frozen in boiling liquid nitrogen (−196 °C) and only 3:1 when freezing was in dry ice (−76 °C). It would seem therefore that the initial freezing procedure is the most critical stage in preparing material for X-ray microanalysis of diffusible ions.

The principle of cutting ultrathin frozen sections on to dry knives has been described in some detail (Appleton, 1974*b*). Where experience with a specific tissue is limited it is wise to start sectioning at a temperature which is obviously too high (−60 °C, for example) and then to lower the temperature progressively until sectioning becomes difficult or erratic. The optimum point is the lowest temperature at which true sections will cut. The optimum temperature may vary considerably (mouse pancreas sections at −76 °C, while kidney sections best at −82 °C) or differences may be slight. In a study of elemental changes in the mantle epithelium of active and regulating snails (*Otala lactea*) which had been deprived of water for several weeks, the epithelium from the active cells had an optimum sectioning temperature of

−72 °C while that from the hibernating (regulating) snail had an optimum temperature of −74 °C (Appleton & Newell, 1977*a*, 1977*b*). Such difference may appear trivial but it is important to pay attention to all the details if one is to have any confidence in the preparative procedures.

Serial sections can be routinely cut at a setting of 70 nm upwards. With experience it is possible to cut 'dry' frozen sections at 30 nm, but for X-ray microanalysis and autoradiography there is insufficient material in sections below 100 nm for adequate signal-to-noise ratios. In practice, sections are usually cut at a setting of 130–140 nm but, because interference colours are not seen when sections are cut on to dry knives, it is probable that the sections are thicker than the microtome setting suggests. Measurements of frozen sections (Appleton, 1974*b*) suggested that 100 nm sections were within 10% of the microtome setting. Those measurements were made on the freeze-dried sections and since the water content of the original sections was at least 60–70% it is probable that the thickness of the sections in the hydrated state was probably nearer to 300 nm. However, after freeze-drying, sections are certainly thin enough to allow penetration of electrons and the sensitivity and spatial resolution is good enough to localize elements to subcellular organelles.

Ultrathin frozen sections will be cut at every stroke of the microtome provided conditions for sectioning are right – i.e. temperature, cutting speed, knife angle, quality of knife. It is also important that the temperature of the specimen, knife, and air have reached a stable point; sectioning will be erratic and difficult if the temperature of the air, knife, and specimen temperatures fluctuates by more than one or two degrees. Left to their 'own devices' sections curl into rolls on the knife edge but can be teased flat with an eye-lash probe or more easily pulled into ribbons of sections using a weak vacuum (Appleton 1974*b*, 1977*a*). A weak vacuum (produced by a vibrator pump) is applied to the leading sections through a flattened syringe needle (Fig. 5.7) and the vacuum device is withdrawn slowly backwards as each section is cut (Fig. 5.7*b*, *c*). Provided sections are being cut right through the block face the vacuum will pull a flat ribbon of serial sections which have a glassy transparent appearance rather like 'cling-film' used for wrapping food. The sections are flat, with minimum compression; compression can be judged by comparing the axes of the sections which were cut from a square block-face. The strength of the vacuum should be adjustable through a valve and will be different for each tissue. Vacuum devices can be constructed easily and fitted to attachment systems although the space available is limited. The cryostat described in Figs. 5.1–5.6 is fitted with a fully controllable vacuum device: the cryostat was designed in collaboration with Slee Medical Equipment and is available from that company.*

* Slee Medical Equipment Ltd., Lanier Works, Hither Green Lane, London SE13 6QD.

5.2.4 COLLECTION OF SECTIONS

The collection of sections is comparatively simple when ribbons of sections have been cut. A piece of PVC insulating tape is placed on the back face of the knife before the knife is placed in the cryostat. This acts as a shelf against which the grid can rest. A formvar-coated nickel grid can be placed on the

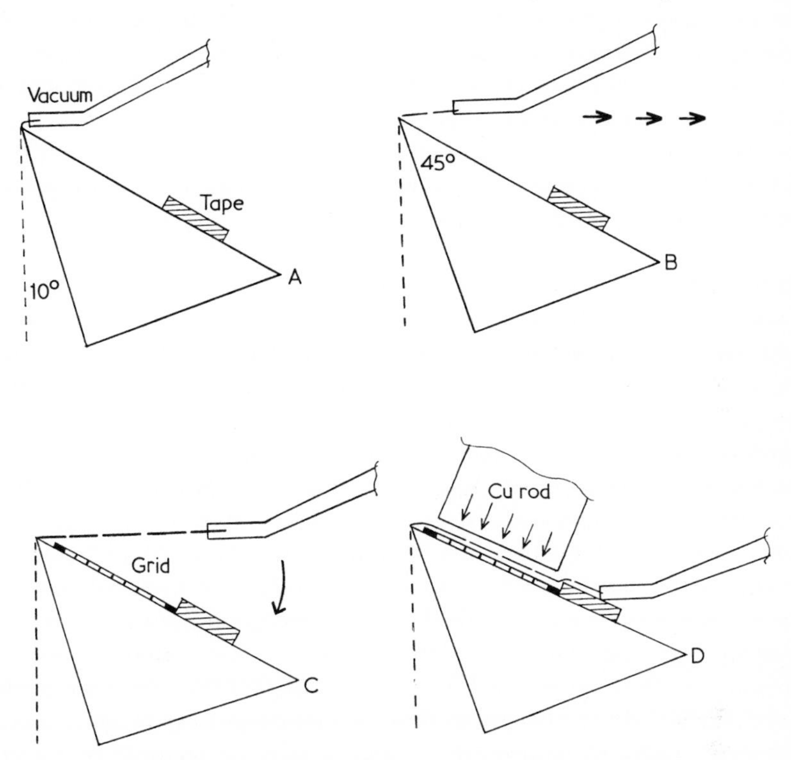

Fig. 5.7 Diagram illustrating the method of pulling a ribbon of dry sections away from the knife edge. Once a section has been caught by the vacuum, the vacuum device is moved progressively back until a ribbon, long enough to cover a grid, is formed. The formvar-coated grid can be in position before section starts or can be slipped under the ribbon of sections after they have been cut – the sections should not be pressed into contact with the grid until the grid is really cold. Only light pressure with the polished copper rod is necessary to ensure that the ribbon of sections is really in contact with the formvar membrane. (After Appleton, 1977*a*).

knife (just behind the cutting edge) before sections are cut or can be slipped underneath the ribbon of sections after cutting; it is important that the grid is cold before sections are allowed to come into contact with it. The ribbon of sections can be lowered through the micromanipulator (Figs. 5.5–5.7), so that the sections lie across the grid; they are then pressed lightly into contact with the formvar-coated grid using a cold, flat polished, copper rod of 4 mm diameter (Christensen, 1971). The pressure is used to press the sections into contact with the grid rather than to flatten the sections. The sections should already be flat and without folds. The grid with ribbon of sections can then be moved to a separate area of the cryostat for freeze drying.

Nickel grids are used for three main reasons: (1) the formvar membranes do not adhere to copper grids at the low temperatures used; (2) grids which are accidently dropped in the cryostat can be picked up more easily with a magnetic probe; (3) there is less overlap of X-ray lines with nickel grids, and it is possible to look at the involvement of copper in biological systems provided, of course, that the copper background due to specimen rods, stages, etc. has been eliminated. A specimen rod has been built for the EMMA-4 in this laboratory so that the grids are held in graphite inserts; metal background peaks have been eliminated.

5.2.5 FREEZE DRYING OF SECTIONS

Frozen sections can be examined in the frozen hydrated state where adequate cold stages and transfer mechanisms have been built (Echlin 1971; Bacaner, Broadhurst, Hutchinson & Lilley, 1973; Gupta, Hall, Maddrell & Moreton, 1976): X-ray microanalysis of diffusible elements has been reported in hydrated frozen sections (Bacaner, Broadhurst, Hutchinson & Lilley, 1973; Sauberman & Echlin, 1976; Gupta, Hall, Madrell & Moreton, 1976). While the initial results seem encouraging, the background of the X-ray spectra is comparatively high due to scattering of ice within the sections. Preliminary results in this laboratory with X-ray microanalysis of hydrated, ultrathin frozen sections at -140 °C suggest that the background scattering is unacceptably high and that the morphology is difficult to interpret. Bacaner, Broadhurst, Hutchinson and Lilley (1973) indicated that 'there are striking morphological differences in the microstructure of deep-frozen, unfixed, unstained, hydrated muscle cells compared to those prepared by the conventional techniques of fixation, dehydration, embedding, and staining. . . .' They go on to suggest that a failure to observe thick and thin filaments, mitochondria, and sarcoplasmic reticulum could be due to insufficient contrast in the hydrated frozen sections. Sjöström and Thornell (1975), however, were able to correlate the ultrastructural detail in freeze-dried, ultrathin frozen sections with the morphology seen in material which had been fixed and embedded. Bacaner (personal communication) suggested that the morphology seen in hydrated frozen sections might be the natural situation and that the conventional morphological interpretations may be based on artefacts. It is clear that more experience of examining hydrated frozen sections at low temperatures is needed before we can condemn the existing evidence from fixed and embedded material. It would, however, be surprising if completely new information was not eventually obtained from techniques using frozen sections.

There are considerable advantages to be gained by using *freeze-dried* ultrathin frozen sections (Appleton & Newell, 1977*a*, 1977*b*). The morphology of the cells is sufficiently clear to be able to relate the structures seen in ultrathin frozen sections with those seen in conventionally prepared material after fixation, dehydration, and embedding. The good resolution and contrast (Plates 5.1–5.9) is partly due to the initial thickness (*ca.* 200–300 nm) and partly due to the removal of water by freeze drying. The 'removal of mass' by freeze drying also means that there is less matter to scatter electrons around the focused beam of electrons (the probe), and consequently leads to an improvement in analytical resolution with good peak-to-background ratios.

It is not possible to provide precise details for freeze drying since the water content of tissues and cells varies quite considerably. However, in general, good preservation is achieved by freeze drying the tissues in the cryostat at the same temperature at which they were cut and at atmospheric pressure (Appleton, 1974*b*, 1977*a*). Rapid freeze drying under vacuum causes gross disruption of the cells (Plate 5.10) while slow freeze drying of adjacent sections at atmospheric pressure provides good morphology (Plates 5.1 and 5.2). The lace appearance in Plate 5.3 is due to freeze drying artefacts which are probably caused by excessive surface tension forces rather than ice crystal artefacts from the initial freezing process. Note also that a grossly underfocused image (Plate 5.2) 'shows' more detail than the true focus image of Plate 5.1. Contrast can be further improved by using scanning transmission electron microscopy – STEM (see Section 5.3.2). The rate of freeze drying in frozen sections can be controlled by regulating the temperature at which they are dried: lowering the temperature reduces the rate of drying. Sections should be dried for at least three hours before they are brought up to ambient temperature. It is most important that sections do not melt; they are therefore allowed to dry in the cryostat for several hours, and usually overnight.

After being dried, the sections are warmed up to ambient temperature in the cryostat by allowing the cryostat chamber and microtome to warm up to −25 °C. The volume of the cryostat is large, and the mass of metal in the microtome is sufficient to ensure that the 'warming-up' process is slow and gentle. Tissues may suffer from thermal shock if the rate of warming is rapid.

The sections are brought to ambient temperature by placing the grids in capsules which have been flushed in dry nitrogen. The capsules in turn are placed in a cannister (also flushed with dry nitrogen), which is allowed to equilibrate to ambient temperature inside a desiccator. There is then no possibility that moisture can condense on the freeze-dried sections. The sections can then be coated with a heavy layer of carbon under vacuum. A layer of evaporated carbon (20–30 nm thick) will protect the sections against water. Sections analysed in the electron microscope show that the carbon layer is sufficiently hydrophobic to protect the sections against moisture and that they can be placed section-side down on to filter paper which has been saturated with water without diffusible material being lost (Baker & Appleton, 1976).

5.3 Electron microscopy of freeze-dried frozen sections

5.3.1 TRANSMISSION ELECTRON MICROSCOPY (TEM)

Transmission electron micrographs of freeze-dried, ultrathin frozen sections show considerable ultrastructural morphology which can be interpreted with some confidence provided one is familiar with the morphology of fixed and embedded material. Staining is unnecessary and would cause gross redistribution of physiologically active ions. Contrast, however, is often the reverse of what one might expect; the condensed chromatin of the nuclei usually appears dense after fixation, embedding, and staining but in some nuclei (e.g. kidney tubules) it is the uncondensed chromatin which is the denser in freeze-dried frozen sections (Baker & Appleton, 1976). It is not yet clear why unstained freeze-dried material shows differences in contrast. It may be due to the presence of certain ions or combinations of ions within structures, or it could be due the absence of a universal 'matrix' such as water or an embedding plastic. Red blood cells, for example, invariably appear as very dense structures in freeze-dried frozen sections but are barely visible in an unstained section from Araldite-embedded material viewed under identical conditions, and contrast is presumably due to differences between RBC in vacuum (in the frozen section) and RBC in Araldite (in the embedded specimen). Since there is no such contrast seen at the edge of the Araldite alone, it would seem that the contrast seen in the red blood cells of the freeze-dried frozen sections may be due to additional 'substances' or combinations of substances present in the frozen section which are absent when fixation, dehydration, and embedding are introduced. The presence of additional material in the unextracted (fixed and embedded) material may also account for the apparent 'absence' of membranes in TEM pictures of frozen sections. Various possibilities have been suggested for the lack of membrane images in freeze-dried frozen sections (Appleton, 1974*b*), the most likely of which is that the electron density of the material on either side of the membrane is the same as the material within the membrane. This would render the membranes invisible. It is unlikely that membranes have been destroyed by the freezing process since material which has been frozen, thawed, fixed, embedded, sectioned, and stained shows membranes with little or no damage provided that the optimum freezing and thawing conditions for a tissue are observed (Trump, 1969), and membrane delineation is clearly seen in freeze-etched material (Smith & Aldrich, 1971). Membrane delineation is occasionally seen in TEM images of freeze-dried frozen sections (when the thickness is low – 70 nm) as a single or double array of granules 8–9 nm in diameter (Appleton, 1974*b*), and these may represent a part of the membrane substructure. Scanning electron microscopy (SEM) and scanning transmission electron microscopy may help considerably where difficulties in interpreting freeze-dried frozen sections occur.

5.3.2 SCANNING ELECTRON MICROSCOPY (SEM) AND SCANNING TRANSMISSION ELECTRON MICROSCOPY (STEM)

Attachments are now available for most modern high resolution election microscopes (TEM) which allow one to examine thin sections by scanning electron microscopy (SEM), and scanning transmission electron microscopy (STEM) in the same instrument. There is considerable advantage in using STEM because the contrast and brightness of the CRT (similar to a TV) display can be adjusted electronically – hence one has considerable latitude – and also thicker sections can be examined than in the conventional TEM. In the STEM mode of the JEM 100C, sections as thick as 1 μm can be examined without difficulty. The great advantage of this new breed of electron microscopes for cryo-ultramicrotomy is that one can now compare the images (obtained by TEM/SEM/STEM) of exactly the same cell (Plates 5.4 and 5.5), so that difficulties associated with one kind of imaging can be offset by comparing three different kinds of image.

Plate 5.4 shows the STEM image of an ultrathin frozen section of mouse spleen. Three cells are clearly visible in the centre of the micrograph but it is difficult to tell where the cell boundaries lie. However, these boundaries are fairly easily visible in the SEM image (Plate 5.5) of the surface of the freeze-dried frozen section taken at the same angle of tilt. Having seen the boundaries on the SEM image it is possible to trace the same boundaries on the STEM image.

The SEM micrograph (Plate 5.5) also shows that the surface of the freeze-dried frozen section has an irregular profile due to the differential 'drying-down' of the structures which results from their varying water contents. The dense parts of the nucleus (Plate 5.4) seen in STEM correspond to the 'high' parts of the nucleus as seen in SEM (Plate 5.5). Similarly the cytoplasm is not a homogeneous layer, presumably because the water content of the various areas of the cytoplasm also varies considerably. The effect of removing water (by freeze drying) on the section is illustrated in three-dimensional terms in Fig. 5.8. The various components of the cell freeze dry 'down' to different heights mainly because of their different water contents. The differences in surface profile may have important geometrical significance in X-ray microanalysis of freeze-dried frozen sections for several reasons:

(1) Structures may shield the X-rays from the detector: it is therefore important to know the relationship of the detector to the structures as seen

in SEM. Analysis of area 'A' in Fig. 5.8 could be very different from that in area 'B'. Such differences could well be due to genuine differences or could be a direct consequence of the angle of 'take-off' for the X-rays (i.e. the path of the X-rays to the detector). In Fig. 5.8 the 'high' structure between 'A' and 'B' could mask the X-rays from 'A' reaching the detector if the position of the detector was similar to that indicated by the arrow.

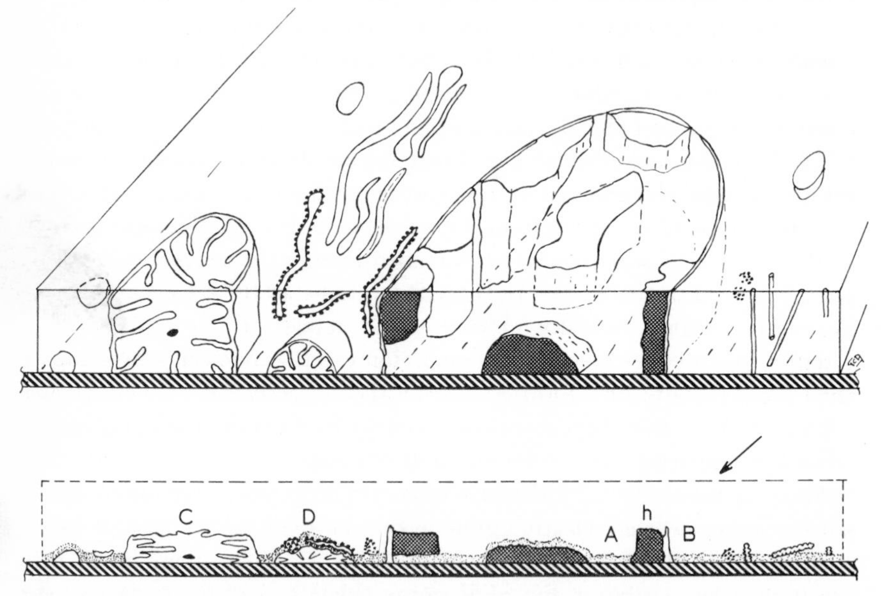

Fig. 5.8 Three dimensional notional diagram to illustrate the freeze drying 'effect' seen in Plate 5.4. Note that when the water has been removed the section collapses down according to the original water content, and that sampling errors may cause serious variances in the analytical results (see discussion in text, Section 5.3.2). Arrow indicates suggested angle of detector, then analysis from area (A) could be influenced by or shielded from the high structure (h) between (A) and (B). Analysis from (B) should show no such interference problem. Analysis over the mitochondrion (C) might show differences from that of mitochondrion (D) although the transmission images may appear similar. SEM of the surface might show differences between (C) and (D).

(2) The 'high' structure 'h' between 'A' and 'B' could itself be contributing towards the X-rays reaching the detector because electrons scattered by 'A' may hit 'h' and generate X-rays which reach the detector together with the X-rays from 'A', whereas no such interference is likely if area 'B' is analysed and the angle of take-off is as indicated by the arrow.

(3) The TEM and STEM images of 'C' and 'D' may appear similar although the three-dimensional reconstruction shows that they are not. It is possible that the image from 'C' may be sharper and denser than 'D' because there

are no overlying structures in 'C'. It is, however, probable that with experience a critical examination of SEM images of the surface using secondary and backscattered electrons may show a clear distinction. Errors in analysis of areas such as 'C' and 'D' are, of course, sampling errors and will become less significant in ultrathin sections than in sections of 1 or 2 μm. Errors of sampling must otherwise be reduced by analysing large numbers of areas.

(4) It may be possible to combine surface profile measurements with X-ray analytical methods to calculate the water content of the areas being analysed. It would then be possible to express the quantitated data in terms which are biologically significant, i.e. mM/l.

The importance of correcting for water content of the subcellular structures will be discussed in greater detail in Section 5.4.2.

5.4 X-ray microanalysis of ultrathin freeze-dried frozen sections

5.4.1 SECTION THICKNESS AND SENSITIVITY

Clearly the optimum section thickness will be governed by the need for clear morphological detail with minimal overlap problems (see Fig. 5.8 and Section 5.3.2), and the necessity for retaining sufficient biological material for significant X-ray detection. Where the material of interest is in the form of a mineralized deposit the section thickness can be legitimately reduced to about 50 nm. However, the majority of biological interests are likely to be in relation to physiologically active elements such as sodium, magnesium, silicon, phosphorus, sulphur, chlorine, potassium, calcium, iron, copper, and zinc. These elements may be present in small amounts and it may be necessary to vary the section thickness according to the element of predominant interest although it would be foolish to ignore the others. As I have already indicated earlier in this chapter, the solution to a particular biological question often lies in the most unexpected elements and so it is advisable to start any investigation with sections which are likely to afford a result. In my experience, sections which have a thickness of 100 nm when they are freeze dried (therefore about 200–300 thick hydrated) are a good starting point. Significant levels of the elements listed above can be detected in such sections using a probe size of 100 nm, a beam current of 4.0 nA at 60 kV accelerating voltage, and 100 s counting time. However, elementary biological knowledge will show that if an element such as potassium is high, then sodium will be low, or maybe below the levels of detection. Where an electron microscope (with X-ray detector of course) is available with the scanning attachments, X-ray mapping or line scans using thick (1 μm) frozen sections may be invaluable in providing preliminary information about the 'gross' sites of activity. Finer studies using static probes or small area scans can then be undertaken on thinner sections to determine the subcellular activity in more detail.

It is difficult to give any precise figures on sensitivity except in the few instances where the answer is already known, e.g. red blood cells. Iron can be detected with a significance greater than 2:1 (Peak: Background) in a probe of 100 nm diameter with a freeze-dried section thickness of 100 nm. Sodium can be similarly detected in a section of red blood cells where the known level is below 0.023%.

Sensitivity depends on the ratio of peak to background and so it is important to keep the background as low as possible and eliminate as far as one can the effects of heavy metals in the area of the specimen. Graphite or aluminium specimen holders, and grids of material with a low atomic number reduce the background and hence increase the sensitivity of detection. Detectors with a large surface area compatible with good peak discrimination should be used and fitted with good collimation, and where possible sections should be analysed without the need for tilting. These and other theoretical points will have been covered in Chapters 2 and 3.

As a rough guide it can be said that levels of elements as low as 1–5 mM per litre can be detected in ultrathin, freeze-dried, frozen sections under normal conditions for EMMA-4 (copper specimen rod; nickel grids; room temperature). This can be improved by a factor of 3 by using graphite inserts in the specimen rod and by a factor of between 2 and 5 by analysing at low temperatures (i.e. −140 °C). Low temperature analysis increases the sensitivity by: (1) reducing the background; and (2) increasing the peak counts for some elements which are 'unstable' even at low beam currents (see Fig. 5.9) (Appleton, 1977*a*). The cold stage developed for EMMA-4 improves the sensitivity considerably but because good conduction of heat is required it utilizes phosphor bronze and so it is impossible to analyse copper at low temperatures. It is not clear at this time why analysis at low temperatures should reduce the background by such a significant amount, although it is possible that it is partly due to a lowering of contamination and a consequent reduction in scattering of electrons. The contribution of the lowering of the background to the increase in sensitivity is seen in Fig. 5.9 – the hatched area of the histogram is due to the lowering of the background and the stippled area to an increase in peak height as a result of low temperature analysis. It is not surprising that there should be an increase in peak height for elements such as chlorine and potassium since these are known to be unstable under the electron beam (Chandler & Battersby, 1976). The apparent instability of phosphorus and calcium is, however, unexpected, and may reflect the dual 'biological state' of these elements – i.e. partly bound and partly labile. It does however emphasize the need for great care in the X-ray microanalysis of naturally occurring elements and that nothing can be taken for granted where biological systems are being investigated.

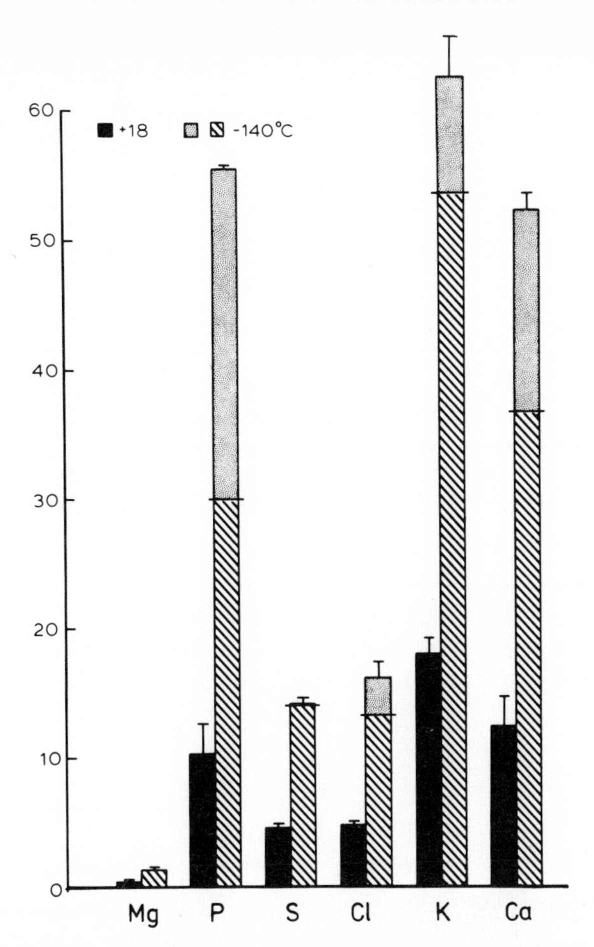

Fig. 5.9 Diagram illustrating the increase obtained by analysing frozen sections (freeze-dried) at −140 °C. The hatched areas indicate the contribution of the reduced background to the overall increase in sensitivity while the stippled areas indicate the contribution of increased peak height which results from low temperature analysis. See text for discussion (Section 5.4.1). It is not clear why low temperatures should reduce the background of X-ray microanalysis.

5.4.2 EXPRESSING AND INTERPRETING THE DATA

The various ways in which the raw data acquired as a result of X-ray microanalysis can be handled or quantified have been reviewed and discussed in detail elsewhere in this book. The intention here is to consider the different ways in which data from X-ray microanalysis of naturally occurring elements can be used, and what corrections may be needed to express the data in biologically significant form. Perhaps the best way of illustrating these points is to consider a few specific examples.

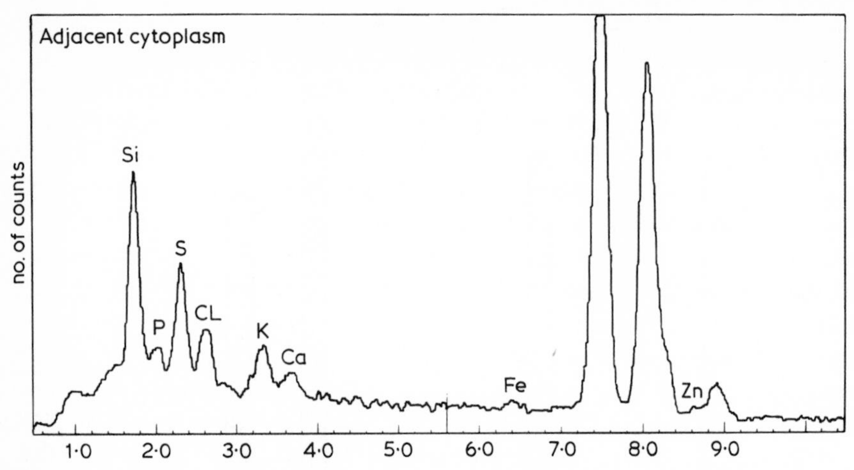

Fig. 5.10 Spectrum obtained after analysis of mitochondrion from lymphocyte as in Plate 5.7.

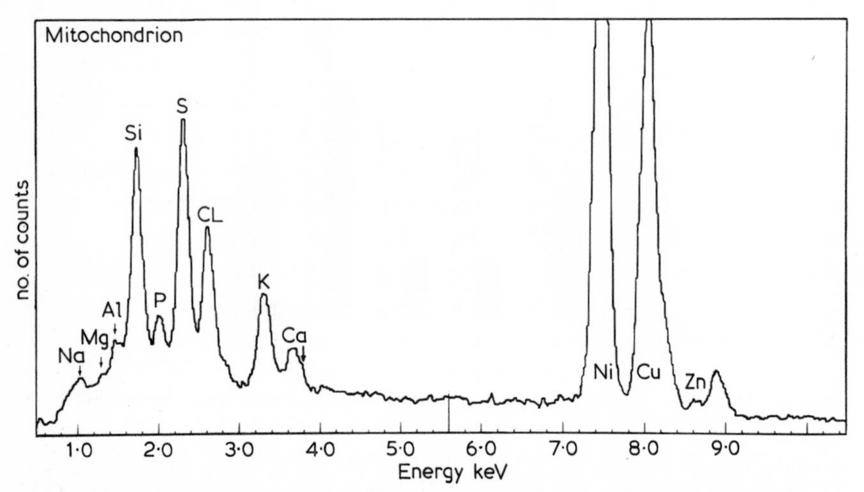

Fig. 5.11 Spectrum from ground cytoplasm adjacent to mitochondrion in Fig. 5.10.

Comparison of spectra

A large number of biological questions can be answered by comparing the results of X-ray microanalysis in one area with those from another, and where the differences are quite obvious, a simple comparison of the X-ray spectra obtained may be quite sufficient. Simple measurements over, for example, crystalline inclusions and over non-crystalline adjacent matrix may provide perfectly adequate answers, or the results can be quantified by comparison with standards of known comparison (Hall, Anderson & Appleton, 1973).

However, the differences in concentrations of naturally occurring elements may be too subtle for such a simple comparison and the inexperienced observer can be misled because the nature of the background is not always apparent to the naked eye. In the following examples the analytical data was obtained from freeze-dried ultrathin frozen sections of mouse spleen analysed at 60 kV; 4.0 nA beam current; 100 s counting time; spot size 120 nm. Plates 5.4, 5.5, 5.6, and 5.7 illustrate the morphology of the sections used. X-ray microanalysis of red blood cells (RBCs) was compared with that from four areas of adjacent lymphocytes – (a) light areas of the nucleus; (b) dark areas of the nucleus; (c) mitochondria; (d) ground cytoplasm adjacent to the mitochondria. In all cases the results were based on at least twelve measurements of each area. Two spectra are shown in Fig. 5.10 and 5.11, and results are expressed by mass in Fig. 5.12. More practical applications of the X-ray microanalysis of RBCs in spleen are shown in Figs. 5.16–5.19.

Figures 5.10 and 5.11 show the spectra obtained over two areas of a lymphocyte mitochondrion and adjacent cytoplasm. Comparison of the two spectra shows certain definite differences between some elements in the two areas; for example it is clear that the sulphur and chlorine levels in the mitochondrion (Fig. 5.10) are higher than in the adjacent cytoplasm (Fig. 5.11), but it would be difficult to be as definite about the phosphorus or calcium levels. Difficulties in such comparisons are partly due to the fact that the human eye finds it difficult to integrate the peaks and make the necessary background corrections, and partly due to the fact that these are single measurements which may reflect errors in sampling. Confidence in one's ability to 'eye-ball' the results are obviously greater when the spectra are very different – for example in Figs. 5.16 and 5.17. However, even here, the differences in the levels of the background may be misleading to all but the most experienced eye. Greater confidence and measures of variability can only be obtained by expressing the results in numerical form. Most analytical systems fitted to electron microscopes have facilities for extracting the numerical data from the spectra. For a thorough analysis of the results, computerized systems with storage, X–Y plotting, and teletype facilities are necessary. Photographic recordings of spectra from CRT displays are seldom sufficient for any but the simplest comparisons.

Comparison by mass fractions (or relative mass fractions)

One of the most useful ways of expressing the results of analysis of naturally occurring elements is by comparing the relative mass fractions measured in one area with those found in another. This expression is obtained by expressing the peak counts ($P-b$) as a ratio of the mass. It is not even necessary to know the thickness of the section so long as there is a measurement taken which is related to mass. In practice this is the continuum measurement obtained by measuring the counts in a wide window set in an area of the spectrum where there are no elemental peaks. In this laboratory the continuum (or white count, W) is routinely set between 11.50 and 14.50 keV (20 eV/ channel). If the mass of the area being analysed increases then the continuum count in that window increases and vice versa. Then:

The *Relative* Mass Fraction = (Peak − Background) ÷ Continuum

$$RMF = \frac{(P-b)}{W} \quad (1)$$

$$\text{The \% } \textit{Relative} \text{ Mass Fraction} = \frac{(P-b)}{W} \cdot 100 = \%RMF \quad (2)$$

The RMF and % RMF values will of course be dependent on the width of the window used to obtain the continuum (W) count but provide a valuable means of comparing the relative concentration of an element in one area with that in another. Such comparisons are valid provided that:

(1) the width of the continuum window is kept constant;

(2) factors which affect the continuum measurement in the region of the specimen are standardized – i.e. material of specimen holder, grid, and tilt (if any).

The correction for mass means that variations in spot size and differences in section thickness have been taken into account provided that the other analytical variables (accelerating voltage, beam current, counting time, dead time) are also standardized.

Figure 5.12 is a comparison between the % RMF of various elements in the five areas of the spleen. Each block represents the mean of at least twelve measurements.

Comparison of relative mass fractions between one area and another is a useful means of expressing the analytical data for the biologist. Occasionally it may be necessary to compare actual mass fractions (MF) or % mass fractions (%MF). The quantitated mass fraction is obtained by comparing the relative mass fractions for the specimen with that obtained from a standard of similar thickness and of known composition. The specimen and standard should be analysed under identical conditions (Hall, Anderson & Appleton, 1973; and Chapter 2 in this book). It is not necessary to know the thickness of the standard or its concentration; the formula provided by Hall, Anderson and Appleton (1973) also takes into account the atomic weights and atomic numbers of the elements involved. Other quantification programmes are available which make the necessary corrections to enable elemental concentrations to be compared (see Chapter 2).

Comparisons of mass fractions (and relative mass fractions) are of particular value where sections of fixed and embedded material are used and where there is little change in the biological material after sectioning has taken place although there may be some loss material during analysis (Hall & Gupta,

1974). Freeze drying of frozen sections removes water from the biological material before analysis and so relative concentrations expressed as mass fractions (relative or actual) may not have true biological significance since the results are expressed as mass fractions DRY WEIGHT and the material

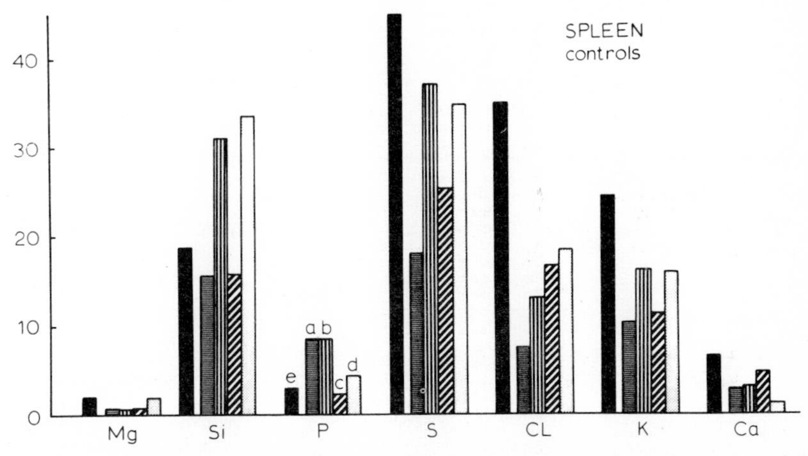

Fig. 5.12 Analytical 'finger-print' for RBCs and four areas of lymphocyte from frozen section of mouse spleen (see plate 5.7). Each block is the mean of at least 12 measurements. Note that differences can be observed between RBCs and lymphocytes and between different areas of lymphocytes. Changes as a result of experiment procedures may influence the characteristic elemental finger-print of cells (see Fig. 5.18 and 5.19). (e) RBC, (a) light area of nucleus, (b) dark area of nucleus, (c) mitochondria, (d) adjacent cytoplasm.

when it was biologically active, living and dynamic, was WET. Expressing the data by mass will therefore tend to smooth out any differences which existed. It may be more realistic to express the data by volume or correct the mass fractions obtained so as to express them as mass fractions WET WEIGHT.

Relative concentrations by volume

In a recent study (Appleton & Newell, 1977*a*, 1977*b*), X-ray microanalytical measurements were made over epithelial cells from the mantle tissue of the snail *Otala lactea*. The relative concentration of various elements was measured by analysing the epithelial cells in frozen sections from normal, actively crawling snails, and snails which had been deprived of food and water for over three weeks (Plates 5.8 and 5.9). Measurements were made from the apical region to the basal nuclear region, and expressed by volume (i.e. $P-b$). The areas of analysis in each of the two groups of sections were very small and so variations in section thickness within one section could be ignored. Continuum measurements (11.50 to 14.50 keV window setting) showed that there was no significant difference between the thickness of the sections from the normal and regulating (water and food deprived) snails. Measurements over 30 areas of each section gave a continuum reading of 3387±231 for the controls, and 3382±94 for the regulating animals analysed under identical conditions. It is clear from these figures that the volumetric measurements

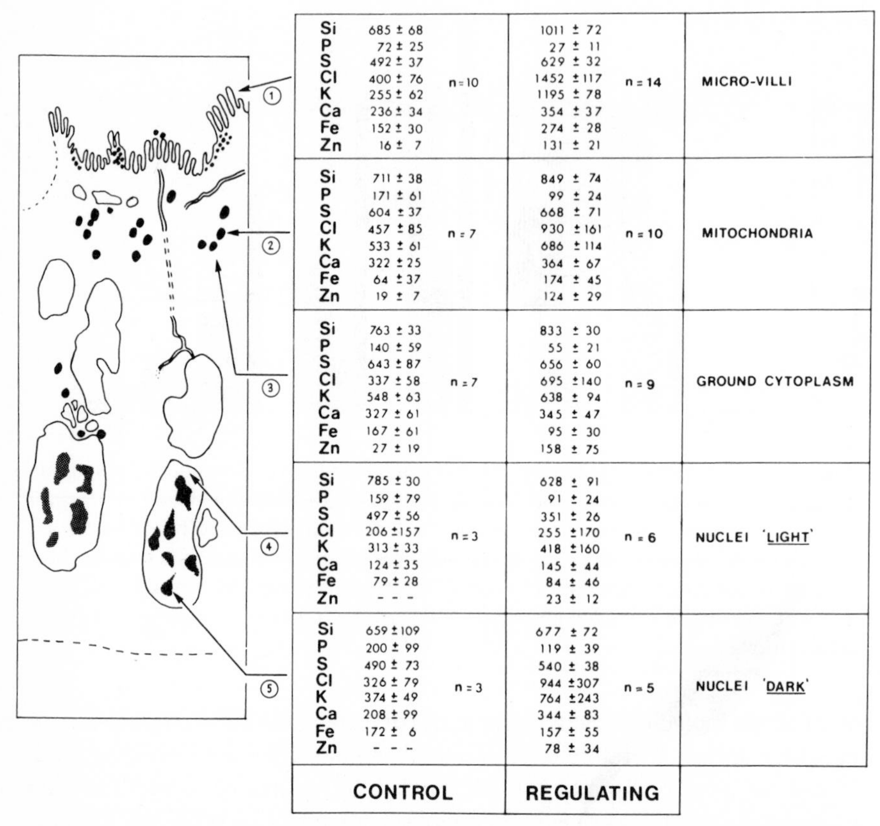

Area	Element	Control	Control n	Regulating	Regulating n
① MICRO-VILLI	Si	685 ± 68	n = 10	1011 ± 72	n = 14
	P	72 ± 25		27 ± 11	
	S	492 ± 37		629 ± 32	
	Cl	400 ± 76		1452 ± 117	
	K	255 ± 62		1195 ± 78	
	Ca	236 ± 34		354 ± 37	
	Fe	152 ± 30		274 ± 28	
	Zn	16 ± 7		131 ± 21	
② MITOCHONDRIA	Si	711 ± 38	n = 7	849 ± 74	n = 10
	P	171 ± 61		99 ± 24	
	S	604 ± 37		668 ± 71	
	Cl	457 ± 85		930 ± 161	
	K	533 ± 61		686 ± 114	
	Ca	322 ± 25		364 ± 67	
	Fe	64 ± 37		174 ± 45	
	Zn	19 ± 7		124 ± 29	
③ GROUND CYTOPLASM	Si	763 ± 33	n = 7	833 ± 30	n = 9
	P	140 ± 59		55 ± 21	
	S	643 ± 87		656 ± 60	
	Cl	337 ± 58		695 ± 140	
	K	548 ± 63		638 ± 94	
	Ca	327 ± 61		345 ± 47	
	Fe	167 ± 61		95 ± 30	
	Zn	27 ± 19		158 ± 75	
④ NUCLEI 'LIGHT'	Si	785 ± 30	n = 3	628 ± 91	n = 6
	P	159 ± 79		91 ± 24	
	S	497 ± 56		351 ± 26	
	Cl	206 ± 157		255 ± 170	
	K	313 ± 33		418 ± 160	
	Ca	124 ± 35		145 ± 44	
	Fe	79 ± 28		84 ± 46	
	Zn	- - -		23 ± 12	
⑤ NUCLEI 'DARK'	Si	659 ± 109	n = 3	677 ± 72	n = 5
	P	200 ± 99		119 ± 39	
	S	490 ± 73		540 ± 38	
	Cl	326 ± 79		944 ± 307	
	K	374 ± 49		764 ± 243	
	Ca	208 ± 99		344 ± 83	
	Fe	172 ± 6		157 ± 55	
	Zn	- - -		78 ± 34	

Fig. 5.13 Relative concentrations of various elements detected in areas of Plate 5.8. Concentrations expressed by volume (i.e. $P-b$). See Section 5.4.2 for discussion. Note particularly the chlorine, potassium, iron, and zinc figures. The gradients for Cl and K are shown in Fig. 5.14. The significance of these elemental changes have been discussed elsewhere (Appleton & Newell, 1977*a*, 1977*b*) and are concerned with regulating the loss of water from the apical surface.

$(P-b)$ can be used to compare the relative concentrations of elements between different areas of both control and hibernating cells: the results are shown in Fig. 5.13 and 5.14. Corrections for any difference in section thickness would need to be made if the continuum measurements show a difference from one section to another; spectra are always stored so that data can be processed in different ways, and continuum measurements are always taken.

Current results (Koepsell, Nicholson, Kriz & Höhling, 1974) suggest that analytical results expressed by volume have more biological significance than those expressed by mass in data obtained from freeze-dried, frozen sections, although it is obviously necessary to work towards expressing quantitative data corrected for the water content of each area analysed.

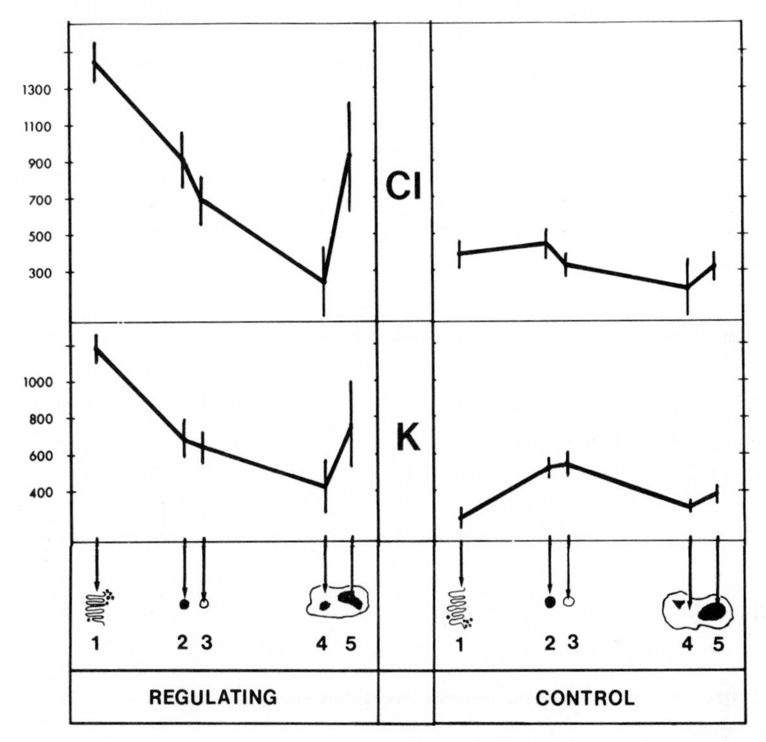

Fig. 5.14 Diagram showing that a gradient of Cl and K can be measured in the regulating epithelium of the hibernating snail *Otala lactea*. The numerical data are from Fig. 5.13 (after Appleton & Newell, 1977*a*, 1977*b*).

Corrections for water content

Any correction for water content will need to be made for each area being analysed because the water content of that area will depend to a large extent on its metabolic state at the moment of freezing. Preliminary results suggest that there may be a relationship between the volumetric measurement ($P-b$) and the mass fraction (or % mass fraction) of an element in solution provided measurements have also been made on a 'biological standard' of known water content. The red blood cells (RBCs) in the spleen sections (Plates 5.6 and 5.7) have been used as biological standards with a water content of 65% to construct the histogram in Fig. 5.15. The scales for the quantitated % mass

Table 5.1 Mass fractions for chlorine quantified and corrected for water content

Cell area, etc.	% Mass fraction DRY*	Water content† (%)	% Mass fraction WET‡
RBCs	0.64±0.05	65	0.22±0.02
Lymphocyte A§	0.15±0.04	74	0.04±0.01
B	0.24±0.02	71	0.07±0.01
C	0.32±0.07	65	0.11±0.03
D	0.33±0.10	79	0.07±0.02

* % MF calculated according to Hall, Anderson and Appleton (1973), and using $CaCl_2$ as standard; † water content calculated for areas of lymphocytes from Fig. 5.15 and using RBCs as biological standard of known water content; ‡ % MF WET calculated from formula (4), below; § (A) light areas of nucleus, (B) dark areas of nucleus, (C) mitochondria, (D) adjacent cytoplasm to mitochondria.

fraction and the volumetric counts ($P-b$) for chlorine in the RBCs was drawn so that the difference in the heights of the solid and stippled blocks represents the water content of the RBCs (i.e. 65%). The histogram for areas (A), (B), (C), and (D) of the lymphocytes were then drawn to the same scale and the water content for each area was calculated from the difference in the histograms. The mM concentration was calculated from the mass fractions and water contents by using the formula below (Appleton & Steward, 1977):

$$\text{Chloride concentration (mM)} = \left(\frac{\%\text{ MF Dry}.(100-\%\text{ water}).10^4}{\text{Atomic weight}.\%\text{ water}}\right) \quad (3)$$

For most purposes a comparison between the mass fractions (WET) would be sufficient:

$$\%\text{ Mass fraction WET} = \frac{\%\text{ MF Dry}.(100-\%\text{ water})}{100} \quad (4)$$

See Table 5.1.

The water content for the areas for the lymphocytes (Fig. 5.15) agree with their suggested profiles as seen in the SEM (see Plate 5.5). It may be that measurements of surface profiles (as in Plate 5.5) could be used to calculate water contents for areas under analysis, particularly if this information can be combined with the analytical information already discussed. Considerable work will be needed to justify these claims but the importance of obtaining such corrections can be seen in Fig. 5.15. The mass fractions for chlorine show no significant difference between areas C and D when expressed as dry weight; when expressed in volumetric terms ($P-b$) or as mM concentrations the two areas do show significant differences in chloride concentration. Differences only become significant when the water content is taken into account. The

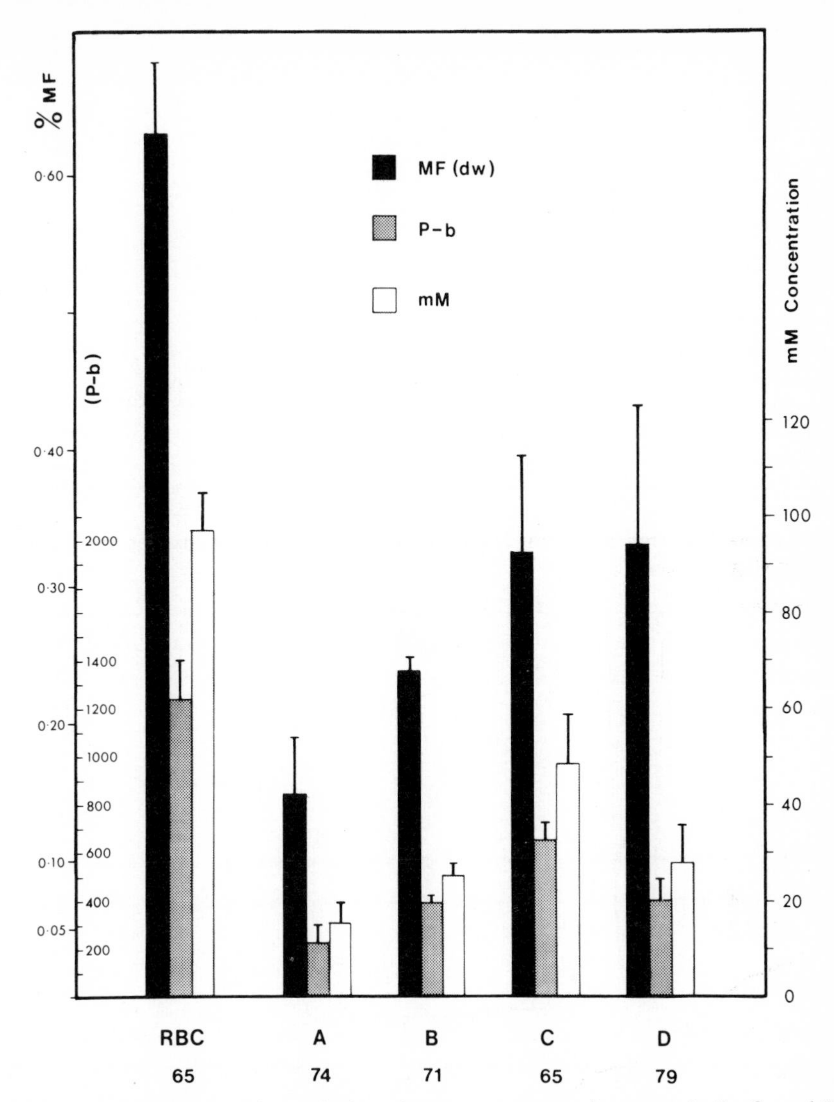

Fig. 5.15 Diagram used to calculate the water contents for areas A, B, C, and D of the lymphocyte. The %MF and $(P-b)$ scales for chlorine in the spleen RBCs (Plate 5.7) were constructed so that the difference in the heights of the solid block and the stippled block reflected the water content (65%) of RBCs. The values for A, B, C, and D were plotted on the same scale and the relative water contents were calculated from the measured differences in the respective solid and stippled blocks. The mM concentrations (right hand scale) were calculated from the formula (Section 5.4.3). Note that there is no significant difference in the %MF dry weight for C and D, and that the concentration only shows significance (C and D) when the water content is taken into account ($P-b$ or mM concentrations). Concentrations expressed as $(P-b)$ are by volume rather than mass and so have more biological significance.

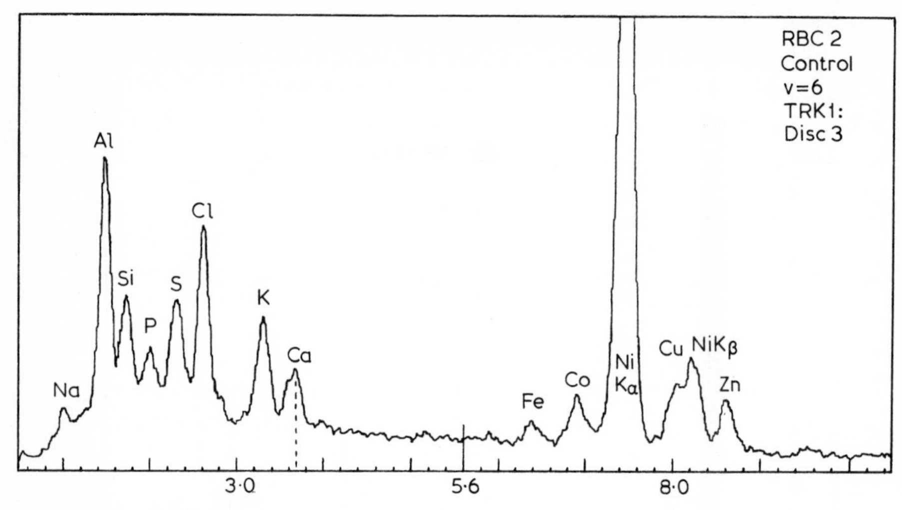

Fig. 5.16 Analytical spectrum from control red blood cell. Cobalt is an impurity in the grid material.

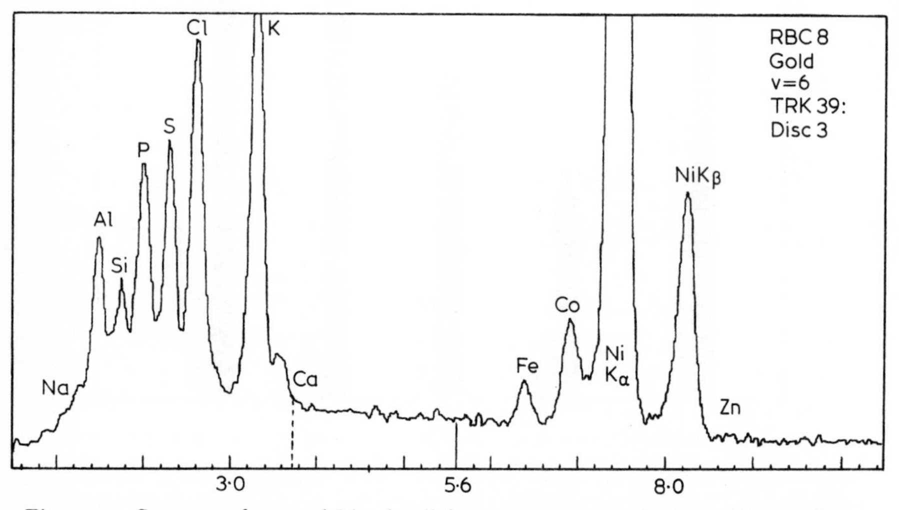

Fig. 5.17 Spectrum from red blood cell from mouse injected with gold as sodium aurothiomalate (Myocrisin) (see text). Note the elevated potassium and phosphorus levels and the reduction in copper and zinc. See Fig. 5.18 for comparison of relative mass fractions.

emphasis for the biologist must be to present his data in biologically significant terms. Percentage mass fractions (dry) obviously mean something to the physicist but mean very little to the biologist when the tissue under investigation may have such varying water contents. X-ray microanalysis of frozen sections in the hydrated state may provide some of the answers since mass fractions from the hydrated state would immediately be in terms of 'wet' weight. However the problems of contrast, interpretation, and relatively high background may limit their widespread use for some years to come. The background caused by the presence of the ice may mask detection of elements

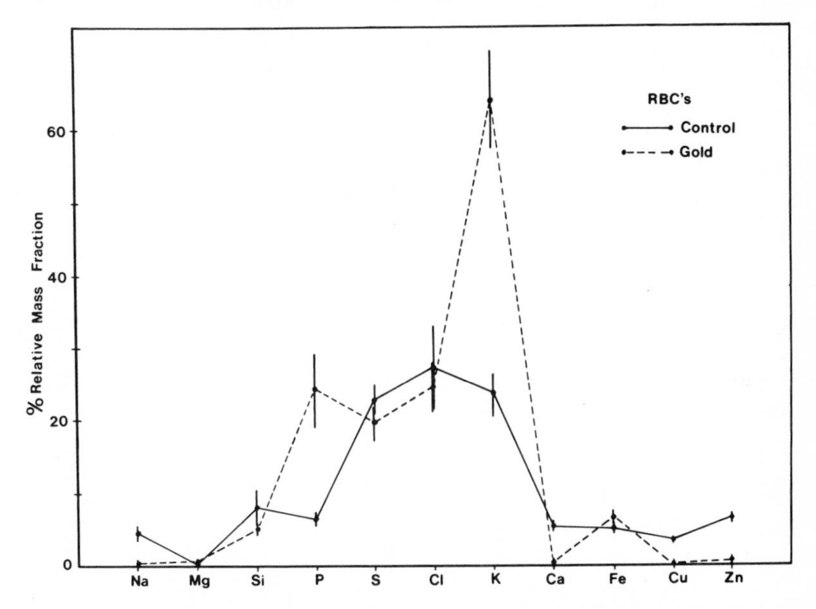

Fig. 5.18 Changes in the elemental 'finger-print' of cells (RBCs) after gold treatment. Such changes may have significance in using X-ray microanalysis of frozen sections for pharmacology and toxicology of drugs. (Appleton, 1977*b*, 1977*c*)

present in low concentrations. The high sensitivity offered by the freeze-dried frozen section makes it possible to study a wide range of applications, but comparison of the results of X-ray microanalytical data with those from complementary techniques (e.g. micro-electrode measurements) will only be possible when the X-ray microanalyst can express his data in common terms, e.g. mM or meq/litre.

5.5 Future contributions of cryo-ultramicrotomy

The techniques of cutting ultrathin frozen sections have already shown that unfixed and unstained material can be examined in the electron microscope and that, provided adequate care is taken to ensure that there is no possibility of diffusion or loss of diffusible elements, these sections can be used for analytical studies such as autoradiography and X-ray microanalysis. These two techniques together provide the biologist, pharmacologist and physiologist with a unique opportunity to study some of the most fundamental processes of life at the subcellular level.

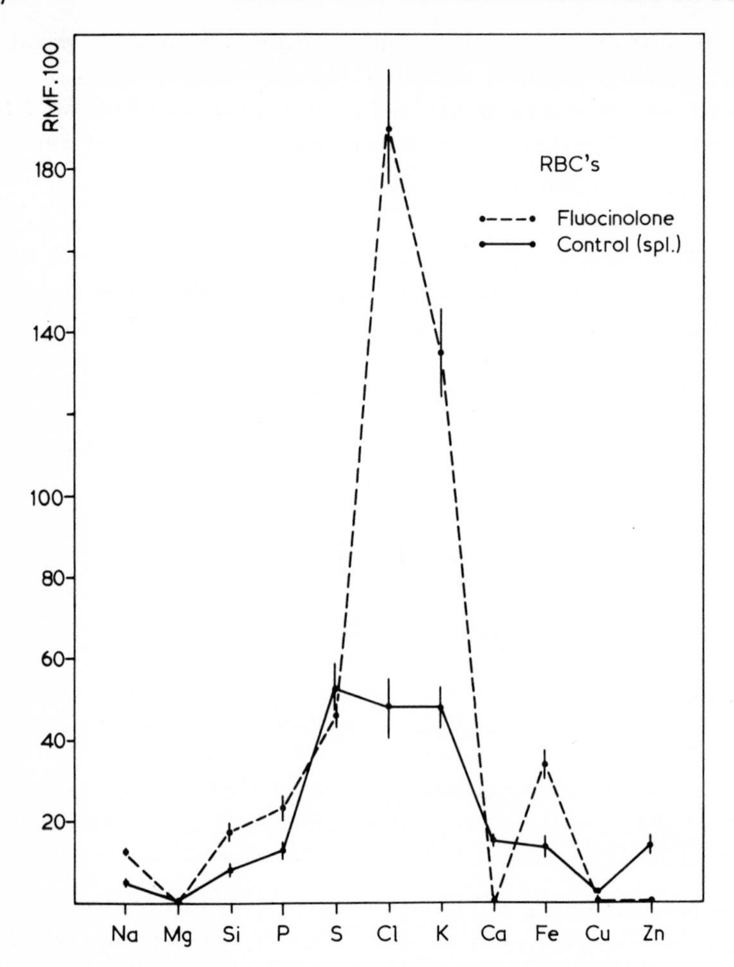

Fig. 5.19 Changes in the elemental 'finger-print' of RBCs after treatment with an anti-inflamatory agent (fluocinolone). Note the increase in the % relative mass fractions of chlorine and potassium with a similar reduction in calcium, copper, and zinc as with gold treatment (Fig. 5.18).

In a recent paper Appleton, Shute and Baker (1976) said:

'The "gross" sites of drug action are usually fairly well known in terms of organ or tissue distribution, and sometimes can be localized to specifically identifiable cells. What is not known is how the various drugs affect the over-all chemistry of the cell at the level of its individual components. How for instance, does an analgesic (pain-killing) drug affect the chemistry of the nerve cells? Does it change the ionic composition of the mitochondria in the axons? Does it change the permeability of the membranes? At what point can we detect toxic effects within the cell? And perhaps more important, why do drugs have side-effects? Do they, for instance, alter the chemistry of cells which are not direct targets? Autoradiography after the application of labelled drugs will indicate the target areas; X-ray microanalysis will show the changes in the ionic concentrations of those cells and their constituents.'

It is probable that the normal elemental composition of cells can be characterized in the form of a kind of 'elemental fingerprint' and that once this has been determined for different cell types it should be possible to detect deviations from that 'norm' by observing changes in one or more of the characterized elements (Appleton, 1977*c*). Already there are indications that this approach is feasible and is likely to yield important results provided that the sensitivity is maximized by care in preparation, that care is taken to reduce the background from instrumental effects, and that the preparative techniques allow us to look at each element detected with equal confidence.

A preliminary study on the study of anti-inflammatory agents on the elemental 'fingerprint' of red blood cells (RBCs) has shown that it is possible to detect such changes. Blood cells from mice which had been injected with 2.0 mg Myocrisin (sodium autothiomalate) intramuscularly and killed 4 days later showed elevated phosphorus and potassium levels, and lowered sodium, calcium, copper, and zinc levels (Figs. 5.16–5.18); yet there was no trace of gold (derived from the aurothiomalate) within the RBCs (Appleton, 1977*b*, 1977*c*).

In a different study the elemental fingerprint of RBCs from mice injected with 1.0 mg of fluocinolone (a topical anti-inflamatory agent) show similar (Fig. 5.19) differences and in addition showed an elevation of chlorine, iron, and sodium with the same marked reduction in calcium, copper, and zinc. It is not yet clear whether the calcium levels seen in the normal RBCs are true, since it is normally assumed that there is no calcium in RBCs. However, it is not unlikely that X-ray microanalysis of naturally occurring elements will fill in many of the gaps in our knowledge as well as proving some clearly held 'truths' to be erroneous.

Cryo-ultramicrotomy allows the biologist to use X-ray microanalysis in the study of diffusible naturally occurring elements. It requires care in all stages of preparation, a basic knowledge of the instrumentation, a small but essential understanding of the physics involved, but above all a critical and enthusiastic honesty at all stages in the techniques involved.

5.6 References

AHKONG, Q. F., FISHER, D., TAMPION, W. & LUCY, J. A. (1975) Mechanisms of Cell fusion. *Nature*, **253**, 194–195.

APPLETON, T. C. (1967) Storage of frozen materials for cryostat sectioning and soluble-compound autoradiography. *Journal of Microscopy*, **87**, 489–492.

APPLETON, T. C. (1972) 'Dry' ultrathin frozen sections for electron microscopy and X-ray microanalysis; the cryostat approach. *Micron*, **3**, 101–105.

APPLETON, T. C. (1973) X-ray microanalysis of diffusible electrolytes in ultrathin frozen sections. *Journal of Physiology*, **233**, 15–17P.

APPLETON, T. C. (1974*a*) Cryo-ultramicrotomy, possible applications in Cytochemistry. In *Electron Microscopy and Cytochemistry*, ed. Wisse, E., Daems, W. Th., Molenaar, I. & van Duijn, P. Pp. 229–242. North Holland Publishing Co.

APPLETON, T. C. (1974*b*) A cryostat approach to ultrathin frozen sections for electron microscopy; a morphological and X-ray analytical study. *Journal of Microscopy*, **100**, 49–74.

APPLETON, T. C. (1977*a*) The use of ultrathin frozen sections for X-ray microanalysis of diffusible elements. In *Analytical and Quantitative Methods in Microscopy*, ed. Meek, G. A. and Elder, H. Y. Pp. 247–268. Cambridge: Cambridge University Press.

APPLETON, T. C. (1977*b*) Subcellular toxicology; an X-ray microanalytical approach. *Acta Pharmacologica et Toxicologica*, **141**, 12–13.

APPLETON, T. C. (1977*c*) Subcellular toxicology; an X-ray microanalytical approach. *Acta Pharmacologica et Toxicologica*, **141**, 14–15.

APPLETON, T. C. & NEWELL, P. F. (1977*a*) X-ray microanalysis of freeze-dried frozen sections of regulating epithelium from the snail *Otala*. *Nature*, **266**, 854–855.

APPLETON, T. C. & NEWELL, P. F. (1977*b*) X-ray microanalysis of diffusible elements in the regulating epithelium of the snail *Otala lactea*. In *Proceedings of the XXVII International Congress of Physiological Sciences*, Paris, July 1977. In press.

APPLETON, T. C., SHUTE, C. C. D. & BAKER, J. R. J. (1976) Understanding toxicity. *Trends in Biochemical Sciences*, **1**, N88–N89.

APPLETON, T. C. & STEWARD, M. C. (1977) Subcellular localization of physiologically active elements using X-ray microanalysis. In *Proceeding of International conference on Microprobe Analysis in Biology and Medicine*. Munster, September 1977. In press.

BACANER, M., BROADHURST, J., HUTCHINSON, T. & LILLEY, J. (1973) Scanning transmission electron microscope studies of deep-frozen unfixed muscle correlated with spatial localization of intracellular elements by fluorescent X-ray analysis. *Proceedings of the National Academy of Science (USA)*, **70**, 3423–3427.

BAKER, J. R. J. & APPLETON, T. C. (1976) A technique for electron microscopy autoradiography (and X-ray microanalysis) of diffusible substances using freeze-dried fresh frozen sections. *Journal of Microscopy*, **108**, 307–315.

BAUER, H. & SIGARLAKIE, E. (1973) Cytochemistry of ultrathin frozen sections of yeast cells. *Journal of Microscopy*, **99**, 205–218.

BERNHARD, W. (1971) Improved techniques for the preparation of ultrathin frozen sections. *Journal of Cell Biology*, **49**, 731–749.

CHANDLER, J. & BATTERSBY, S. (1976) X-ray microanalysis of ultrathin freeze-dried sections of human sperm cells. *Journal of Microscopy*, **107**, 55–65.

CHRISTENSEN, A. K. (1971) Frozen thin sections of fresh tissue for electron microscopy, with a description of pancreas and liver. *Journal of Cell Biology*, **51**, 772–804.

ECHLIN, P. (1971) The examination of biological material at low temperatures. In *Proceedings of the 4th Scanning Electron Microscope Symposium*, 225–232.

GUPTA, B. L., HALL, T. A., MADDRELL, S. H. P. & MORETON, R. B. (1976) Distribution of ions in a fluid transporting epithelium determined by electron-probe X-ray microanalysis. *Nature*, **264**, 284–287.

HALL, T. A., ANDERSON, H. C. & APPLETON, T. C. (1973) The use of thin sections for X-ray microanalysis in biology. *Journal of Microscopy*, **99**, 177–182.

HALL, T. A. & GUPTA, B. L. (1974). Beam-induced loss of organic mass under electron microprobe conditions. *Journal of Microscopy*, **100**, 177–188.

HODSON, S. & MARSHALL, J. (1970) Ultramicrotomy – a technique for cutting ultrathin frozen sections of unfixed biological tissues for electron microscopy. *Journal of Microscopy*, **91**, 105–117.

HODSON, S. & MARSHALL, J. (1972) Evidence against through-section thawing whilst cutting on the ultracryotome. *Journal of Microscopy*, **95**, 459–466.

HÖHLING, H. J., BARCKHAUS, R. H., KREFTING, E. R. & SHREIBER, J. (1976) Electron microscope microprobe analysis of mineralized collagen fibrils and extra collagenous regions in turkey leg tendons. *Cell & Tissue Research*, **175**, 345–350.

KIRK, R. G. & DOBBS, G. H. (1976) Freeze fracturing with a modified cryo-ultramicrotome to prepare large intact replicas and samples for X-ray microanalysis. *Science Tools*, **23**, 28–31.

KOEPSELL, H., NICHOLSON, W. A. P., KRIZ, W. & HÖHLING, H. J. (1974) Measurements of exponential gradients of sodium and chlorine in the rat kidney medulla using electron microprobe. *Pflugers Archives*, **350**, 167–184.

KUSHIDA, H. (1961) A styrene–methacrylate resin embedding method for ultrathin sectioning. *Journal of Electron Microscopy*, **10**, 16–19.

MCINTYRE, I. A., GILULA, N. B. & KARNOVSKY, M. J. (1974) Cryoprotectant-induced redistribution of intramembraneous particles in mouse lymphocytes. *Journal of Cell Biology*, **60**, 192–203.

MERRYMAN, H. T. (1956) Mechanics of freezing in living cells and tissues. *Science*, **124**, 515–521.

MIZUHIRA, V. (1976) Elemental analysis of biological specimens by electron probe X-ray microanalysis. *Acta Histochemica Cytochemica*, **9**, 69–87.

MOOR, H., KISTLER, J. & MÜLLER, M. (1976) Freezing in a propane jet. *Experientia*, **32**, 805.

NIEDERMEYER, W. & MOOR, H. (1976) The effect of glycerol on the structure of membranes. A freeze-etch study. *Proceedings of the 6th European Congress on Electron Microscopy*, Vol. **2**, 108–109.

PETERS, P. D., YAROM, R., DORMANN, A. & HALL, T. A. (1976) X-ray microanalysis of intracellular zinc: Emma-4 examinations of normal and injured muscle and myocardium. *Journal of Ultrastructure Research*, **57**, 121–132.

PLATTNER, H., SCHMITT-FUMIAN, W. W. & BACHMANN, L. (1973) Cryofixation of single cells by spray freezing. In *Freeze-etching*, ed. Benedetti, E. L. & Favard, P. Pp. 81–100. Paris: Société Française de Microscopie Électronique.

PRYDE, J. A. & JONES, G. O. (1957) Properties of vitreous water. *Nature*, **170**, 685–688.

RIEHLE, U. & HÖCHLI, M. (1973) The theory and technique of high pressure freezing. In *Freeze-etching*, ed. Benedetti, E. L. & Favard, P. Pp. 31–61. Paris: Société Française de Microscopie Électronique.

SAUBERMAN, A. J. & ECHLIN, P. (1976) The preparation, examination, and analysis of frozen hydrated sections by scanning transmission electron microscopy and X-ray microanalysis. *Journal of Microscopy*, **105**, 155–191.

SJÖSTRÖM, M., THORNELL, L. E. & CEDERGREN, E. (1973) Ultramicrotomy in the study of fine structure of myofilaments. *Journal of Microscopy*, **99**, 193–204.

SJÖSTRÖM, M. & THORNELL, L. E. (1975) Preparing sections of skeletal muscle for transmission electron analytical microscopy (TEAM) of diffusible elements. *Journal of Microscopy*, **103**, 101–112.

SMITH, D. J. & ALDRICH, H. C. (1971) Membrane systems of freeze-etched striated muscle. *Tissue and Cell*, **3**, 261–281.

SPRIGGS, T. L. B. & WYNNE-EVANS, D. (1976) Observations on the production of frozen-dried thin sections for electron microscopy using unfixed fresh liver, fast-frozen without cryoprotectants. *Journal of Microscopy*, **107**, 35–46.

SPURR, A. R. (1969) A low viscosity epoxy resin embedding medium for electron microscopy. *Journal of Ultrastructure Research*, **26**, 31–43.

TRUMP, B. F. (1969) Effects of freezing and thawing on cells and tissues. In *Autoradiography of Diffusible Substances*, Stumpf, W. E. & Roth, L. J. Pp. 211–240. New York: Academic Press.

TSUCHIYA, T. (1976) Electron microscopy and electron probe analysis of Ca-binding sites in the cilia of *Paramecium caudatum*. *Experientia*, **32**, 1176–1177.

YAROM, R. & CHANDLER, J. A. (1974) Electron probe microanalysis of skeletal muscle. *Journal of Histochemistry and Cytochemistry*, **22**, 147–154.

YAROM, R., PETERS, P. D. & HALL, T. A. (1974) Effect of glutaraldehyde+Urea embedding on intracellular ionic elements. *Journal of Ultrastructure Research*, **49**, 405–418.

YAROM, R., MAUNDER, C. A., SCRIPPS, M., HALL, T. A. & DUBOWITZ, V. (1975) A simplified method of specimen preparation for X-ray microanalysis of muscle and blood cells. *Histochemistry*, **45**, 49–59.

Acknowledgements

The work described in this chapter was generously supported by grants from the Medical Research Council, and CIBA-Geigy Pharmaceuticals.

The cold-stage for the EMMA-4 microscope was designed and constructed by J. H. Lucas, Fine Mechanics, Hillside, Rickling, Saffron Walden, Essex, under a grant from the Royal Society, Paul Instrument Fund.

6

I. D. BOWEN AND T. A. RYDER

The application of X-ray microanalysis to histochemistry

6.1 Introduction

By definition, histochemistry involves the localization of chemicals or chemical reactions within cells, tissues, or cell products, and since X-ray microanalysis is a technique whereby the identity and distribution of elements within a source can be determined, a close interrelationship has naturally grown between the two disciplines. X-ray microanalysis can be employed to study the specificity of histochemical reactions and the composition of their products; in this context, microanalysis can establish the validity of accepted techniques (Ryder & Bowen, 1974). Since X-ray microanalysis, interfaced with microscopy, also releases the histochemist from the constraint of being dependent on a 'visual' reaction product, new techniques can be evolved based simply on elemental detection (Bowen, Ryder & Downing, 1976). Such techniques can also take advantage of new physical preparation procedures such as freeze drying and ultracryotomy.

In any scientific endeavour the results achieved are only as good as the validity of the techniques employed; X-ray microanalysts should, therefore, be wary of employing histochemical methods initially designed for optical microscopy and which may not meet the stringent requirements of cytochemistry at electron microscope level. Problems encountered at this level relate to the specificity and sensitivity of the histochemical test and to the resolving capabilities of the analytical system. At an ultrastructural level, problems relating to movement or translocation of elements and compounds become particularly acute, due to the fine compartmentalized nature of biological material involving membrane barriers. Thus certain elements may be selectively retained by plasma and organelle membranes. Such semipermeable membranes may also retard the entry and penetration of precipitating compounds during specimen preparation. Preparative procedures all produce their own artefacts (see Chapter 4). Each histochemical test should, therefore, have its efficacy and specificity fundamentally examined before the final application to a biological problem is made.

All types of X-ray microanalytical instruments have been employed in histochemical studies. These include the X-ray microprobe analyser, and transmission and scanning electron microscopes fitted with X-ray microanalysers. More recently, scanning electron microscopes using the transmission mode, and transmission electron microscopes using the scanning mode (STEM) have also been used in conjunction with analysers. X-ray analysers may be either wavelength-dispersive crystal spectrometers, or energy-dispersive analysers. The principles of X-ray microanalysis and the instrumentation used are reviewed in Chapters 2 and 3.

The histochemical advantages of the various kinds of analysis can be summarized as follows. Point analysis, although time consuming, is capable of detecting lower concentrations of elements and presents quantitative possibilities. Line scan analysis is useful in that it provides a visual indication of elemental distribution which can be superimposed on a morphological image. X-ray images relate histochemical information directly to topography or surface features. Both line scan analyses and X-ray images relate to only one element at a time whereas point analyses using energy-dispersive systems give a display of a broad energy spectrum and have the advantage of presenting the over-all chemical composition as well as the element or elements of interest. Energy-dispersive systems require the operation of a window to select for particular elements during line scan analysis or production of X-ray images. Wavelength-dispersive analysis, which selects only one element depending on the crystal chosen (see Chapters 2 and 3), is thought to be more sensitive than energy-dispersive systems in terms of spectral resolution. In this sense, wavelength-dispersive systems display a higher histochemical resolution.

An important consideration when investigating the possibility of using X-ray microanalysis in a cytochemical study is the detection limit of the system, which will in turn determine the minimum amount of reaction product which must be present to be detectable. This limit will depend not only on the nature of the specimen but to a large degree on the instrumentation used. Russ (1974*b*) has indicated that under favourable conditions the minimum concentration of an element in an organic matrix that can be detected is 100–500 ppm. Cytochemists who manipulate their experimental system to produce the required amount of reaction product, e.g. by extending incubation times or altering temperature in the case of enzyme cytochemical studies, are unlikely to find the minimum detection limits a problem, although care must be taken here not to lose the spatial specificity of the reaction. Workers dealing with products that cannot be 'built up' may face a more difficult task.

The ultimate aim in most histochemical and cytochemical studies is to identify and locate the chosen element in the specimen. The spatial resolution

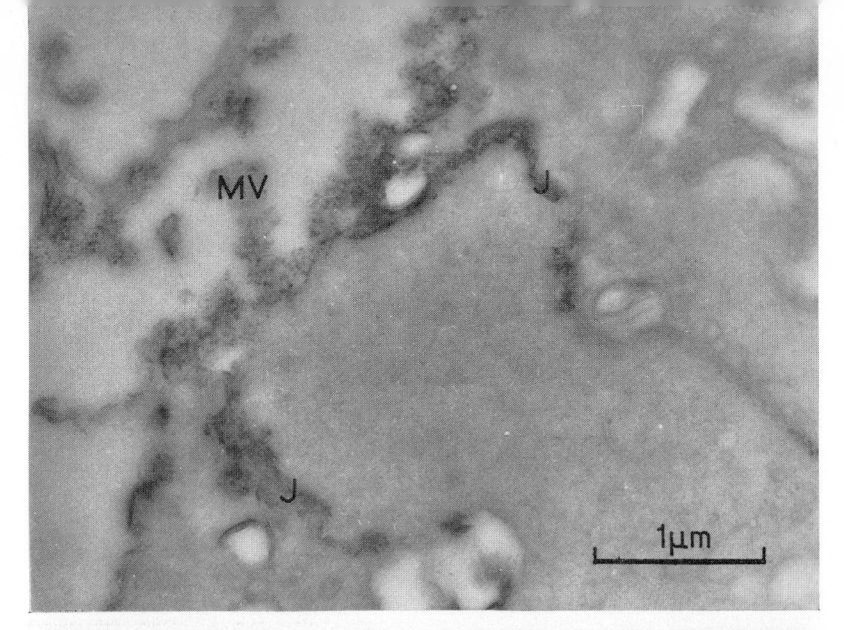

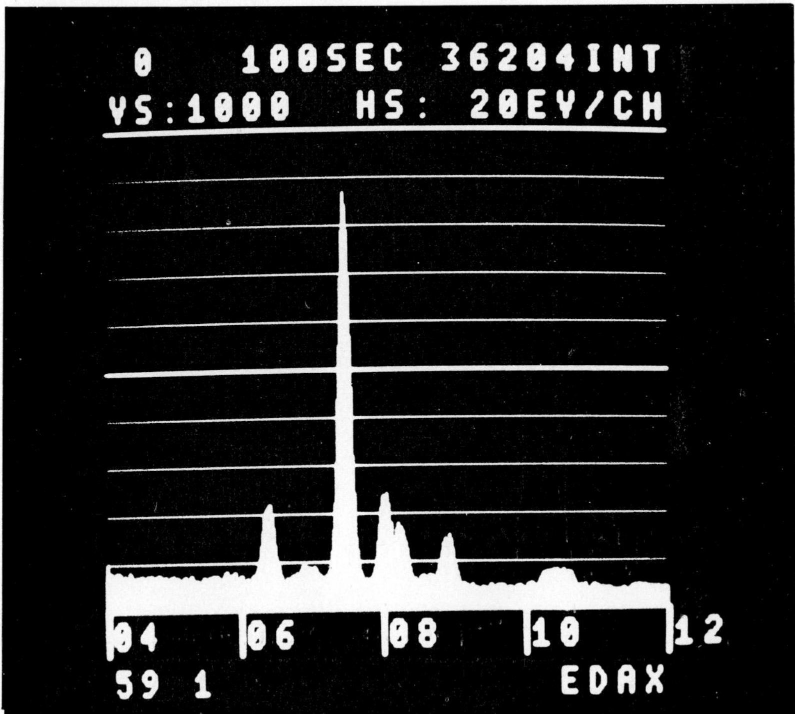

Plate 6.1 Section through part of the foot epithelial cells of the slug *Agriolimax reticulatus*. The slug has been allowed to 'walk' for 30 min at room temperature over a substratum impregnated with 1000 ppm copper sulphate. Uptake of copper has been demonstrated, at electron microscope level, by precipitation with potassium ferrocyanide. Electron-opaque copper ferrocyanide can be seen around the microvilli (MV) and the paracellular junctions (J). ($\times$ 24 000)

Plate 6.2 The energy spectrum obtained on analysing the electron opaque product present in a cell junction (Plate 6.1). The peaks represent: Iron, $K\alpha$ 6.398 keV and $K\beta_2$ 7.057 keV; Nickel, $K\alpha$ 7.471 keV; Copper, $K\alpha$ 8.040 keV and $K\beta_1$ 8.904 keV; Osmium, $L\alpha$ 8.910 keV and $L\beta_1$ 10.35 keV. The nickel is derived from the grid and the osmium from the fixative.

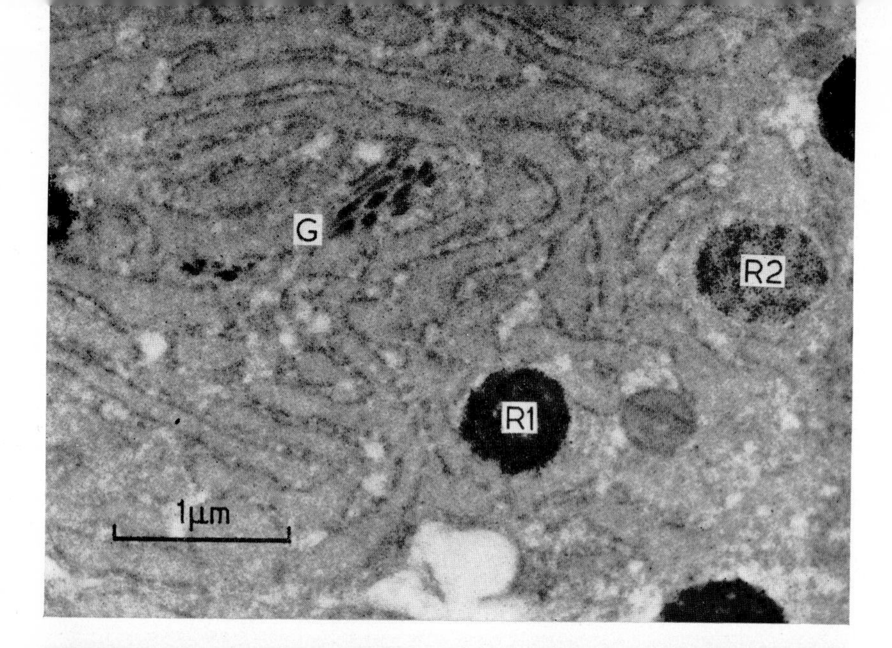

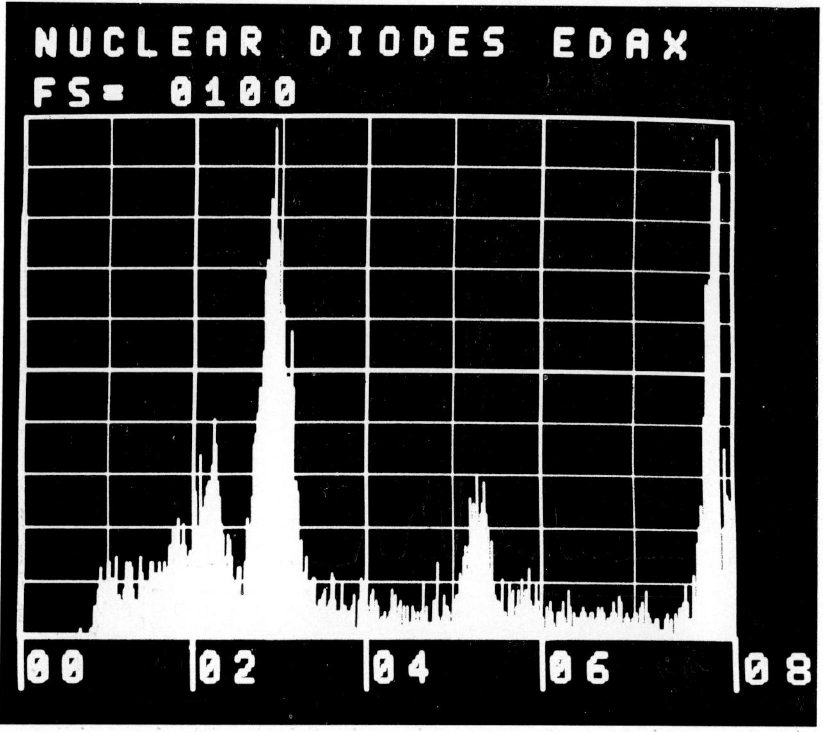

Plate 6.3 Section through part of a cyanophil gland cell from a planarian worm exposed to the chromic acid oxidation-silver technique. Dense silver granules, depicting the site of polysaccharide, are deposited in two types of reticulate granules (R1) and (R2) and in the Golgi apparatus (G). (×25 000)

Plate 6.4 An energy spectrum obtained on analysis of the reticulate granule (R1) (Plate 6.3). The Lα peak for silver can be seen at 2.98 keV. The Kα peak at 2.307 is sulphur from the Araldite embedding medium and the chromium Kα peak at 5.4 keV is an instrumental artefact.

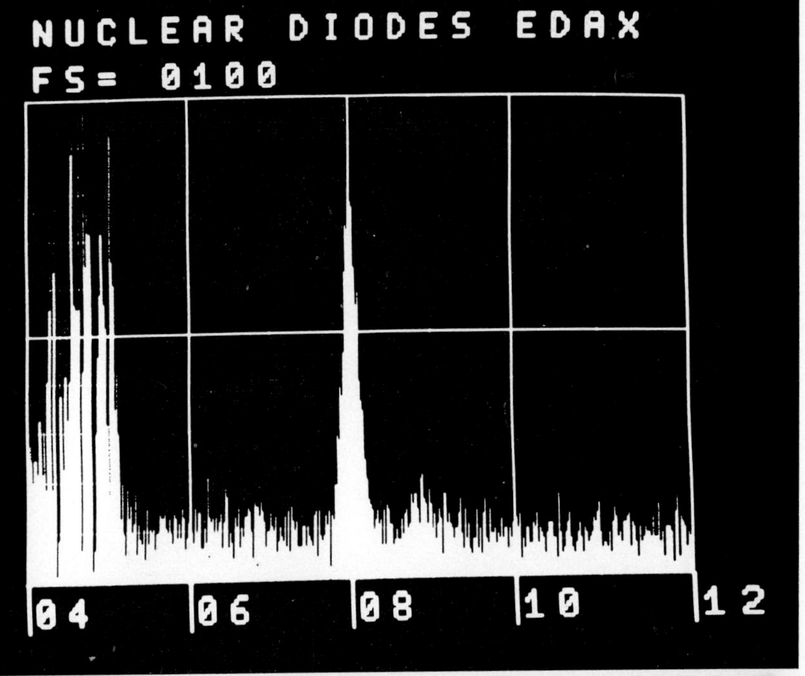

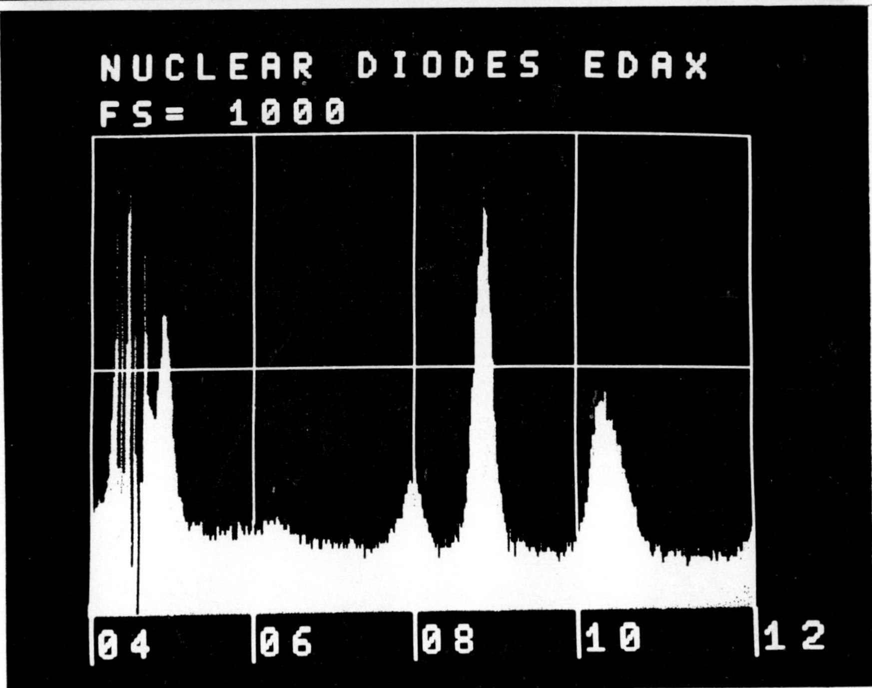

Plate 6.5 An energy spectrum derived from a small droplet of 1 % Alcian blue sprayed on to a carbon film supported by a titanium grid. The Kα line for copper (obtained from the Alcian blue) can be seen at 8.040 keV.

Plate 6.6 An energy spectrum obtained on analysis of a mucous droplet on the surface epithelium of a planarian worm. The tissue has been exposed to the Alcian blue test for acid mucopolysaccharide. The characteristic Kα line for copper can be seen at 8.040 keV and the Kβ_1 line at 8.904. The Kβ peak for titanium derived fom the grid can be seen at 4.931 keV. Lα and Lβ_1 peaks for osmium can be seen at 8.910 and 10.354 keV respectively.

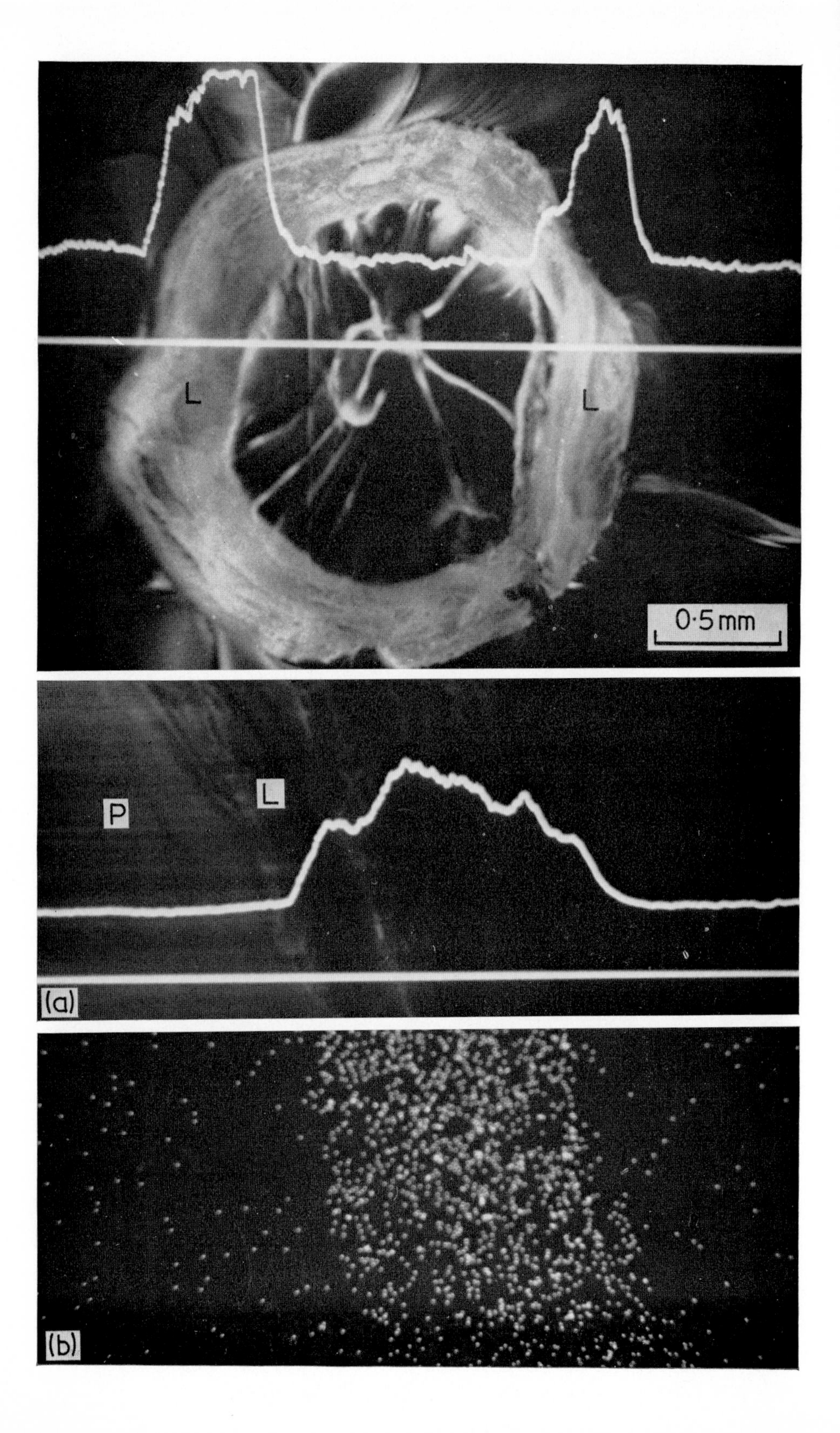
L
L
0·5 mm
P
L
(a)
(b)

Plate 6.7 A stereoscan image of the cut surface of a fractured slug egg (previously exposed to a solution of 100 ppm copper sulphate for 4 h at 18 °C). The lower line shows the line scanned by the electron beam, the upper trace represents the number of counts for copper along that line (full scale $= 3 \times 10^2$ counts). The peaks indicate that copper from the molluscicide is limited to the outer layers of the egg (L). (×40)

Plate 6.8*a* Part of a specimen similar to that shown in Fig. 6.7 depicting the egg shell (L) and perivitelline jelly (P) at a higher magnification. Again the line scan indicates that all the detectable copper is retained in the outer layers. (×270)

Plate 6.8*b* An X-ray pattern for copper obtained from the specimen shown in Fig. 6.8*a*. All the detectable copper is retained in a band representing the egg shell. Some background counts are seen outside the shell. (×270)

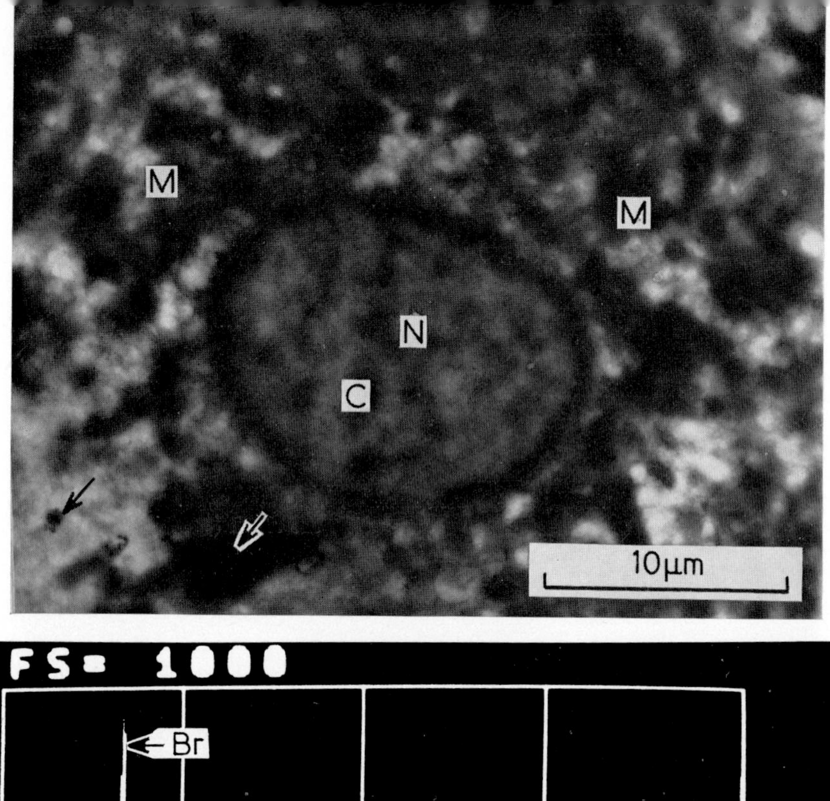

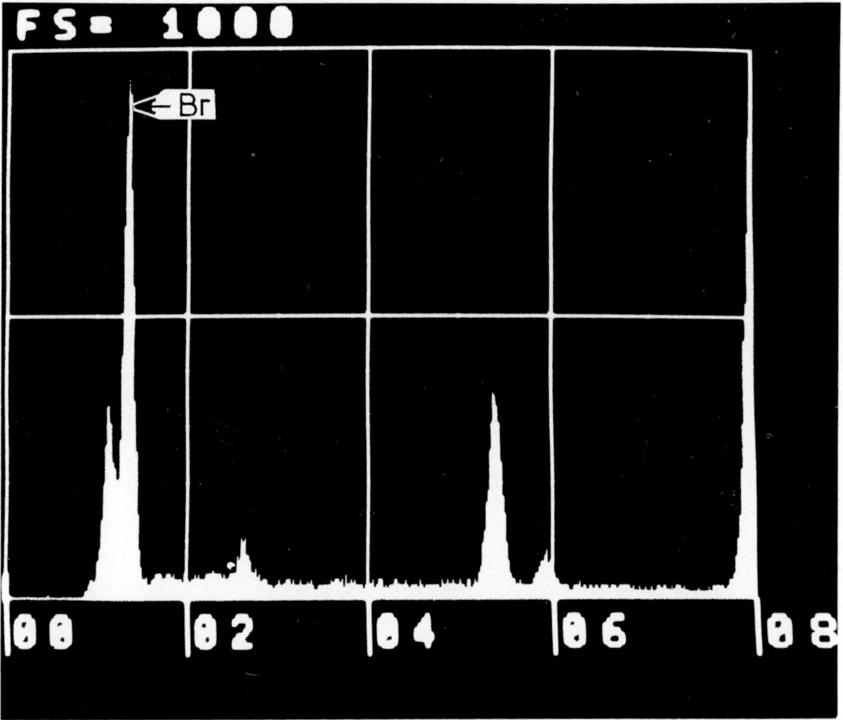

Plate 6.9 Part of an ultrathin cryo-section from rat liver tissue exposed to the simultaneous coupling azo dye test for acid phosphatase activity (Bowen, Ryder & Downing, 1976). Dense azo dye reaction product can be seen in the pericanalicular region (arrows). The nucleus (N), chromatin (C), and mitochondria (M) of a liver cell can also be seen. (× 3400).

Plate 6.10 The energy spectrum obtained from X-ray microanalysis of the azo dye reaction product depicted in Plate 6.9. Note a substantial peak for bromine (Br) at 1.5 keV. The chromium peak at 5.4 keV is an instrumental artefact.

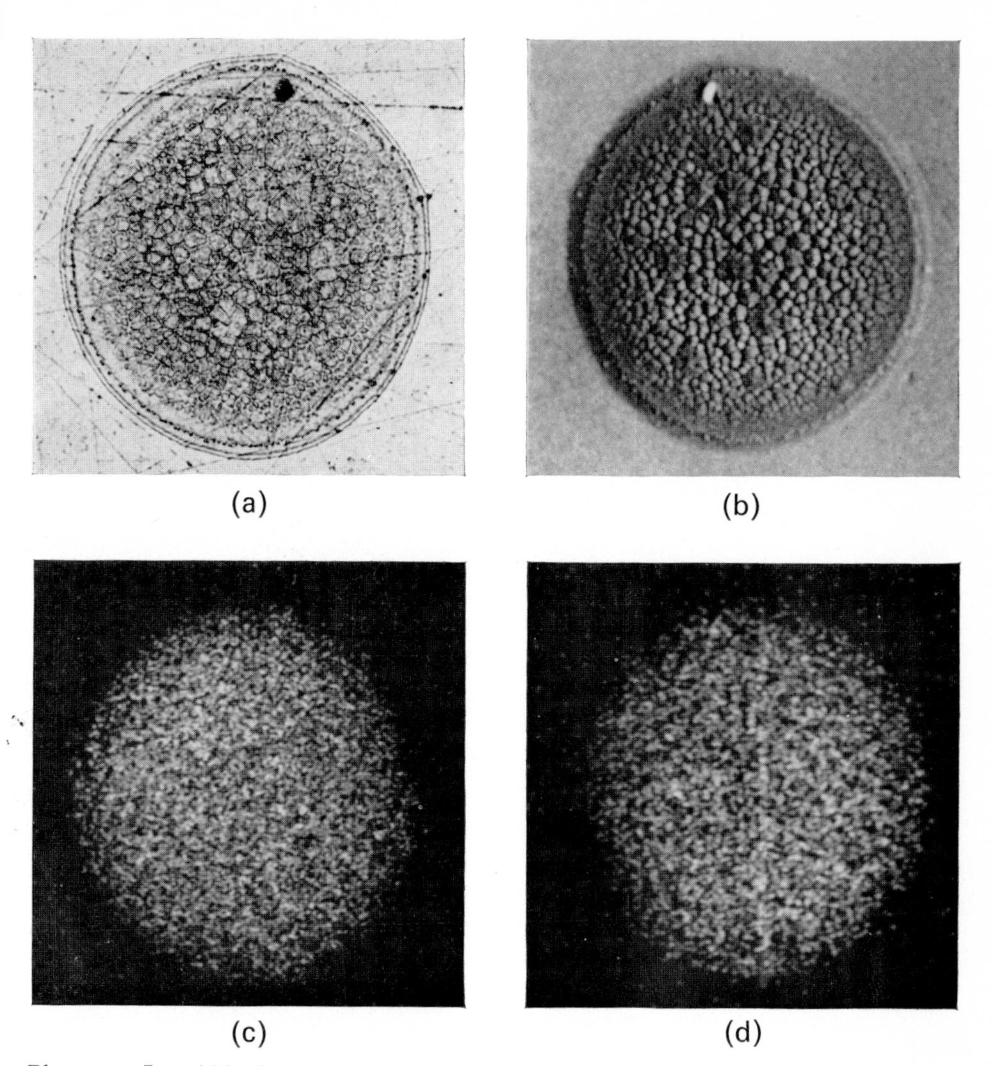

Plate 7.1 Lyophilized renal tubular fluid sample obtained from *Necturus* prepared for electron probe microanalysis. Spot diameter approximately 120 μm. (*a*) Optical image; (*b*) electron image; (*c*) X-ray image of sodium; (*d*) X-ray image of chlorine (from Garland, Hopkins, Henderson, Haworth & Chester Jones, 1973).

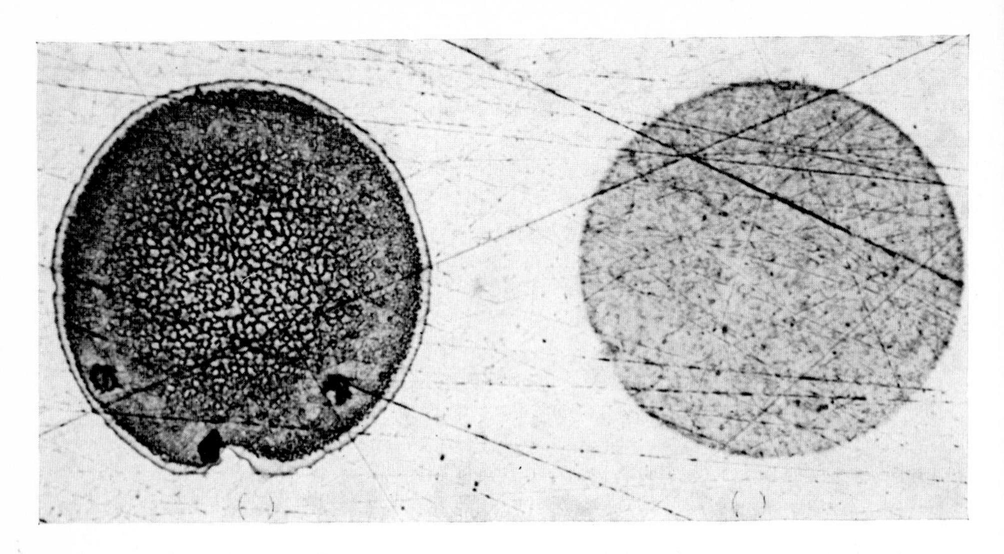

Plate 7.2 Lyophilized renal tubular fluid sample obtained from *Necturus* prepared for electron probe microanalysis. At *left* sample itself, 120 μm in diameter, and at *right* carbon contamination mark on aluminium substrate built up by a defocused electron beam over a period of about 20 minutes (from Garland, Hopkins, Henderson, Haworth & Chester Jones, 1973).

of the analytical system is, therefore, of importance. For a given homogeneous specimen and accelerating voltage, the spatial resolution, i.e. the volume from which X-rays are generated, will depend on specimen thickness. The beam spread in an ultrathin section may be regarded as negligible and thus the beam diameter represents the spatial resolution of the analysis system. In thick sections or bulk specimens, the electrons scatter and the X-rays emanate from a cone shape, which is somewhat larger at the base than the beam diameter. With such samples high resolution, subcellular histochemistry is impossible.

From the foregoing statements it would appear that to achieve ultimate resolution, ultrathin specimens and very small beam diameters are required. Unfortunately, there are complicating factors; sections thinner than 100 nm may be unstable under the intense beam, and because the excited volume is small analysing times must usually be extended. This can be disadvantageous because of the increase in background counts produced. Extended analysis times also increase the possibility of specimen damage. In many histochemical studies where concentrations of marker elements are low, the ultimate limit of resolution of which the analytical system is capable is unlikely to be achieved.

The translocation or movement of chemical components within the biological material under investigation is a critical consideration for histochemists. Some of the leaching effects of fixatives are known, and the various preparative techniques dealt with in Chapter 4 remove or may even add various elements to the cells and tissues. Loss and translocation during examination under the electron beam deserves urgent attention. A specimen subjected to an electron beam in the electron microscope is altered both physically and chemically, the extent of which is still largely unknown (Stenn & Bahr, 1970). There is a loss of mass under the beam (Hall & Gupta, 1974), and independent reports by Morgan, Davies and Erasmus (1975), and Roinel (1975) have indicated that chlorine is particularly susceptible. Of greater importance to the histochemist, is the possible translocation of elements due to irradiation which may under some circumstances result in a false localization.

Histochemistry in its broadest sense includes the study of the distribution of elements in biological material and, as such, direct X-ray microanalysis of biological material such as is presented in Chapters 3, 4 and 5, represents histochemical technique. In this review we will confine ourselves to a stricter definition of histochemistry, where specific reactions are employed to generate a product within the cell. Virtually all histochemistry prior to the development of X-ray microanalysis fell into this category of 'reactive histochemistry'. For practical purposes, therefore, our definition excludes naturally occurring elements unless their presence is detected or highlighted through some additional chemical manipulation.

6.2 Preparation of cells and tissues for histochemistry and X-ray microanalysis

6.2.1 PRESERVATION AND FIXATION

It is essential to histochemical studies that the morphology and fine structural relationships of the specimen be preserved whilst maintaining, as far as is possible, the original *in vivo* distribution of chemical components. Such ideals are never fully realized for all components using any one method of fixation. Broadly, preservation and fixation can be achieved either by physical means which include freezing, freeze drying, air drying, and heat fixation, or by chemical means which include treatment with aldehydes and/or metallic fixatives such as osmium tetroxide, potassium permanganate, or mercuric chloride.

From a theoretical point of view, the precipitation techniques which are employed to insolubilize or immobilize elements in tissues prior to analysis could also be regarded as fixatives, since they primarily function to 'fix' chemical elements *in situ*. The ideal method of fixation will depend on the type of histochemical test being undertaken. One cannot generalize in this area, but precipitation agents such as pyroantimonate and oxalate are often included in a routine biological fixative such as osmium tetroxide in order to fix specific ions for subsequent analysis. Routine aldehyde fixation may be adequate in certain instances, e.g. fixation of resistant enzymes such as acid phosphatase or morphological fixation prior to staining procedures. Nevertheless, organic or aqueous chemical fixatives do tend to remove elements from tissues and cells (Morgan & Bellamy, 1973). On occasion this can be used to histochemical advantage. Naturally occurring chlorine can be washed out of tissues by buffer rinse and fixation, thus making possible the localization of covalently linked product-associated chlorine (Ryder & Bowen, 1974). Physical methods of fixation represent an attempt to circumvent this problem of solvent action. In the case of enzyme histochemistry a certain degree of fixation may also prove necessary to stabilize proteins in order to avoid enzyme diffusion. Such fixation may remove co-factors or ions essential for enzymatic activity, and these may have to be replaced in the incubating medium.

Freezing, freeze drying, and freeze-dried substitution methods have been established for some time (Gersh, Vergara & Rossi, 1960), and from a histochemical point of view appear to have several advantages. Preservation of enzyme activity for enzyme histochemistry, absence of solvent action, and maintenance of true electrolyte picture are among the advantages claimed. Recently, however, the translocation effects of freezing have been demonstrated, although some of these are actually associated with thawing and can be avoided by maintaining the biological material in a frozen state throughout, even in the microscope. Such an approach may also necessitate storage of the biological material for prolonged periods in liquid nitrogen. Various procedures can minimize ice crystal growth; from a histochemical point of view the incorporation of a cryo-protective such as polyvinyl pyrrolidone (PVP) is an useful precaution.

The use of ultrathin cryo-sections for X-ray microanalysis has histochemical advantages. Azo dye and indoxyl reaction products can be microanalytically detected in ultrathin cryosections, and organic solvents can be excluded from the preparative procedures (Ryder & Bowen, 1974). Azo dye histochemistry of cryostat sections is a generally accepted technique for obtaining a relatively true picture of acid phosphatase activity. It is, therefore, very useful to develop this type of approach at electron microscope and X-ray microanalytical levels.

In a recent review Coleman and Terepka (1974) have emphasized the advantages of rapid drying as a means of heat fixation. The approach appears to be most suitable for single cells and isolated organelles. Coleman, Nilsson, Warner and Batt (1972, 1973*a*, and 1973*b*) reported successful heat fixation of protozoa where the cells retained recognizable morphology and normal distribution of diffusible elements. The histochemical efficacy of this approach has not been examined, but as it involves heat denaturation it would be of little value for enzymatic localization.

6.2.2 EMBEDDING

Embedding in media routinely used for electron microscopy often influences the chemical content of tissue specimens (e.g. Appleton, 1972; Davies & Erasmus, 1973; Weavers, 1973). During fixation some elements are removed or partially removed and others may be introduced, e.g. osmium. Embedding media introduce their own chemical artefacts. As long as the media do not affect the distribution of relevant histochemical reactants or products, however, they can be used in the content of X-ray microanalysis. Araldite may still be used as an embedding medium when analysing lead derived from the Gomori acid phosphatase test, although it introduces sulphur and chlorine. In the case of other reactants, it may prove necessary to exclude Araldite or similar resins. Freezing, both for preservation, fixation, and as a supporting procedure, then becomes invaluable; under such circumstances infiltration of the tissue with a cryo-protective such as PVP results in a supportive as well as a cryo-protective action. Other supporting methods have evolved in relation to ultracryotomy, e.g. cross-linked albumin (Kuhlmann & Viron, 1972), fibrin (Rowe, 1972), and gelatin encapsulation (Baur & Sigarlakie, 1973).

6.2.3 SECTIONING AND PRETREATMENT

The technique of ultracryotomy makes available for electron microscopy and analysis, thin frozen sections which can be examined in the frozen hydrated state or after they have been freeze dried. The principles of ultracryotomy are described elsewhere (Bernhard & Viron, 1971; Appleton, 1972; Davies & Erasmus, 1973), and in Chapter 5. The use of fresh or fixed frozen sections vastly increases the potential number of chemical components that can be analysed in the electron microscope; this is particularly true in relation to enzyme histochemistry due to the preservation of enzymatic activity in frozen sections.

A range of histochemical techniques have been applied to ultrathin cryo-sections. The demonstration of enzymatic activity has been established by Leduc, Bernhard, Holt and Tranzer (1967), Holt (1972), and Baur and Sigarlakie (1973). Zotikov and Bernhard (1970) have localized the activity of nucleases, and Couteaux and Delaitre (1972) have demonstrated cholin-esterases in ultrathin frozen sections. Leung and Babai (1974) reported the detection of 5-nucleotidase in thin frozen sections, and Yokota and Nagata (1974*a* and 1974*b*) presented an immunoferritin method for demonstrating urate oxidase and catalase respectively in ultrathin cryo-sections. Kuhlmann and Miller (1971) have developed methods for the ultrastructural localization of anti-enzyme antibodies in the ultrathin frozen sections.

With regard to non-enzyme histochemistry, Babai and Bernhard (1971), and Babai (1972) described the cytochemical localization of polysaccharides in thin frozen sections using phosphotungstic acid, and Puvion and Bernhard (1975) have described the ribonucleoprotein components of liver nuclei in ultrathin cryo-sections. Christensen and Paavlova (1972) introduced the possibility of using fresh ultrathin frozen sections in autoradiography, and Bernier, Iglesias and Simard (1972), Geuskens (1972), and Simard (1972*a*) demonstrated DNA in such sections by means of tritiated actinomycin. Simard (1972*b*) presented a useful method for the autoradiography of diffusible substances incorporating (^{125}I). Hellström (1973) described the localization of catecholamines in ultra-thin sections using formaldehyde-induced fluorescence. Painter, Tokuyasu and Singer (1973) introduced a technique of direct immunoferritin staining for the localization of intracellular antigens in ultrathin cryo-sections, and Fournier-Laflèche, Chang, Bénichou and Ryter (1975) described the immuno-labelling of frozen ultrathin sections of bacteria.

Some techniques have been developed for the X-ray microanalysis of enzymatically reacted frozen sections. Engel, Resnick and Martin (1968) described an electron probe analysis method suitable for cryostat sections containing the products of ATPase activity. Beeuwkes and Rosen (1975) also looking at ATPase activity employed 10 μm cryostat sections in their X-ray microanalytical study. Reaction products containing label atoms such as chlorine and bromine enzymatically deposited within ultrathin frozen sections can be examined by means of X-ray microanalysis (Ryder & Bowen, 1974;

Bowen, Ryder & Downing, 1976). These last studies revealed the fine structural localization of acid phosphatase. Indeed, a new field of enzyme-based X-ray microanalysis is now becoming established (Bowen & Ryder, 1977).

In relation to cryo-techniques, tissues may be treated with histochemical tests either before freezing and sectioning, or after ultracryo-sectioning. In the former case, thin slivers of fresh or fixed tissue are cut on a tissue chopper, incubated, and subsequently quenched in liquid nitrogen. In the latter case, tissues are exposed, after cryo-sectioning, to the histochemical test for a limited period under optimal conditions.

Although 'wet knife' cryo-techniques may be capable of producing morphologically superior sections, care must be taken not to use solvents in the knife bath which might affect the localization of histochemical reactants and products. In general, from a histochemical point of view, we would strongly recommend cutting and picking up cryo-sections in the dry state.

It is important to realize that ultracryotomy may cause the movement or translocation of naturally occurring, unbound ions, due to melting at the cutting edge, although Hodson and Marshall (1972) have indicated that there is no evidence for through-section thawing whilst cutting. It would appear unlikely that translocation of insoluble histochemical products occurs, unless small particles are physically carried by the cutting edge of the knife. This is a possibility when the final reaction product contains heavy crystalline deposits, as with the Gomori test for acid phosphatase (Bowen & Downing, unpublished observations). Finally, support of the sections by a suitably inert material is obviously important when attempting to detect particular elements within histochemical products. In this context, titanium or Teflon grids and graphite grid holders are invaluable.

Sectioning is obviously not a necessary prelude to scanning for surface associated histochemical products in the scanning electron microscope. It may, however, be necessary to freeze fracture in a simple way under liquid nitrogen in order to determine the uptake of elements by internal structures (Ryder & Bowen, 1977*a*). Similarly, histochemically reacted bulk specimens could be fractured before X-ray microanalysis of reactants and/or products, by means of point analysis, line scan analysis, or X-ray imaging. Since scanning electron microscopes can be fitted with a transmission stage, relatively thick (0.5–4 μm) frozen or freeze-dried sections of biological material could be examined for histochemical products. This could be an useful low-resolution preliminary to more exhaustive TEM analysis.

6.3 X-ray microanalysis of endogenous substances

Endogenous materials include all the naturally occurring and metabolic components of cells, tissues, and their products. We shall confine ourselves here to the reactive histochemical techniques employed in order to demonstrate such components.

6.3.1 PRECIPITATION TECHNIQUES

Precipitation methods include all procedures that result in the chemical immobilization *in situ*, of elements to be analysed. The aim is that precipitation should prevent the displacement and redistribution or translocation of ions during the preparation of the biological specimen for microscopy. Precipitation techniques for diffusible substances have been reviewed recently by Läuchli (1975). Cytochemically, many precipitation techniques have evolved as attempts to produce a visible dense deposit when examined in the conventional electron microscope. Several methods have, therefore, employed heavy metals. Komnick (1962) used organic silver salts for the localization of chloride ions. Tandler and Solari (1969) have demonstrated the distribution of phosphate in cells by precipitating the ions with lead.

To date by far the most widely used precipitation technique is the potassium pyroantimonate method (Komnick, 1962; Garfield, Henderson & Daniel, 1972; Simpson & Spicer, 1975) for subcellular localization of various cations including Ca^{2+}, Mg^{2+}, Na^{+}, K^{+}, H^{+}, and NH_3^{+}. This technique has been adapted for X-ray microanalysis (Tandler, Libanati & Sanchis, 1970).

Calcium has been precipitated by means of oxalate incorporation (Constantin, Franzini-Armstrong & Podolsky, 1965) and this histochemical technique has been adapted for X-ray microanalysis by several workers (Coleman, DeWitt, Batt & Terepka 1970; Oschman & Wall, 1972; Yarom & Chandler, 1974). Calcium has been localized by the silver impregnation technique of V. Kossa. The action of this histological stain probably depends on the precipitation of silver by phosphate or carbonate anions of insoluble calcium salts (Gardner & Hall, 1969).

A precipitation method for the fine structural and X-ray microanalytical demonstration of copper in rats exposed to an intraperitonial injection of copper sulphate (0.5%) has been described by Scheuer, Thorpe and Marriott (1967) who adapted the histochemical method of Timm. Copper was converted into the sulphide by treating sections of liver from the injected animals with 0.5% ammonium sulphide. Silver was subsequently deposited from silver nitrate at the site of sulphide formation. In our laboratory, copper has been insolublized with potassium ferrocyanide (Ryder & Bowen, 1977*b*) and also through chelation with diethyldithiocarbamate (Ryder & Bowen, 1977*a*). The copper containing product has been traced in slug foot and slug eggs by means of X-ray microanalysis (see p. 198).

Precipitation of final reaction product is an important part of many enzymatic histochemical techniques. The application of X-ray microanalysis to enzyme histochemistry is an expanding field and is dealt with separately later (see p. 198). Most precipitations in this context employ metal salts of lead, strontium, cobalt, and occasionally copper for the formation of insoluble final reaction products.

(Pyro) antimonate methods

An antimonate–osmium method was originally developed by Komnick (1962) to study the intracellular distribution of Na^+. The histochemical basis of the method has been subsequently challenged by several authors. Garfield, Henderson and Daniel (1972), in a constructively critical paper, justifiably questioned the utility of pyroantimonate for the demonstration of tissue sodium on both quantitative and qualitative grounds. It is now understood that several factors govern the precipitation of antimonate and these involve not only cations such as Ca^{2+}, Mg^{2+}, Zn^{2+}, Fe^{2+}, Mn^{2+}, and H^+ but also NH_3^+-rich histones. The histochemical picture has been neatly summarized recently by Simpson and Spicer (1975). They state that, 'care must be taken in the interpretation of results' and point out that variables such as temperature, buffer, and rinse time may dramatically alter precipitate density and distribution.

The application of X-ray microanalysis to tissues exposed to pyroantimonate, enables the elemental content of the final precipitate to be determined. To this extent, X-ray microanalysis removes the 'non-specificity' of the original histochemical technique; nevertheless, the question of translocation of elements remains an open one.

Lane and Martin (1969) indicated that identification of the cation is advisable before designating sites of precipitation as loci of high sodium content, and they employed X-ray microanalysis to this end. Using buffered 2% potassium pyroantimonate in 1% osmium tetroxide at pH 6.9, they produced precipitates in the lamina propria and along the apical plasma membrane of mouse vas deferens *in vitro*. Under the conditions described they claimed that electron-dense precipitate deposits contained largely sodium and antimony, although low levels of calcium and magnesium were also present.

Tandler, Libanati and Sanchis (1970), in an interesting paper described the use of potassium pyroantimonate as a fixative as well as a precipitating agent. They employed X-ray microanalysis to determine the elemental content of the resultant precipitate and argued that their technique reflected the *in vivo* localization of cations in growing maize roots, broad bean embryos, and rat liver. They reported the intracellular precipitation of calcium, magnesium, and sodium in the nucleus and nucleolus as well as in the cytoplasm.

Tisher, Weavers and Cirksena (1972) also employed X-ray microanalysis to elucidate the cation content of precipitated pyroantimonate complexes.

Precipitates were found on the intercellular side of plasma membranes and interchromatin regions of cell nuclei of rat kidney exposed to 2% phosphate buffered potassium pyroantimonate at pH 7.2 to 7.4. Standard micropuncture technique was employed to perfuse single proximal tubules *in vivo*, calcium, magnesium, and potassium ions, in addition to sodium, being demonstrated. One critical point to emerge from this study was the heavy concentration of potassium and antimony detected in unrinsed tissue, presumably arising from non-specific precipitation of potassium pyroantimonate.

The fact that X-ray microanalysis precisely defines the elemental content of antimonate precipitates has eventually led to the utilization of the antimonate technique for localizing elements other than sodium. Thus, the fine structural distribution of calcium in dog myocardial tissue has been described by Yarom, Peters, Scripps and Rogel (1974). They employed the electron microscope X-ray analyser (EMMA 4) and reported that calcium could be detected in dog myocardial tissue in varying concentrations, depending on the fixative employed; osmium fixation showing the highest retention of calcium. Other interesting data emerging from this study indicated that the antimonate precipitate was not always stable and was subject to alterations during and after processing. Thus precipitate was lost in buffer, alcohol, and when uranyl acetate '*en bloc*' staining was employed. Yarom, Maunder, Scripps, Hall and Dubowitz (1975) also found calcium loss during dehydration after the cation had been precipitated by pyroantimonate. Hales, Luzio, Chandler and Herman (1974) have reported the localization of calcium in the smooth endoplasmic reticulum of isolated rat fat cells employing 2% potassium pyroantimonate during fixation, following by staining in veronal-buffered uranyl acetate.

Yarom and Chandler (1974) investigated the calcium distribution in resting frog sartorius muscle fixed by the pyroantimonate technique. They penetrated the tissue with ethyleneglycoltetraacetate to chelate calcium, and with tris-chloride to replace sodium. They reported significant concentrations of calcium in the nuclei, muscle triads, thin filaments, and the sarcolemma. The concentration of calcium appeared to be decreased by ethyleneglycoltetraacetate pretreatment, and slightly increased by tris-chloride pretreatment. Van Steveninck, Van Steveninck, Peters and Hall (1976) have demonstrated that barley roots fixed in osmium tetroxide containing potassium pyroantimonate showed the presence of antimony containing deposits at the root surface associated with the mucilaginous sheath. The deposits contained iron and phosphorus in constant proportions. In the root cells, deposits contained osmium and antimony with occasional epidermal vacuoles containing some iron. Iron was also located in nuclear deposits in endodermal cells.

Oxalate methods

These methods have been used generally to immobilize calcium. The histochemical technique was described by Constantin, Franzini-Armstrong and Podolsky (1965) who localized calcium accumulating structures, at fine structural level, in striated muscle fibres. The technique consisted of the application of 10 mM sodium oxalate as a perfusate and also as a component in the buffered glutaraldehyde and osmium tetroxide fixatives. The authors concluded that electron-opaque material, probably calcium oxalate, accumulated in the terminal sacs of the sarcoplasmic reticulum. Galle (1967*a*, 1967*b*) has studied calcification of the kidneys in the rat. Among the various agents he injected intraperitoneally were sodium oxalate and sodium phosphate. The oxalate produced intracellular crystals in the tubular cells, and X-ray microprobe analysis showed that the crystals were formed through precipitation with calcium. Similarly, phosphate ions formed smaller deposits at the same cytoplasmic sites. A constant calcium/phosphorus ratio was demonstrated by microprobe analysis.

Coleman, DeWitt, Batt and Terepka (1970), in a combined electron probe and electron microscope analysis, have demonstrated that only certain ectodermal cells of the chick chorioallantoic membrane contained concentrations of calcium. They employed a 1% solution of sodium oxalate incorporated into 6% acrolein fixative to precipitate the calcium ions. This initial report was followed by more extensive studies (Coleman & Terepka, 1972*a*, 1972*b*) where sodium oxalate was also employed in the buffer wash and in the post-fixation dehydration procedures. Coleman and Terepka (1972*a*) assessed the preparative procedures especially in relation to electron probe investigation. The authors claimed that the method employed preserved fine structure, prevented selective loss of calcium, and resulted in only minor losses of total calcium. Resolution problems and preparative artefacts were also discussed. Coleman and Terepka (1972*b*) presented the biological significance of their results. They reported the occurrence of calcium at specific sites within 'capillary-covering' cells of the chick chorioallantoic membrane, and presented evidence that the calcium identified represents that involved in active transport. They proposed that sequestered calcium might be contained within endocytotic vesicles involved in transcellular transport.

Plattner and Fuchs (1975) investigating calcium triggering of the exocytosis of 'trichocysts' in *Paramecium* at fine structural level, immobilized calcium ions by the oxalate method of Constantin, Franzini-Armstrong and Podolsky (1965), and also by glutaraldehyde fixation as described by Oschman and Wall (1972). Precipitates were demonstrated in some resting trichocysts and regularly in discharging trichocysts. In fixed material, deposits occurred on the trichocyst membrane and on the inner lamellar sheath extending into secretory material. Energy-dispersive X-ray microanalysis indicated the presence of both calcium and phosphorus in the deposits. Some deposits contained calcium; phosphorus and sulphur were also obtained on cilia and ciliary granule plaques. Cells containing calcium-storing vacuoles were rich in calcium and sulphur, but contained no phosphorus.

Methods for precipitating copper

Timm (1960) described a method for the direct demonstration of copper in the electron microscope in which copper was converted into the sulphide and silver was then deposited from silver nitrate at the site of the sulphide. Scheuer, Thorpe and Marriott (1967), studying the liver of copper treated rats, used the same basic method and analysed the subcellular granules formed in EMMA with a view to demonstrating the occurrence together of copper and silver. Silver was in fact shown to be present, although copper appeared to have been lost during preparation procedures.

In this laboratory Ryder and Bowen (1977*b*), investigating the uptake of copper molluscicides by the slug pest *Agriolimax reticulatus*, have precipitated copper with potassium ferrocyanide in the manner described by Mendel and Bradley (1905). The potassium ferrocyanide was incorporated in the prefix before routine preparation for electron microscopy. The product proved to be sufficiently dense for observation in the electron microscope (Plate 6.1). The copper content of the precipitate was confirmed using energy-dispersive X-ray microanalysis (Plate 6.2). The technique demonstrated a paracellular uptake of copper by slug foot epithelia. A chelating rather than a precipitating method is also presented for uptake studies on copper molluscicides (see p. 198).

6.3.2 STAINED ENDOGENOUS MATERIAL

Included in this section are techniques where morphological or histological components have been stained in the classical sense and where X-ray microanalysis has assisted in the localization of such stained entities. The staining mechanism may be specific and may be based on a well understood chemical principles. More often than not, however, staining reactions are not specific in the sense that more than one chemical component is localized. The mode of action of many biological dyes and staining reagents is not known. By breaking down the composition of stains and stained material into its component elements, X-ray microanalysis may assist in unravelling some of these problems.

X-ray microanalysis can usefully be employed to identify particular elements within histologically deposited stains. Sims and Marshall (1966) reported the histological localization of nucleic acids in the rat spinal cord by means of the gallocyanin chrome alum method of Berube, Powers, Kerkay and Clark (1966). Optical microscopy demonstrated that only nuclear chromatin, nucleoli, and Nissl substance had stained. They showed by means of X-ray microanalysis that the chromium component of the dye corresponded to the distribution of the staining pattern. They also pointed out the possibility of obtaining quantitative data.

Another application based on the use of a chromium-containing stain was described by Wood (1974, 1975). This technique uses potassium dichromate to react with an isoquinoline derivative of a glutaraldehyde–biogenic amine complex. The test was reported to be sufficiently sensitive to pin-point chromium-located biogenic amine at fine structural level. X-ray microanalysis combined with electron microscopy was used to localize chromium intraneuronally in the median eminence and arcuate nucleus of cat brain.

Lever, Santer, Lu and Presley (1977) have published an electron probe X-ray microanalysis study of small granulated cells in rat sympathetic ganglia after sequential aldehyde and dichromate treatment. They indicated that 'chromium binds to the Schiff monobase formed by glutaraldehyde and noradrenaline during fixation'. They concluded from the distribution of chromium, estimated by X-ray microanalysis, that noradrenaline was present in granules of the type II sympathetic small granulated cell as well as in adrenomedullary NA cells. Chromium was not detected in adrenomedullar adrenaline cells nor in two other sympathetic small granulated cell types.

Cosslett and Switsur (1963) indicated that staining reactions could be employed to concentrate an element 'without' altering its distribution; this would be of use in a situation where the initial concentration of an element was below that detectable by means of X-ray microanalysis. The case which they cited as an example was that described by Lever and Duncumb (1961) who increased the iron emission from rat duodenal epithelia by means of the Prussian blue reaction. Ghadially, Oryschak, Ailsby and Mehta (1974) studied the distribution of iron deposits in chondrocytes of haemarthrotic articular cartilage by means of X-ray microanalysis correlated with electron microscope observations. Parallel histochemical investigations using the Prussian blue and Turnbull blue reaction for iron were aso carried out. The authors concluded that haemosiderin deposits occurred in chondrocytes from haemarthrotic joints.

Silver salts have been used to demarcate morphological features. De Bruijn, Von Mourik and Bosveld (1974) successfully used silver nitrate and silver proteinate to define the aortic endothelial cell borders in the scanning electron microscope. Energy-dispersive analysis confirmed that silver was present along the cell borders. Silver proteinate was the preferred stain.

Bodian silver proteinate (Protargol) is commonly used to stain morphological components of protozoa, particularly ciliates. Zagon, Vavra and Steele (1970) have described the microprobe analysis of Protargol stain deposition in two protozoa, *Carchesium polypinum* and *Gregorina rhyparobiae*. A Protargol staining technique involving the use of silver proteinate in the presence of copper, followed by a gold chloride toner, is described. The electron microprobe analyser was used to determine the elemental composition of the stain and its morphological distribution.

Bowen, Ryder and Winters (1975) demonstrated oxidizable mucosubstances employing ammoniacal silver nitrate after chromic acid oxidation in the planarian worm *Polycelis tenuis*. Silver was deposited on secretory granules and Golgi complexes. Energy-dispersive X-ray microanalysis was used to confirm this and also to assist in differentiating between normally dense morphological features and those containing precipitated silver (Plates 6.3 and 6.4).

Gardner and Hall (1969) have submitted the Von Kossa silver impregnation technique to X-ray microanalysis in an attempt to verify the method as an indirect technique for the location of insoluble tissue calcium.

Ohkura, Iwatsuki, Asai, Watanabe and Fukouka (1973) have used X-ray analysis to demonstrate copper in material stained with Alcian blue. Bowen, Ryder and Winters (1975) demonstrated that a 1% solution of the acid mucopolysaccharide stain Alcian blue contained detectable quantities of copper (Plate 6.5). A similar peak for copper was obtained by means of energy-dispersive X-ray microanalysis of mucous droplets found on the surface of a planarian worm stained with 1% Alcian blue (Plate 6.6).

In a recent short review of microanalysis and cytochemistry, Martoja, Szöllösi and Truchet (1975) described the X-ray microanalysis of several stains including phosphotungstic acid, silicotungstic acid, hexachloroplatinic acid, and ferrocyanide. The paper lists the specific rat tissue components which take up these stains under various conditions.

The methods currently available for the X-ray microanalysis of endogenous substances are summarized in Table 6.1.

6.4 X-ray microanalysis of exogenous substances

The use of X-ray microanalysis to follow the uptake and fate of deliberately or accidentally introduced exogenous material is dealt with in Chapters 3 and 4. In this section we shall direct our attention to those investigations that have relied on some basic histochemical principle or technique to bring about final localization of material of exogenous origin.

6.4.1 LABELLING TECHNIQUES

In a recent review, Chandler (1975) discussed the theoretical possibilities of applying X-ray microanalysis to various immunohistochemical techniques. Clarke, Salsbury and Willoughby (1970) specifically described the application of electron probe microanalysis and electron microscopy to the study of the

Table 6.1 Methods available for the X-ray microanalysis of endogenous material.

Method	*Elements detected*	*Cells and tissues*	*Author*
Pyroantimonate	Na, Ca, Mg	Mouse vas deferens	Lane and Martin, 1969
Pyroantimonate	Na, Ca, Mg	Maize roots, bean embryos, rat kidney	Tisher *et al.*, 1972
Pyroantimonate	Ca	Rat isolated fat cells	Hales *et al.*, 1974
Pyroantimonate	Ca	Resting frog sartorius muscle	Yarom and Chandler, 1974
Pyroantimonate	Ca	Dog myocardial tissue	Yarom *et al.*, 1974
Pyroantimonate	Ca, Zn, Cu	Rat muscle, human blood	Yarom *et al.*, 1975
Pyroantimonate	Fe, P	Barley root tip and sheath	Van Steveninick *et al.*, 1976
Oxalate and phosphate	Ca	Rat kidney	Galle, 1967*b*
Oxalate	Ca	Chick chorioallantoic membrane	Coleman *et al.*, 1970
Oxalate	Ca	Intestinal membranes	Oschman and Wall, 1972
Oxalate	Ca	Chick chorioallantoic membrane	Coleman and Terepka, 1972*a*, *b*
Oxalate	Ca	*Paramecium aurelia*	Plattner and Fuchs, 1975
Copper (silver precipitation)	Ag	Rat liver	Scheuer *et al.*, 1967
Copper (diethyldithiocarbamate)	Cu	Slug eggs	Ryder and Bowen, 1977*a*
Copper (ferrocyanide)	Cu	Slug foot epithelium	Ryder and Bowen, 1977*b*
Nucleic acids (gallocyanin–chrome alum)	Cr	Rat spinal cord	Sims and Marshall, 1966
Biogenic amine (dichromate test)	Cr	Cat adrenal medulla	Wood, 1974
Biogenic amine (dichromate test)	Cr	Cat adrenal medulla monkey brain	Wood, 1975
Biogenic amine (dichromate test)	Cr	Rat adrenal medulla and celiac-mesenteric sympathetic ganglia	Lever *et al.*, 1977
Cell border demarcation	Ag	Aortic endothelial cells	de Bruijn *et al.*, 1974
Protargol staining	Cu, Ag, Au	*Carchesium polypinum* *Gregarina rhyparobiae*	Zagon *et al.*, 1970
Histological staining (including Alcian blue)	Cu	Dog synovial membrane	Ohkura *et al.*, 1973
Oxidizable polysaccharides (Alcian blue and ammoniacal silver)	Cu, Ag	Planarian gland cells	Bowen *et al.*, 1975
Stains (phosphotungstic, silicotungstic, hexachlorotungstic acids and ferrocyanide)	W, Si, Pt, Fe	Rat tissues	Martoja *et al.*, 1975

transfer of antigenic material. In this investigation a purified protein derivative was labelled with iodine and subsequent cell preparations were analysed for iodine distribution using EMMA 4. Salsbury and Clarke (1972) further publicized the method as illustrative of how X-ray microanalysis can be applied to immunological problems. They established that it was possible to trace the passage of antigenic material from one cell to another by labelling the antigen with iodine. The results suggest that the iodine-labelled purified protein derivative could pass from sensitized macrophages to lymphocytes via intercellular processes.

6.4.2 UPTAKE STUDIES

The work of Galle (1967*a*, 1967*b*) and others has established the general utility of X-ray microanalysis in uptake studies. Under certain circumstances, simple physical processing followed by straight X-ray microanalysis may not be entirely adequate to preserve the localization of the exogenous material. It may first be necessary to immobilize the exogenous material by histochemical means. The histochemical tests applied may be of a precipitative nature more or less identical to those already described.

The uptake of drugs (Erasmus, 1974) and/or pesticides could thus be established and followed directly if the substance contained a distinctive marker element, or by suitably combining histochemical tests with X-ray microanalysis. One such application was described by Ryder and Bowen (1977*a*) who employed the chelating agent diethyldithiocarbamate to precipitate out copper taken up from copper sulphate molluscicide by eggs of the slug *Agriolimax reticulatus* (M). Eggs, after chelation, were quenched in liquid nitrogen, fractured, and the freeze-dried halves were analysed in the scanning electron microscope. Sites of the copper-rich chelate were then established by means of energy-dispersive and wavelength-dispersive X-ray microanalysis. Line scans (Plates 6.7 and 6.8*a*) and X-ray distribution patterns (Plate 6.8*b*) may be compared.

Copper was localized in the perivitelline membranes where it appeared to be initially retained. In this particular instance, an identical localization of the copper was achieved in eggs not exposed to the chelate.

6.5 X-ray microanalysis and enzyme histochemistry

6.5.1 PRINCIPLES

Since enzymes are biological catalysts, in general they have been localized in cells and tissues by employing the reactions which they catalyse to ensure deposition of a reaction product or products more or less at the site of the enzyme's activity. As proteins, enzymes can also be localized directly by immunochemical means. M. R. J. Morgan (personal communication) has suggested that enzymes might be localized by using labelled specific inhibitors which would attach to the active sites.

If the reaction which the enzyme catalyses is employed as a means for its localization, the first requirement is that the cells or tissues be exposed to a reaction mixture containing an appropriate substrate. The enzyme then catalyses the transformation of the substrate into a product or products. In most cases the primary reaction product is treated in some way, so as to make it immobile, stable, and detectable.

The methods evolved for the detection of enzymatic activity in cells, tissues, and tissue sections at the level of the optical microscope have been reviewed by Burstone (1962), and Pearse (1968, 1972). These methods depend largely on the formation of an insoluble coloured final reaction product. Different methods have had to be developed for electron microscopy which depend on the deposition of insoluble, electron-opaque products (Hayat, 1973). The advent of X-ray microanalysis has further enabled the histochemist and electron cytochemist to analyse reaction products in terms of their elemental content and by these means study either the course of the reaction in elemental terms or the final site of deposition of the product elements. In general, X-ray microanalysis is able to detect elements of atomic number eleven and above (i.e. above sodium), although some energy-dispersive systems have been developed to analyse elements of lower atomic number. Essentially this means that enzyme cytochemists need not confine themselves to the use of electron-opaque heavy metal markers (Ryder & Bowen, 1973, 1974).

Most enzyme cytochemical studies employing X-ray microanalysis have been carried out on hydrolases, although Weavers (1974) has published an investigation on succinate dehydrogenase, and Rosen and Koffman (1972) have described a method for detecting carbonic anhydrase. In many respects, histochemical reactions for hydrolytic enzymes provide an ideal testing ground for analytical electron microscopy. Incubating conditions can be arranged so that maximal concentration of reaction product is achieved without destroying the specificity of the reaction. A wide spectrum of well worked methods can be found in the histochemical literature. The coupling, conversion, and chelation methods that are available can be used as a means of introducing appropriate label or marker elements. With respect to coupling, markers could be introduced in two ways: either during incubation (simultaneous capture or coupling), or subsequent to incubation (post-coupling, or more correctly post-incubation coupling).

Simultaneous coupling is the principle used in the established metal salt precipitation techniques such as those employed to demonstrate sites of acid and alkaline phosphatase activity. Simultaneous coupling is employed in many azo dye methods where diazonium salts are used to precipitate naphtholic reaction products. Naphthyl compounds have also been used in post-coupling reactions (Fishman, Goldman & DeLellis, 1967; Smith & Fishman, 1968).

The development of ultracryotomy is of particular significance to the enzymatic application of X-ray microanalysis (Bowen, Ryder & Downing, 1976) since it allows one to use a wider range of azo dye, indoxyl, and azo–indoxyl methods than would be usefully applicable using routine electron microscopic preparative technique. The use of cryo-sections means that the routine organic solvents and preparative media associated with electron microscopy can be avoided. This makes available for analysis reaction products which, although soluble in organic media, are insoluble and immobile in aqueous media.

6.5.2 METHODS

One of the earliest applications of X-ray microanalysis to enzyme cytochemistry was described by Hale (1962) who identified the primary, secondary, and final reaction products of alkaline phosphatase activity in sections of rat kidney by means of scanning X-ray emission analysis. The histochemical method employed was the classical Gomori–Takamatsu reaction. Alkaline phosphatase within tissue sections was permitted to catalyse the hydrolysis of a phosphate ester in the presence of calcium. Calcium phosphate, the primary reaction product, was deposited at the sites of enzymatic activity. To identify the reaction product at the level of the optical microscope, the primary product must be further converted into cobalt phosphate and sulphided to give a visible final reaction product of cobalt sulphide. By means of X-ray microanalysis, Hale demonstrated that conversion of the primary reaction product to cobalt sulphide was inefficient in that a certain amount of calcium phosphate remained unconverted.

It should be realized that conversion to cobalt sulphide is not necessary for demonstration of the enzyme's activity by means of X-ray microanalysis, and that Hale used the full technique in order to achieve a greater understanding of the reactions involved. He demonstrated that both the calcium and the phosphorus of the primary reaction product was not naturally occurring in that both were absent from, or at least undetectable in, the controls. The work neatly showed that X-ray microanalysis could be used as a technique for studying enzyme histochemical reactions and their products.

Weavers (1974) also studied the value of X-ray microanalysis in critically assessing cytochemical reactions. He investigated the ferricyanide reaction for the demonstration of succinate dehydrogenase activity. This involved ferricyanide reduction and simultaneous coupling with Cu^{+}. Dehydrogenase activity was localized at various sites on mouse liver mitochondrial cristal membranes, and also in the intracristal spaces.

Previous to this investigation it had been thought that ferrocyanide was produced from ferricyanide at the sites of reductase activity. It was assumed that ferrocyanide was simultaneously captured by copper ions to be precipitated as copper ferrocyanide. Had this been so the final reaction product should contain copper and iron in the stochiometric ratio of 2.276:1. This was shown by Weavers not to be the case. He demonstrated that although the reaction product always had the same appearance, its composition varied. In particular, variations in the copper/iron ratio were noted with different incubation times. It was concluded that the primary reaction product was not simultaneously captured by copper ions and that the histochemical reactions involved were more complex than previously thought.

Several authors have employed X-ray microanalysis as a chemical check for enzymatically released reaction products. Goldfischer and Moskal (1966), whilst studying the distribution of copper in the liver of patients suffering from Wilsons' disease, demonstrated by means of X-ray microanalysis, that the copper occurred in the same organelles that contained lead phosphate reaction product, deposited by the Gomori method for localizing acid phosphatase. They used β-glycerophosphate and cytidine-5'-monophosphate as substrates. Engel, Resnick and Martin (1968) demonstrated the suitability of electron probe analysis for the detection of the invisible products of ATPase activity in cryostat sections of human skeletal muscle. The histochemical test they used was based on the Gomori principle as modified for ATPase by Drews and Engel (1966). Operation of the electron microscope in the scanning mode revealed concentrations of calcium and phosphorus (derived from primary reaction product) in certain muscle fibres. This finding was subsequently confirmed by means of routine optical histochemistry.

Beeuwkes and Rosen (1975) have described the optical localization and X-ray microanalysis of renal sodium–potassium dependent adenosine triphosphatase in tissue sections from human, rabbit, and rat kidneys. The histochemical method involved the hydrolysis of *p*-nitrophenyl phosphate in alkaline medium containing dimethyl sulfoxide. The products of the histochemical reactions were sequentially examined by electron probe microanalysis. The authors identified the primary reaction product as being a mixture containing $KMgPO_4$ and $Mg_3(PO_4)_2$. Subsequent conversion into a visible cobalt sulphide salt was found to be linear. The results indicated that the highest levels of enzymatic activity were to be found in the thick ascending limbs and distal convoluted sections of the renal tubule. The enzyme reaction was potassium dependent, and was inhibited by ouabain as is typical of microsomal Na–K–ATPase. The authors also presented the measurement of primary reaction product phosphorus by means of electron probe analysis, as a direct method of quantitation of enzymatic activity in tubule segments.

Using energy-dispersive X-ray microanalysis, Ryder and Bowen (1974) were able to confirm that the reaction product of acid phosphatase activity, demonstrated by a modification of the Gomori (1952) lead salt method, contained lead and phosphorus. The elemental composition of the enzyme's reaction product was compared with a chemical standard of lead phosphate. So-called 'non-specific' nuclear staining prevalent in this method, was also characterized as containing lead and phosphorus. In this case, therefore, true reaction product and 'non-specific stain' were shown to have the same composition.

Heavy-metal salt methods in general produce dense crystalline products which are readily visible in the electron microscope. X-ray microanalysis proves to be particularly useful in localizing azo dye, indoxyl, or azo–indoxyl reaction products which on the whole are not easily visible and may on occasion be confused with naturally occurring dense components. Holt and Sheldon (1972), employing an energy-dispersive X-ray analysis system, confirmed the presence of osmium in the osmiophilic reaction product generated by reacting osmium tetroxide with an azo–indoxyl compound deposited as a result of esterase activity in rat kidney. The indoxyl method used was that introduced by Holt and Hicks (1966). Lysosomes which contained reaction product showed an intense peak for osmium, whereas other structures showed little osmiophilia.

Lewis (1972), using a previously developed electron microscope method for demonstrating cholinesterase activity (Lewis & Shute, 1966, 1969), was able to localize the copper-containing final reaction product in sections of rat brain examined in EMMA 4.

A few enzyme cytochemical techniques have been specifically designed to take advantage of X-ray microanalysis (Ryder & Bowen, 1973, 1974; Bowen, Ryder & Downing, 1976). Ryder and Bowen (1974) coupled enzymatically released naphthyl ASTR with the active diazotate of 2,5-dichloroaniline in rat liver. The resultant azo dye contained three covalently linked chlorine atoms which proved detectable in ultrathin cryo-sections by means of energy-dispersive X-ray microanalysis. The same principle was employed by Bowen, Ryder and Downing (1976) to establish an insoluble brominated azo dye at the sites of acid phosphatase activity. The reactions involved are illustrated in Fig. 6.1. The method devised involves the simultaneous coupling of naphthyl AS BI, enzymatically released from the substrate naphthyl AS BI phosphoric acid, with diazotized 2,5-dibromoaniline. This produces an insoluble red azo dye containing three covalently bound bromine atoms. Ultrathin cryo-sections of rat liver were employed and a lysosomal localization of acid phosphatase was achieved by detecting the bromine present in the final reaction product (Plate 6.10). Fine crystals of reaction product were also frequently visible (Plate 6.9).

Stabilization of the diazotized 2,5-dibromoaniline with zinc chloride is advisable since under reaction conditions at pH 5.0 the diazotate tends to precipitate out as the inert isodiazotate, which also contains bromine.

The method provides a technique whereby cryostat based azo dye histochemistry can be tested at electron microscope level. The red colour of the

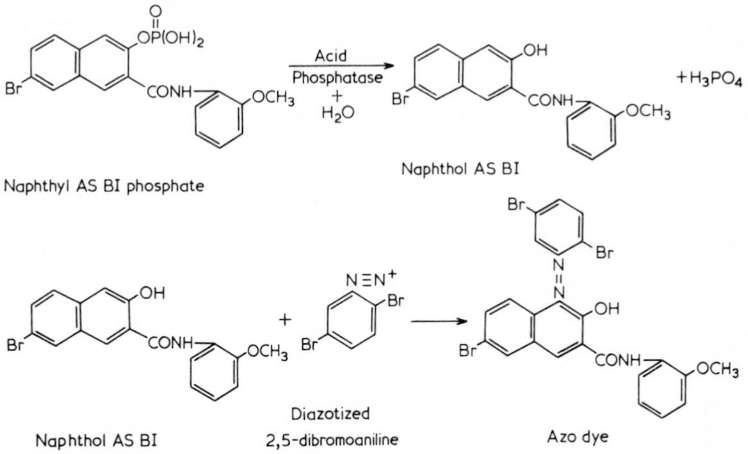

Naphthyl AS BI phosphate — Acid Phosphatase + H_2O → Naphthol AS BI + H_3PO_4

Naphthol AS BI + Diazotized 2,5-dibromoaniline → Azo dye

Fig. 6.1 An azo dye reaction for the localization of acid phosphatase. The substrate Naphthyl AS BI phosphate is hydrolysed by the enzyme acid phosphatase to give the primary reaction product naphthol AS BI. This is coupled with 2,5-dibromoaniline to give a red azo dye which is insoluble in aqueous media.

azo dye means that the reaction product can be checked optically. Indeed, the technique compares favourably with established methods at the levels of the optical microscope, provided a simultaneous capture is employed. The use of the scanning electron microscope combined with electron probe analysis in histochemistry has been recently reviewed by Makita (1974). The results of applying a test for alkaline phosphatase (Makita & Sandborn, 1969) are presented, and reports of the localization of ATPase are also given. Makita proposes that in addition to phosphatases, other enzymatic reactions such as uranyl ferrocyanide in the carnitine acetyl transferase test, copper ferrocyanide in cholinesterase and succinate dehydrogenase tests, and gold thiocholine phosphate or gold sulphate in the case of acetylcholinesterase can be studied in the analytical scanning electron microscope.

Rosen (1972), and Rosen and Koffman (1972) have presented an X-ray microanalytical method for detecting carbonic anhydrase activity in relation to surface topography, using the scanning electron microscope. Mucosal carbonic anhydrase activity in turtle bladder was demonstrated in topographically identifiable cells. These cells were shown to contain the final reaction product, cobalt sulphide.

Table 6.2 Methods for detecting enzymatically released reaction products.

Enzyme localized	*Element detected*	*Cells and tissues*	*Author*
Alkaline phosphatase	Ca, P, Co, S	Rat kidney	Hale, 1962
Alkaline phosphatase	Pb, P	Rat duodenum	Makita, 1974
Acid phosphatase	Pb	Human liver	Goldfischer and Moskal, 1966
Acid phosphatase	Pb, P	*Polycelis tenuis* gland cells	Ryder and Bowen, 1973, 1974
Acid phosphatase	Cl	Rat liver	Ryder and Bowen, 1974
Acid phosphatase	Br	Rat liver	Bowen *et al.*, 1976
ATPase	Ca, P	Human skeletal muscle	Engel *et al.*, 1968
ATPase	Mg, P, Co, K	Human, rabbit and rat kidney	Beeuwkes and Rosen, 1975
Esterase	Os	Rat kidney	Holt and Sheldon, 1972
Cholinesterase	Cu	Rat brain	Lewis, 1972
Succinate dehydrogenase	Cu, Fe	Isolated mouse liver mitochondria	Weavers, 1974
Carbonic anhydrase	Co	Turtle bladder mucosa	Rosen, 1972

A list of the methods currently available, involving the X-ray microanalysis of enzymatically released reaction products, is summarized in Table 6.2.

6.6 Concluding remarks

In a biological context, X-ray microanalysis is a relatively young technique and the histochemical application of X-ray microanalysis is a recent development, dating from the mid-sixties. As X-ray microanalysis, combined with electron microscopy, becomes generally available in research laboratories, the volume of research in this area will doubtless expand.

To date, few histochemical techniques have been investigated fully by means of X-ray microanalysis and the vast potential for new methods has hardly been touched. Some possibilities for future development have been discussed by Chandler (1975), and Makita (1974).

The advent of 'ultrathin' cryo-technique has added a new dimension, allowing as it does the possibility of investigating naturally occurring elements with minimal preparative interference. Nevertheless, investigators must still be on their guard, new techniques are inevitably accompanied by new artefacts.

Currently, accurate methods of quantitation are being worked out for X-ray microanalysis (see Chapters 2 and 3). As yet, however, the major and most successful histochemical contribution has been of a qualitative nature.

There is plenty of scope for future development from the point of view of purely qualitative histochemistry. The (pyro)antimonate precipitation methods have shown that histochemical techniques, fallen into disrepute and disuse due to apparent lack of specificity, can be looked at again more favourably with X-ray microanalysis which provides a precise assessment of the various products formed. Histochemical methods that suffer other drawbacks such as lack of resolution or translocation are, however, not going to be improved by the application of X-ray microanalysis. Indeed, X-ray microanalysis may well introduce its own translocation problems.

The area holding the most immediate promise would seem to be the development of new qualitative histochemical methods with X-ray microanalysis primarily in mind. Unfortunately, analysis has been used too often simply as an adjunct or after-thought rather than as a central tool. The possibility of obtaining accurate and meaningful quantitative data can be held out as a long-term aim in a histochemical context.

X-ray microanalysis, combined with electron microscopy, has a major role to play in elucidating the composition of living matter. It is now hoped that biologists in general, and histochemists in particular, will realize fully the advantages provided by this new approach.

6.7 References

APPLETON, T. C. (1972) 'Dry' ultrathin frozen sections for electron microscopy and X-ray microanalysis: The cryostat approach. *Micron*, **3**, 101–105.

BABAI, F. & BERNHARD, W. (1971) Détection cytochimique par l'acide phosphotungstique de certains polysaccharides sur coupes à congélation. *Journal of Ultrastructure Research*, **30**, 642–663.

BABAI, F. (1972) Mise en évidence de polysaccarides sur coupes à congélation ultrafines. *Journal de Microscopie et de Biologie Cellulaire*, **13**, 147.

BAUR, H. & SIGARLAKIE, E. (1973) Cytochemistry on ultrathin frozen sections of yeast cells. Localization of acid and alkaline phosphatase. *Journal of Microscopy*, **99**, 205–218.

BEEUWKES, R. & ROSEN, S. (1975) Renal sodium–potassium adenosine triphosphatase optical localization and X-ray microanalysis. *Journal of Histochemistry and Cytochemistry*, **23**, 828–839.

BERNHARD, W. & VIRON, A. (1971) Improved techniques for the preparation of ultrathin frozen sections. *Journal of Cell Biology*, **49**, 731–746.

BERNIER, R., IGLESIAS, R. & SIMARD, R. (1972) Detection of DNA by tritiated actinomycin on ultrathin frozen sections. *Journal of Cell Biology*, **53**, 798–808.

BERUBE, G. R., POWERS, M. M., KERKAY, J. & CLARK, G. (1966) The gallocyanin–chrome alum stain, influence of methods of preparation on its activity and separation of active staining compound. *Stain Technology*, **41**, 73–81.

BOWEN, I. D. & RYDER, T. A. (1977) The application of X-ray microanalysis to enzyme cytochemistry. In *Electron Microscopy of Enzymes, Principles and Methods*, ed. Hayat, M. A. Vol. 5, Ch. 7, New York: Van Nostrand Reinhold.

BOWEN, I. D., RYDER, T. A. & DOWNING, N. L. (1976) An X-ray microanalytical azo dye technique for the localization of acid phosphatase activity. *Histochemistry*, **49**, 43–50.

BOWEN, I. D., RYDER, T. A. & WINTERS, C. (1975) The distribution of oxidizable mucosubstances and polysaccharides in the planarian *Polycelis tenuis* Iijima. *Cell and Tissue Research*, **161**, 263–275.

DE BRUIJN, W. C., VON MOURIK, W. & BOSVELD, I. D. (1974) Cell border demarcation in the scanning electron microscope by silver stains. *Journal of Cell Science*, **16**, 221–239.

BURSTONE, M. S. (1962) *Enzyme Histochemistry and its Application to the Study of Neoplasms*. New York and London: Academic Press.

CHANDLER, J. A. (1975) Electron probe X-ray microanalysis in cytochemistry. In *Techniques of Biochemical and Biophysical Morphology*, ed. Glick D. & Rosenbaum, R. M. Vol. 2, p. 307. New York: John Wiley & Sons, Inc.

CHRISTENSEN, A. K. & PAAVLOVA, L. G. (1972) Frozen thin sections of fresh frozen tissue, and the possibility of their use for autoradiography. *Journal de Microscopie et de Biologie Cellulaire*, **13**, 148–150.

CLARKE, J. A., SALSBURY, A. J. & WILLOUGHBY, D. A. (1970) Application of electron probe microanalysis and electron microscopy to the transfer of antigenic material. *Nature*, **227**, 69–71.

COLEMAN, J. R., DE WITT, S. M., BATT, P. & TEREPKA, A. R. (1970) Electron probe analysis of calcium distribution during active transport in chick chorioallantoic membrane. *Experimental Cell Research*, **63**, 216–220.

COLEMAN, J. R., NILSSON, J. R., WARNER, R. R. & BATT, P. (1972) Qualitative and quantitative electron probe analysis of cytoplasmic granules in *Tetrahymena pyriformis*. *Experimental Cell Research*, **74**, 207–219.

COLEMAN, J. R., NILSSON, J. R., WARNER, R. R. & BATT, P. (1973*a*) Electron probe analysis of refractive bodies in *Amoeba proteus*. *Experimental Cell Research*, **76**, 31–40.

COLEMAN, J. R., NILSSON, J. R., WARNER, R. R. & BATT, P. (1973*b*) Effects of calcium and strontium on divalent ion content of refractive granules in *Tetrahymena pyriformis*. *Experimental Cell Research*, **80**, 1–9.

COLEMAN, J. R. & TEREPKA, A. R. (1972*a*) Electron probe analysis of the calcium distribution in cells of the embryonic chick chorioallantoic membrane. I. A critical evaluation of techniques. *Journal of Histochemistry and Cytochemistry*, **20**, 401–413.

COLEMAN, J. R. & TEREPKA, A. R. (1972*b*) Electron probe analysis of the calcium distribution in cells of the embryonic chick chorioallantoic membrane. II. Demonstration of intracellular location during active transport. *Journal of Histochemistry and Cytochemistry*, **20**, 414–424.

COLEMAN, J. R. & TEREPKA, A. R. (1974) Preparatory methods for electron probe analysis. In *Principles and Techniques of Electron Microscopy Biological Applications*, ed. Hayat, M. A. Vol. 4, Ch. 8. New York: Van Nostrand Reinhold.

CONSTANTIN, L. L., FRANZINI-ARMSTRONG, C. & PODOLSKY, R. J. (1965) Localization of calcium accumulating structures in striated muscle fibres. *Science*, **147**, 158–160.

COSSLETT, V. E. & SWITSUR, V. R. (1963) Some Biological Applications of the Scanning Microanalyser. In *X-Ray Optics and X-Ray Microanalysis*, ed. Pattee, H. H., Coslett, V. E. & Engström, A. pp. 507–512. New York, Academic Press.

COUTEAUX, R. & DELAITRE, D. (1972) Essais de localisation des cholinestérases sur coupes à congélation ultrafines. *Journal de Microscopie et de Biologie Cellulaire*, **13**, 150–152.

DAVIES, T. W. & ERASMUS, D. A. (1973) Cryo-ultramicrotomy and X-ray microanalysis in the transmission electron microscope. *Science Tools*, **20**, 9–13.

DREWS, G. A. & ENGEL, W. K. (1966) Reversal of the ATPase reaction in muscle fibres by EDTA. *Nature*, **212**, 1551–1553.

ENGEL, W. K., RESNICK, J. S. & MARTIN, E. (1968) The electron probe in enzyme histochemistry. *Journal of Histochemistry and Cytochemistry*, **16**, 273–275.

ERASMUS, D. A. (1974) The application of X-ray analysis in the transmission electron microscope to a study of drug distribution in the parasite *Schistosoma mansoni* (Platyhelminthes). *Journal of Microscopy*, **102**, 59–69.

FISHMAN, W. M., GOLDMAN, S. S. & DE LELLIS, R. (1967) Dual localization of β-glucuronidase in endoplasmic reticulum and in lysosomes. *Nature*, **213**, 457–460.

FOURNIER-LAFLÈCHE, D., CHANG, A., BÉNICHOU, J. C. & RYTER, A. (1975) Immunolabelling of frozen ultrathin sections of bacteria. *Journal de Microscopie et de Biologie Cellulaire*, **23**, 17–28.

GALLE, P. (1967*a*) Microanalyse des inclusions minérales du rein. *Proceedings of the 3rd International Congress of Nephrology, Washington, D.C.*, **2**, 306–319.

GALLE, P. (1967*b*) *Les nephrocalcinoses: Nouvelles dounces d'ultrastructure et de microanalyse. Actualités nephrologiques de l'hospital Necker, pp. 303–315.* Paris: Flammarion.

GARDNER, D. L. & HALL, T. A. (1969) Electron microprobe analysis of sites of silver deposition in avian bone stained by the V. Kóssa technique. *Journal of Pathology and Bacteriology*, **98**, 105–109.

GARFIELD, R. E., HENDERSON, R. M. & DANIEL, E. E. (1972) Evaluation of the pyroantimonate technique for localization of tissue sodium. *Tissue and Cell*, **4**, 575–589.

GERSH, I., VERGARA, J. & ROSSI, G. I. (1960) Use of anhydrous vapours in post-fixation and in staining of reactive groups of proteins in frozen-dried specimens for electron microscopical studies. *Anatomical Record*, **138**, 445–460.

GEUSKENS, M. (1972) Fixation d'actinomycine D tritiée sur la chromatine de coupes à congélation. *Journal de Microscopie et de Biologie Cellulaire*, **13**, 153–154.

GHADIALLY, F. N., ORYSCHAK, A. F., AILSBY, R. L. & MEHTA, P. N. (1974) Electron probe X-ray analysis of Siderosomes in haemarthrotic articular cartilage. *Virchows Archives B. Cell Pathology*, **16**, 43–49.

GOLDFISCHER, S. & MOSKAL, J. (1966) Electron probe microanalysis of liver in Wilson's disease. *American Journal of Pathology*, **48**, 305–315.

GOMORI, G. (1952) *Microscope Histochemistry: Principles and Practice.* Illinois: University of Chicago Press.

HALE, A. J. (1962) Identification of cytochemical reaction products by scanning X-ray emission microanalysis. *Journal of Cell Biology*, **15**, 427–435.

HALES, C. N., LUZIO, J. P., CHANDLER, J. A. & HERMAN, L. (1974) Localization of calcium in the smooth endoplasmic reticulum of rat isolated fat cells. *Journal of Cell Science*, **15**, 1–15.

HALL, T. A. & GUPTA, B. L. (1974) Measurement of mass loss in biological specimens under an electron microbeam. In *Microprobe Analysis as Applied to Cells and Tissues*, ed. Hall, T. A., Echlin, P. & Kaufmann, R. pp. 147. New York: Academic Press.

HAYAT, M. A. (1973) Electron microscopy of enzymes: principles and methods. Vol. I. New York: Van Nostrand Reinhold.

HELLSTRÖM, S. (1973) Formaldehyde-induced fluorescence of biogenic amines in frozen thin sections. *Journal of Ultrastructure Research*, **42**, 396.

HODSON, S. & MARSHALL, J. (1972) Evidence against through section thawing whilst cutting on the ultratome. *Journal of Microscopy*, **95**, 459–465.

HOLT, S. J. (1972) Cytochemical localization of enzymes in ultrathin frozen sections. *Journal de Microscopie et de Biologie Cellulaire*, **13**, 155–156.

HOLT, S. J. & HICKS, R. M. (1966) The importance of osmiophilia in the production of stable azo–indoxyl complexes of high contrast for combined cytochemistry and electron microscopy. *Journal of Cell Biology*, **29**, 361–366.

HOLT, S. J. & SHELDON, F. (1972) Cytochemical applications of energy-dispersive X-ray analysis combined with transmission electron microscopy. *Proceedings of the International Congress of Histochemistry and Cytochemistry, Kyoto*, 279–280.

KOMNICK, H. (1962) Lokalisation von Na^+ und Cl^- in Zellen und Geweben. *Protoplasma*, **55**, 414–418.

KUHLMANN, W. D. & MILLER, H. R. P. (1971) A comparative study of the techniques for ultrastructural localization of antienzyme antibodies. *Journal of Ultrastructure Research*, **35**, 370–385.

KUHLMANN, W. A. & VIRON, A. (1972) Cross linked albumin as supporting matrix in ultrathin cryomicrotomy. *Journal of Ultrastructure Research*, **41**, 385–394.

LANE, B. P. & MARTIN, E. (1969) Electron probe analysis of cationic species in pyroantimonate precipitate in epon-embedded tissue. *Journal of Histochemistry and Cytochemistry*, **17**, 102–106.

LÄUCHLI, A. (1975) Precipitation technique for diffusible substances. *Journal de Microscopie et de Biologie Cellulaire*, **22**, 239–246.

LEDUC, E. H., BERNHARD, W., HOLT, S. J. & TRANZER, J. P. (1967) Ultrathin frozen sections. II. Demonstration of enzymatic activity. *Journal of Cell Biology*, **34**, 773–786.

LEUNG, T. K. & BABAI, F. (1974) Detection ultracytochimique de la 5′-nucléolidase sur les coupes à congélation ultrafines. *Journal de Microscopie et de Biologie Cellulaire*, **21**, 111–118.

LEVER, J. D. & DUNCUMB, P. (1961) The detection of iron in rat duodenal epithelium. In *Electron Microscopy in Anatomy*. ed. Boyd, J. D., Johnsons, F. R. & Lever, J. D. pp. 278. London: Edward Arnold, Ltd.

LEVER, J. D., SANTER, R. M., LU, K. S. & PRESLEY, R. (1977) Electron probe X-ray microanalysis of small granulated cells in rat sympathetic ganglia after sequential aldehyde and dichromate treatment. *Journal of Histochemistry and Cytochemistry*, **25**, 295–299.

LEWIS, P. R. (1972) The application of an analytical electron microscope to enzyme studies of brain tissue. *5th European Congress of Electron Microscopy, Abstracts*, p. 93. London: Institute of Physics.

LEWIS, P. R. & SHUTE, C. C. D. (1966) The distribution of cholinesterase in cholinergic neurons demonstrated with the electron microscope. *Journal of Cell Science*, **1**, 381–390.

LEWIS, P. R. & SHUTE, C. C. D. (1969) An electron microscopic study of cholinesterase distribution in the rat adrenal medulla. *Journal of Microscopy*, **89**, 181–193.

MAKITA, T. (1974) Sections incubated in the histochemical media. In *Principles and Techniques of Scanning Electron Microscopy*, ed. Hayat, M. A. Vol. 2, Ch. 4, pp. 47. New York: Van Nostrand Reinhold.

MAKITA, T. & SANDBORN (1969). Aldehyde fixation and fine structural localization of alkaline phosphatase activity in intestinal epithelial cells. *27th Annual Proceedings EMSA*, p. 280.

MARTOJA, R., SZÖLLÖSI, A. & TRUCHET, M. (1975) Microanalyse et cytochimie. *Journal de Microscopie et de Biologie Cellulaire*, **22**, 247–260.

MENDEL, L. B. & BRADLEY, H. C. (1905) Experimental studies on the physiology of the molluscs. *American Journal of Physiology*, **14**, 313–323.

MORGAN, A. J. & BELLAMY, D. (1973) Microanalysis of elastic fibres of rat aorta. *Age and Ageing*, **2**, 61.

MORGAN, A. J., DAVIES, T. W. & ERASMUS, D. A. (1975) Analysis of droplets from iso-atomic solutions as a means of calibrating a transmission electron analytical microscope (TEAM). *Journal of Microscopy*, **104**, 271–280.

OHKURA, T., IWATSUKI, H., ASAI, T., WATANABE, T. & FUKOUKA, T. (1973) Studies of mucopolysaccharide tissue chemical reaction by energy-dispersive type X-ray detector. *Hitachi Scientific Instrument News*, **16**, 10.

OSCHMAN, L. J. & WALL, B. J. (1972) Calcium binding to intestinal membranes. *Journal of Cell Biology*, **55**, 58–73.

PAINTER, R. G., TOKUYASU, K. T. & SINGER, S. F. (1973) Immunoferritin localization of intracellular antigens: The use of ultracryotomy to obtain ultrathin sections suitable for direct immunoferritin staining. *Proceedings of the National Academy of Science. USA*, **70**, 1649–1653.

PEARSE, A. G. E. (1968) *Histochemistry, Theoretical and Applied.* Vol. 1, 3rd ed. London: J. & A. Churchill.

PEARSE, A. G. E. (1972) *Histochemistry, Theoretical and Applied.* Vol. 2, 3rd ed. London: J. & A. Churchill.

PLATTNER, H. & FUCHS, S. (1975) X-ray microanalysis of calcium binding sites in *Paramecium* with special reference to exocytosis. *Histochemistry*, **45**, 23–47.

PUVION, E. & BERNHARD, W. (1975) Ribonucleoprotein components in liver cell nuclei as visualized by cryo-ultramicrotomy. *Journal of Cell Biology*, **67**, 200–214.

ROINEL, N. (1975) Electron microprobe quantitative analysis of lyophilised 10^{-10} l volume samples. *Journal de Microscopie et de Biologie Cellulaire*, **22**, 261–268.

ROSEN, S. (1972) Localization of carbonic anhydrase in turtle and toad urinary bladder mucosa. *Journal of Histochemistry and Cytochemistry*, **20**, 548–551.

ROSEN, S. & KOFFMAN, D. M. (1972) Carbonic anhydrase containing cells in the turtle bladder mucosa: surface topography and electron probe analysis. In *Thin Section Microanalysis*, ed. Russ, J. C. & Panessa, B. J. p. 41. Raleigh: EDAX Laboratories.

ROWE, A. J. (1972) The possible use of fibrin as a harmless encapsulating medium in cryo-ultramicrotomy. *Journal de Microscopie et de Biologie Cellulaire*, **13**, 163.

RUSS, J. C. (1974*a*) The direct element ratio model for quantitative analysis of thin sections. In *Microprobe Analysis as Applied to Cells and Tissues*, ed. Hall, T., Echlin, P. & Kaufmann, R. p. 269. London: Academic Press.

RUSS, J. C. (1974*b*) X-ray microanalysis in the biological sciences. *Journal of Submicroscopic Cytology*, **6**, 55–80.

RYDER, T. A. & BOWEN, I. D. (1973) The application of X-ray microanalysis to the cytochemical localization of acid phosphatase. *Proceedings of the Royal Microscopical Society*, **8**, 202.

RYDER, T. A. & BOWEN, I. D. (1974) The use of X-ray microanalysis to investigate problems encountered in enzyme cytochemistry. *Journal of Microscopy*, **101**, 143–151.

RYDER, T. A. & BOWEN, I. D. (1977*a*) The use of X-ray microanalysis to demonstrate the uptake of the molluscicide copper sulphate by slug eggs. *Histochemistry*, **52**, 55–60.

RYDER, T. A. & BOWEN, I. D. (1977*b*) The slug foot as a site of uptake of copper molluscicide. *Journal of Invertebrate Pathology*, (in press).

SALSBURY, A. J. & CLARKE, J. A. (1972) Electron probe microanalysis in relation to immunology. *Micron*, **3**, 135–137.

SCHEUER, P. J., THORPE, M. E. C. & MARRIOTT, P. (1967) A method for the demonstration of copper under the electron microscope. *Journal of Histochemistry and Cytochemistry*, **15**, 300–301.

SIMARD, R. (1972*a*) Autoradiography of diffusible substrates (^{125}I). *Journal de Microscopie et de Biologie Cellulaire*, **13**, 164–165.

SIMARD, R. (1972*b*) Detection of DNA by actinomycin D-^{3}H on ultrathin frozen sections. *Journal de Microscopie et de Biologie Cellulaire*, **13**, 165–166.

SIMPSON, J. A. V. & SPICER, S. S. (1975) Selective subcellular localization of cations with variants of the potassium (pyro)antimonate technique. *Journal of Histochemistry and Cytochemistry*, **23**, 575–598.

SIMS, R. T. & MARSHALL, D. J. (1966) Location of nucleic acids by electron probe X-ray microanalysis. *Nature*, **212**, 1359.

SMITH, R. E. & FISHMAN, W. H. (1968) (Acetoxymercuric) aniline-diazotate: a reagent for visualising the naphthyl AS BI product of acid hydrolase action at the level of the light and electron microscope. *Journal of Histochemistry and Cytochemistry*, **17**, 1–22.

STENN, K. & BAHR, G. F. (1970) Specimen damage caused by the beam of the transmission electron microscope. *Journal of Ultrastructure Research*, **31**, 526–550.

TANDLER, C. J. & SOLARI, A. J. (1969) Nucleolar orthophosphate ions. Electron microscope and diffraction studies. *Journal of Cell Biology*, **41**, 91–108.

TANDLER, C. J., LIBANATI, C. M. & SANCHIS, C. A. (1970) The intracellular localization of inorganic cations with potassium pyroantimonate. *Journal of Cell Biology*, **45**, 355–366.

TIMM, F. (1960) The histochemical demonstration of normal heavy metals in the liver. *Histochemie*, **2**, 150–162.

TISHER, C. C., WEAVERS, B. A. & CIRKSENA, W. J. (1972) X-ray microanalysis of pyroantimonate complexes in rat kidney. *American Journal of Pathology*, **69**, 255–264.

VAN STEVENINCK, M. E., VAN STEVENINCK, R. F. M., PETERS, P. D. & HALL, T. A. (1976) X-ray microanalysis of antimonate precipitates in barley roots. *Protoplasma*, **90**, 47–63.

WEAVERS, B. A. (1973) Combined transmission electron microscopy and X-ray microanalysis of ultrathin frozen dried sections – an investigation to determine the normal elemental composition of mammalian tissue. *Journal of Microscopy*, **97**, 331–341.

WEAVERS, B. A. (1974) An X-ray microanalytical study of the ferricyanide reaction for the electron cytochemical demonstration of succinate dehydrogenase activity in isolated mitochondria. *Histochemical Journal*, **6**, 121–131.

WOOD, J. G. (1974) Positive identification of intracellular biogenic amine reaction product with electron microscopic X-ray analysis. *Journal of Histochemistry and Cytochemistry*, **22**, 1060–1067.

WOOD, J. G. (1975) Use of the analytical electron microscope (AEM) in cytochemical studies of the central nervous system. *Histochemistry*, **41**, 233–240.

YAROM, R. & CHANDLER, J. A. (1974) Electron probe microanalysis of skeletal muscle. *Journal of Histochemistry and Cytochemistry*, **22**, 147–154.

YAROM, R., PETERS, D. D., SCRIPPS, M. & ROGEL, S. (1974) Effect of specimen preparation on intracellular myocardial calcium. *Histochemistry*, **38**, 143–153.

YAROM, R., MAUNDER, C., SCRIPPS, M., HALL, T. A. & DUBOWITZ, V. (1975) A simplified method of specimen preparation for X-ray microanalysis of muscle and blood cells. *Histochemistry*, **45**, 49–59.

YOKOTA, S. & NAGATA, T. (1974*a*) Studies on mouse liver urate oxidase. III. Fine localization of urate oxidase in liver cells revealed by means of ultracryotomy – immunoferritin method. *Histochemistry*, **39**, 243–250.

YOKOTA, S. & NAGATA, T. (1974*b*) Ultrastructural localization of catalase on ultracryotomic sections of mouse liver by ferritin-conjugated antibody technique. *Histochemistry*, **40**, 165–175.

ZAGON, I. S., VAVRA, J. & STEELE, I. (1970) Microprobe analysis of protargol stain deposition in two protozoa. *Journal of Histochemistry and Cytochemistry* **18**, 559–564.

ZOTIKOV, L. & BERNHARD, W. (1970) Localisation au Microscope Electronique de l'activité de certaines nucléases dans de coupes à congélation ultrafines. *Journal of Ultrastructure Research*, **30**, 642–663.

7

H. O. GARLAND, J. A. BROWN AND
I. W. HENDERSON

X-ray analysis applied to the study of renal tubular fluid samples

7.1 Introduction

7.1.1 DEVELOPMENT OF THE TECHNIQUE

At its inception the electron microprobe was principally applied to investigations in the general fields of metallurgy, mineralogy, solid state physics, and geology. In the early 1960s X-ray analysis of biological materials, in particular hard tissue sections (e.g. bone), was carried out (see Brooks, Tousimis & Birks, 1962), and in 1967, Ingram and Hogben suggested using the electron microprobe for electrolyte analysis of renal tubular fluids. In their preliminary studies it was concluded that analysis was accurate, sample preparation was simple, and measurement was rapid. Since these earlier studies the technique, somewhat modified and greatly refined, has become a routine procedure for the analysis of electrolytes in renal tubular fluids and has contributed greatly to our understanding of many facets of kidney function. In particular, multi-elemental analysis of single specimens has permitted the elucidation of the delicate interrelationships between electrolyte and water transport across renal tubular epithelia. Such analyses are simply not possible using other more standard analytical procedures.

The following papers are relevant to the development and application of electron probe microanalysis to renal physiology: Cortney (1969); Morel & Roinel (1969); Morel, Roinel & Le Grimellec (1969); Lechene (1970); Le Grimellec & Lechene (1970); Amiel, Kuntziger, Roinel & Morel (1972); Murayama, Morel & Le Grimellec (1972); Agus, Gardner, Beck & Goldberg (1973); Beck & Goldberg (1973*a*, *b*); Beck, Senesky & Goldberg (1973); Garland (1973, 1974); Garland, Henderson & Chester Jones (1973); Garland, Hopkins, Henderson, Haworth & Chester Jones (1973); Le Grimellec, Roinel & Morel (1973*a*, *b*; 1974*a*, *b*); Roinel, Richard, Robin & Morel (1973); de Rouffignac, Morel, Moss & Roinel (1973); Goldberg, Agus & Beck (1974);

Kuntziger, Amiel, Roinel & Morel (1974); Moss, Moriarty & Rankin (1974); Baumann, de Rouffignac, Roinel, Rumrich & Ullrich (1975); Garland, Henderson & Brown (1975); Le Grimellec (1975); Le Grimellec, Poujeol & de Rouffignac (1975); Morel (1975); Roinel (1975); Brown (1976); de Fronzo, Goldberg & Agus (1976); Moss (1976); Shirley, Poujeol & Le Grimellec (1976).

7.1.2 PRINCIPLE OF THE METHOD

The general underlying principles of microprobe analysis have been described in Chapter 2. The present chapter considers the analysis of renal tubular fluids. In principle, a minute known volume of fluid for analysis (less than 1 nl) is deposited on to a suitable substrate, and evaporated to dryness to form a thin, even layer of crystals 1 to 2 μm thick. Once in the probe, the deposit is bombarded totally and evenly by an accelerated beam of electrons, and the emitted X-rays are analysed by one or more X-ray spectrometers. The wavelengths of the X-rays are characteristic of the atoms present, and X-ray intensities proportional to the amount of element in the sample. If a series of known standard solutions of identical volumes is then treated in exactly the same way as the unknowns, the concentration of the various elements in the unknown deposits can be calculated directly.

7.1.3 ADVANTAGES OF THE METHOD

There are several advantages of microprobe analysis over other ultramicroanalytical methods available in this field. The technique permits the simultaneous analysis of four or more elements in a single fluid sample. In theory, the involvement of an electron probe allows the analysis of any element with an atomic number greater than 5 (see, for example, Robertson, 1968). Thus under standard conditions, sodium, potassium, calcium, magnesium, phosphorus, sulphur, and chlorine can be determined in a single specimen, something that is almost impossible using microphotometric and other techniques available. Moreover, several studies have also demonstrated the feasibility of probing elemental iron (in sodium ferrocyanide) (Morel & Roinel, 1969; Morel, Roinel & Le Grimellec, 1969), and elemental cobalt (in cyanocobalamin – vitamin B_{12}) (Garland, Henderson & Chester Jones, 1973; Garland, Hopkins, Henderson, Haworth & Chester Jones, 1973; Garland, Henderson & Brown, 1975). These two substances can be used as glomerular filtration markers and monitors of water movement.

The electron probe can analyse undiluted tubular fluid and plasma samples, thereby overcoming the difficulties and errors introduced by diluting small volumes for some of the microphotometric methods. The size of the sample required for microprobe analysis is extremely small. Fluid volumes of 0.05–0.5 nl are normally used (Morel & Roinel, 1969; Lechene, 1970; Garland, Hopkins, Henderson, Haworth & Chester Jones, 1973; Roinel, 1975), and this is important when only a few nanolitres of tubular fluid are obtained from some nephrons. Such volumes form crystalline deposits about 100 μm in diameter. A large number of samples can therefore be deposited on a relatively small area. Roinel (1975) comments that from a single micropuncture experiment, up to 400 samples can be deposited on an 8×6 mm^2 surface, and adds that there is still room for more!

Finally, the method is extremely sensitive, the microprobe being able to detect as little as 10^{-15} g of most elements (Ingram & Hogben, 1967; Morel & Roinel, 1969).

7.2 Micropuncture techniques

7.2.1 COLLECTION PROCEDURES

The concept of renal micropuncture emanates from the researches of Richards and his co-workers in the 1920s (see Richards, 1938, for a review of the early studies). The process essentially involves the withdrawal of fluid from individual renal tubules so that its composition can be compared with the blood plasma from which it was derived by ultrafiltration. A detailed description of the methodology involved in obtaining fluid samples is outside the scope of this chapter. Collection procedures are based on methods originally described by Richards and Walker (1937), and Walker and Oliver (1941). More recent reviews include those of Windhager (1968), Giebisch (1972), and Gottschalk and Lassiter (1973).

7.2.2 HANDLING OF SAMPLES

The extremely small size of renal tubules means that fluid volumes obtained are in the region of nanolitres (10^{-9} l). Samples are therefore withdrawn and subsequently handled under water-equilibrated mineral oil to prevent an otherwise inevitable spontaneous evaporation. Analysis is usually performed as soon as possible after collection, although the use of a probe allows dried deposits to be stored for some time prior to microanalysis (see below).

7.3 Sample preparation

The preparation of liquid samples for X-ray analysis is extremely critical, and the accuracy to which tubular fluid samples can be analysed depends to a large extent on the actual loading of samples on to a suitable substrate. Any variation in the diameter or thickness of the deposits formed will lead to inaccuracies in subsequent analyses. Consistent results from this analytical method can only be expected from deposits showing a regular distribution of small crystals as seen in Plates 7.1 (*a*, *b*), and 7.2.

7.3.1 TECHNIQUES IN USE

The original techniques for sample preparation, described by Ingram and Hogben (1967), are perhaps the simplest available. The apparatus is shown in Fig. 7.1; liquid samples are transferred on to the surface of a quartz slide using a xylene-filled micropipette. The xylene evaporates completely from

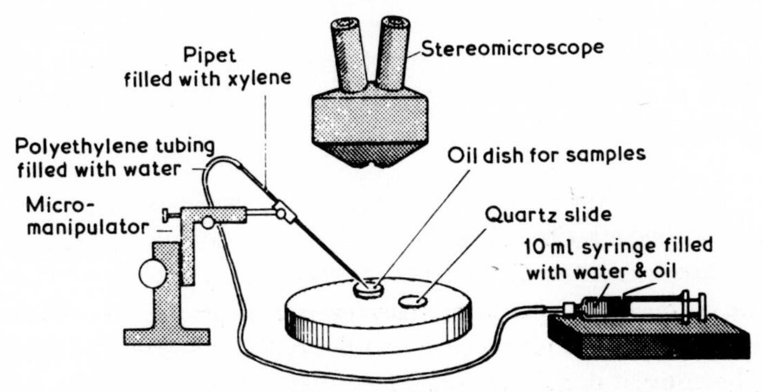

Fig. 7.1 General arrangement for depositing samples prior to electron probe micro-analysis (from Ingram & Hogben, 1967).

the slide, and the dried deposits are then carbon coated to make the slide conductive and thus ground the electron beam. Courtney (1969) slightly modified this original technique to measure simultaneously sodium, potassium, and chlorine levels in renal tubular fluids collected from adrenalectomized rats. The modifications ensured that samples were spread over a similar area after being dried, and involved the initial spotting of a 20% urea solution as a marker for the subsequent positioning of the samples. Final dried deposits were then of similar diameter and all would be covered by a suitably defocused electron beam (see Section 7.4).

Methods described by Morel and Roinel (1969) employed beryllium instead of quartz as a substrate for sample deposition, thus minimizing 'background' counts in the microprobe (elements of low atomic numbers produce fewer counts than those further down the periodic table, e.g. beryllium, 4; carbon, 6; aluminium, 13; silicon, 14). To avoid the spontaneous evaporation of liquid samples, and thus facilitate the formation of smaller crystals, the beryllium was held on a block of cold copper. When all the droplets had been transferred, the block was frozen with solid carbon dioxide, and the samples were freeze-dried below −5 °C. Deposits formed under these conditions were 1–2 μm in thickness and comprised a regular distribution of small crystals. Carbon coating was not necessary.

Lechene (1970) also used beryllium as a substrate, but deposited samples under oil to avoid evaporation. Subsequent studies (Le Grimellec, 1975; Le

Table 7.1 X-ray count rates (counts/30 s) emitted from aluminium and beryllium substrates bombarded at 20 kV with a specimen current of 100 nA and a beam diameter of 150 μm (from Morel & Roinel, 1969).

	X-ray			
Substrate	*Phosphorus* $K\alpha_1$	*Potassium* $K\alpha_1$	*Calcium* $K\alpha_1$	*Iron* $K\alpha_1$
Aluminium	230	95	35	530
Beryllium	190	20	7	150

Grimellec, Roinel & Morel, 1973*a*, *b;* 1974*a*, *b*) also used this technique. The oil was then washed off with xylene and/or chloroform and the liquid droplets were snap-frozen with isopentane cooled to −150 °C (Lechene) or solid carbon dioxide (Le Grimellec, *et al.*). Roinel (1975) recommends the use of the latter, commenting that isopentane cooled with liquid nitrogen may produce explosive bursting of droplets. Samples are then freeze-dried under vacuum at −40 °C; again very small crystals are formed. A further modification, described by de Rouffignac, Morel, Moss and Roinel (1973), and Kuntziger Amiel, Roinel and Morel (1974) again aimed at precluding the formation of large crystals, involves the diluting of more concentrated samples to an osmolarity of around 0.3 Osmol/l before analysis. Roinel (1975) describes the possibility of storing dried samples (in a dust-free environment) when it is not possible to analyse them immediately. The initial drying is then not as critical and can be performed at room temperature. Before analysis, the samples are rehydrated, then snap-frozen, and lyophilized as before.

Aluminium has been used as an alternative substrate to beryllium (Garland, Henderson & Chester Jones, 1973; Garland, Hopkins, Henderson, Haworth & Chester Jones, 1973; and Garland, Henderson & Brown, 1975). Its advantages, including availability, low cost and non-toxicity, however, have to be weighed against inevitably higher backgrounds, which become particularly relevant for elements such as calcium, and potassium. A comparison of the difference in background counts between the two substrates is shown in Table 7.1. In addition, the methods described by the above authors also involved the loading of samples in a cold room at 2–3 °C to avoid spontaneous sample evaporation.

Finally, Moss (1976) has described the use of carbon sample supports instead of beryllium. He also simplifies the sample-freezing and oil-removing stages, combining both operations by submerging the sample block in liquid Arcton 12 (ICI Ltd.) at −20 °C. After about 30 minutes the oil has been dissolved and the frozen droplets are left intact on the substrate. Samples are lyophilized as before.

7.3.2 BASIC EQUIPMENT

Immaterial of the details of the method employed, the required basic equipment and apparatus is similar. Samples are deposited under 25× binocular magnification using pipettes precisely controlled by a micromanipulator. The

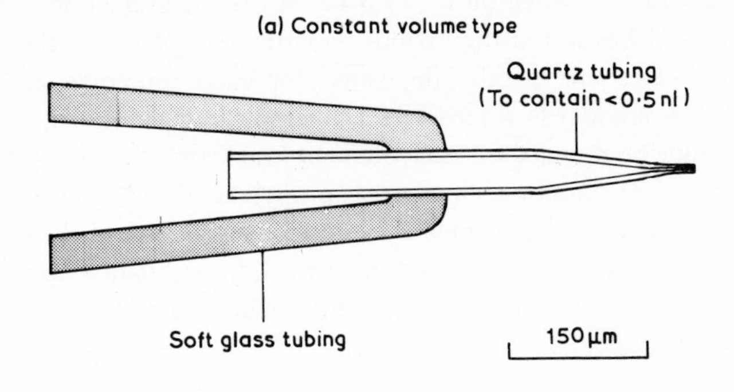

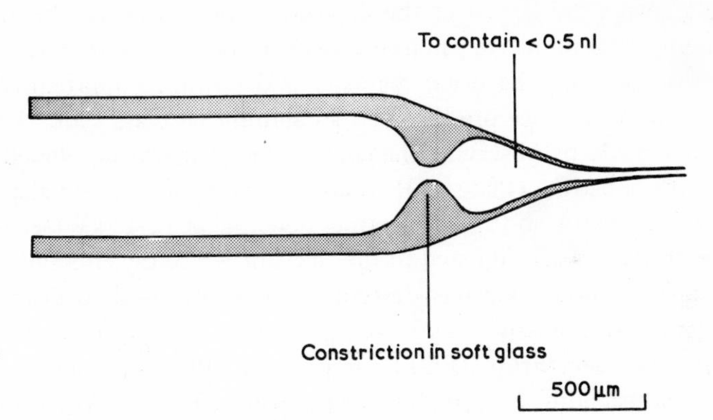

Fig. 7.2 Types of micropipettes for handling sub-nanolitre volumes: (a) constant volume type; (b) constriction type.

pipettes themselves can be of the constriction or constant volume variety (Fig. 7.2). The constriction type are constructed from suitable glass tubing using a microforge of the de Fonbrune type, and the methods described by Prager, Bowman and Vurek (1965) can be used to make constant volume pipettes. Pipette volumes of 0.05–0.5 nl are normally used. Exact volumes can be determined using radioactive solutions while approximate volumes can be calculated from measurements of pipette dimensions. Deposits formed from such volumes comprise a regular distribution of 1–2 μm crystals over an area of 80–150 μm diameter (Plates 7.1 *a*, *b*, and 7.2). Larger volumes produce larger deposits, which may necessitate further defocusing of the electron beam and possibly excessive defocusing of the spectrometers (see Section 7.4). A cold fibre-optic light source provides illumination without contributing to sample evaporation, when oil is not used. A grid scratched on the substrate facilitates identification of the various deposits once placed in the probe. Up to five replicates are usually deposited for each unknown and standard solution. The standards themselves are simple salt solutions, four or five concentrations being used for each element bracketing the biological samples under study. Substances such as urea, propylene glycol, and albumin have been added to standard solutions to increase the viscosity of the droplets and produce better deposits (Garland & Brown, unpublished data). Finally, standard vacuum and freeze-drying equipment is used during the critical lyophilization stage.

7.4 Microprobe techniques

7.4.1 CONDITIONS FOR MICROANALYSIS

Microanalysis is performed using a sufficiently defocused electron beam which completely covers the largest of the deposits to be analysed. This involves a beam diameter of up to 150 μm (see above). Deposit sizes are measured by light microscopy, and the beam diameter is determined approximately by viewing the luminescence produced on an aluminized glass slide. A further check can be made by observing the contamination marks developed on the surface of the substrate (Plate 7.2). Analysis is performed using a specimen current of 200–500 nA and an accelerating potential of 15–20 kV (see below). X-ray spectrometers are adjusted to the relevant Kα X-ray emission line for the elements required. Various detector crystals are used to discriminate certain elements; potassium hydrogen phthalate (KAP) for sodium and magnesium, pentaerythritol (PET) for chlorine, PET or quartz $10\bar{1}1$ for potassium and calcium, PET or KAP for phosphorus and $10\bar{1}1$ for iron and cobalt. Variations exist between laboratories (see Morel & Roinel, 1969; Lechene, 1970; Garland, Hopkins, Henderson, Haworth & Chester Jones, 1973; Roinel, 1975). The lyophilized deposits and the electron beam are normally aligned manually for each sample. In addition, the presence of only two spectrometers in most microprobes necessitates a readjustment of the wavelengths after each pair of elements analysed. Crystals may also have to be changed, depending on the combination of elements probed (see above). Such time-consuming exercises have been minimized by Roinel, Richard, Robin and Morel (1973) who have devised a computer-linked automatic microprobe sample-feeder (Fig. 7.3). The *X*, *Y* and *Z* co-ordinates of each sample are initially fed into the computer memory, and during the analytical cycle, the computer controls the displacements and subsequent analyses. Whether manually or automatically aligned, samples are always analysed for long enough to give good statistical accuracy of counting. Plate 7.1 shows light optical, electron and X-ray images of a typical fluid sample under

Fig. 7.3 Internal top view of the automatic sample feeder for electron probe microanalysis described by Roinel, Richard, Robin and Morel (1973). The specimen stage with two grooves for the substrate blocks is shown internally. Externally two motors for *X* and *Y* displacements are visible.

analysis. Readings are taken for unknowns and standards, and an assessment is made of background radiation by analysing an uncontaminated area of substrate. Calibration curves relating peak count rates to concentration are obtained for all elements under analysis. Elemental concentrations in unknown samples can then be calculated directly. It is important to include a set of standards with each analysis, as probe performance may not be identical from day to day.

7.4.2 POTENTIAL PROBLEMS

A number of studies have considered in some detail the potential problems associated with this technique: reproducibility, sensitivity, specificity, linearity, and possible volatility.

Table 7.2 Reproducibility of sample deposition. Four deposits from one solution were made and serially counted. For potassium, calcium, and cobalt two counts, and the mean value are given. For sodium, deposits were counted four times and the mean and standard error are presented. Overall means and standard errors together with the standard error as a percentage of the mean are also given (from Brown, 1976).

Deposit	*Sodium (cps)*	*Potassium (cps)*	*Calcium (cps)*	*Cobalt (cps)*
1	1533.23±55.23 (4)	53.59, 55.52 } 54.54	36.15, 41.03 } 38.59	76.40, 70.40 } 73.40
2	1500.08±10.75 (4)	49.12, 55.62 } 52.37	35.13, 35.63 } 35.38	65.30, 72.00 } 68.65
3	1399.75±2.83 (4)	46.27, 47.52 } 46.90	37.13, 28.10 } 32.62	59.30, 67.70 } 63.50
4	1602.48±19.02 (4)	53.54, 52.27 } 52.95	35.53, 34.83 } 35.18	64.30, 64.55 } 64.43
Mean value±SE (cps)	1508.89±42.17	51.69±1.66	35.44±1.22	67.50±2.26
SE as % of mean	2.78%	3.21%	3.16%	3.36%

Reproducibility

Assessments of reproducibility give some idea of the accuracy of both sample preparation and microprobing conditions. Analyses of sample replicates have indicated standard errors to be around 2–3% of the mean for sodium, potassium, calcium, phosphorus, iron, and cobalt (Morel & Roinel, 1969; Lechene, 1970; Brown, 1976). Data from Brown (1976) are shown in Table 7.2. For sodium, the mean count rate from four deposits of a single sample did not differ significantly from the mean count rate in any one deposit. Le Grimellec, Roinel and Morel (1973*a*) have further compared samples obtained by recollection micropuncture techniques. Again there is evidence for good reproducibility (Fig. 7.4).

Sensitivity

Morel and Roinel (1969) critically discussed the sensitivity of the method. In practice a focused microprobe of 1 μm^2 gives a sensitivity of approximately 5×10^{-13} M. Sensitivity appears to be a function of sample diameter (Fig. 7.5) and accelerating potential (Fig. 7.6). Moreover, responses to changing kV (accelerating voltage) varies between elements (Fig. 7.6). With increasing kV, count rates for sodium and chlorine decrease, while those for cobalt increase. Count rates for calcium and potassium are optimum at around 15 kV (Morel & Roinel, 1969). Fifteen or 20 kV is thus used to produce optimum sensitivity for the majority of elements; only cobalt analysis appears to benefit from higher values. The linearity and sensitivity of standard solutions arc discussed below.

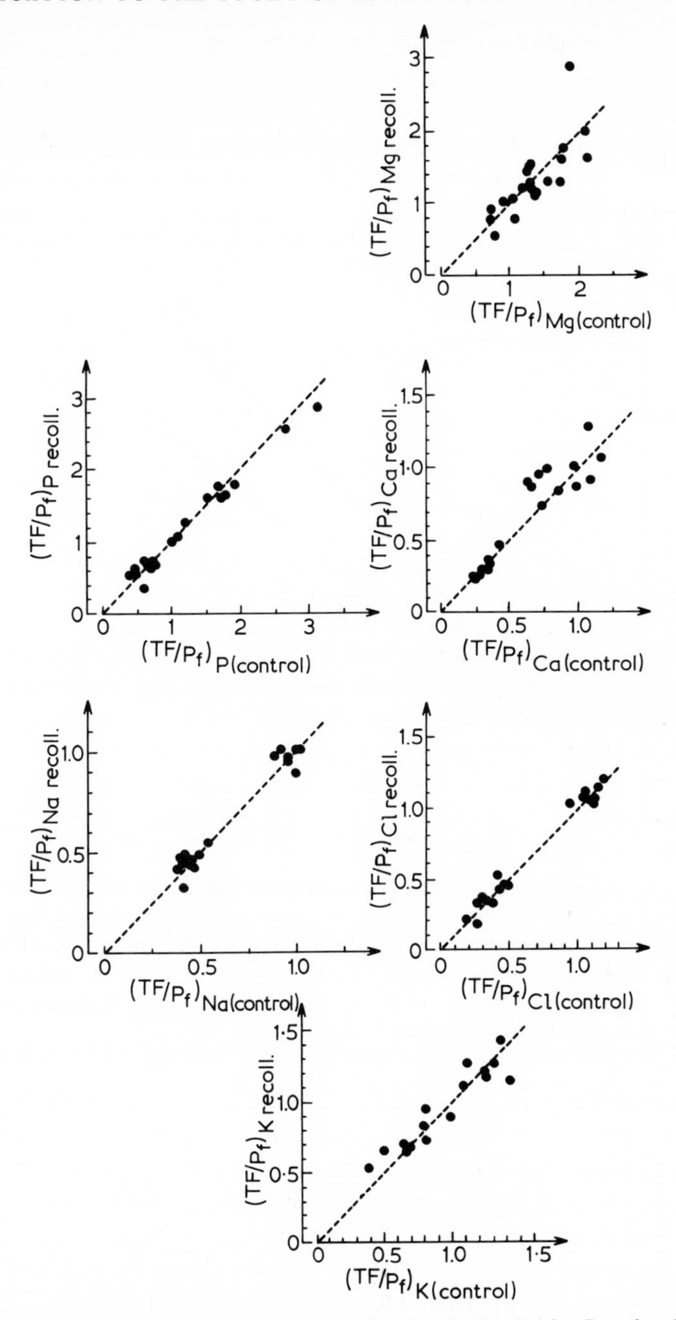

Fig. 7.4 Multielemental analysis of rat renal tubular fluids. Renal tubules were punctured twice during the same experimental period. First collection values (abscissa) are plotted against second (ordinate) (from Le Grimellec, Roinel & Morel, 1973*a*).

Specificity and potential interference between elements

Theoretically, X-rays emitted from a bombarded sample are specific for the various elements present. Detection of and discrimination between the different wavelengths, however, is dependent upon both the quality of the

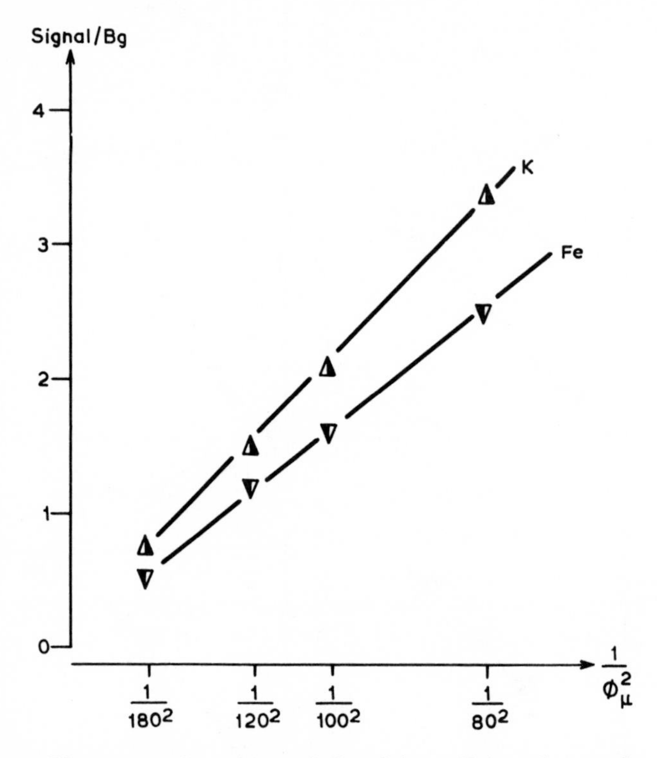

Fig. 7.5. Electron probe microanalysis of iron (Fe) and potassium (K). The relationship between sensitivity (signal: background ratio, S/Bg on ordinate) and sample size (reciprocal of square of the spot diameter, $1/\phi_{\mu}^{2}$ on abscissa). Volume deposited 0.4 nl. (From Morel & Roinel, 1969).

X-ray spectrometers and the conditions under which they are used. Potential interference between elements was investigated in a preliminary study by Garland, Hopkins, Henderson, Haworth and Chester Jones (1973). Separate standard solutions of sodium, potassium, and calcium were analysed in the presence and absence of another element. Results are reproduced in Table 7.3. The consistency of the count rates for a given element in pure and mixed solutions indicates no significant interference between the elements measured at the concentrations employed. Morel and Roinel (1969) also analysed elements with wavelengths relatively close together. The specific case involved phosphorus, with a $K\alpha_1$ line of 6.154 nm, and calcium, which has an adjacent

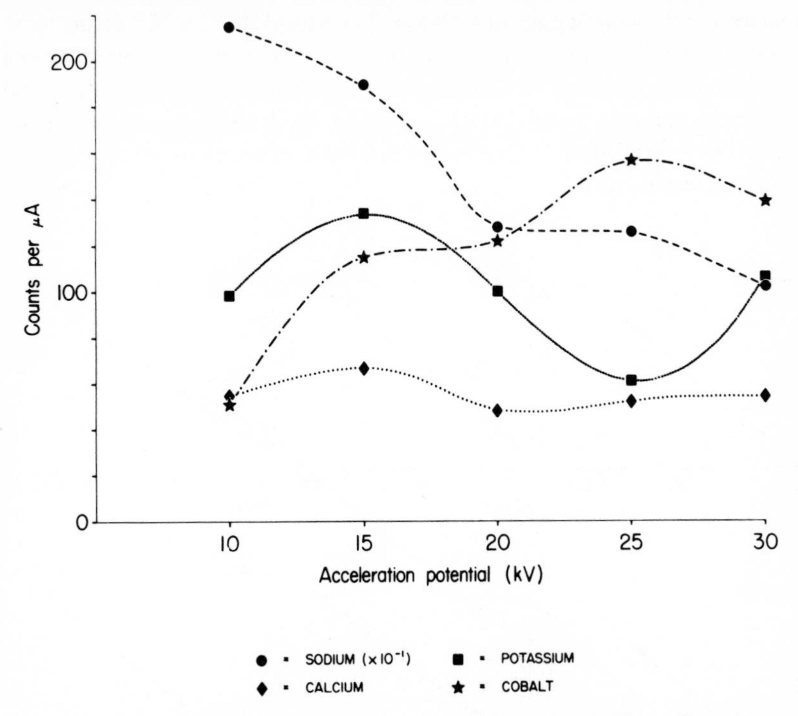

Fig. 7.6 Effects of different kVs on count rates from cobalt (*), sodium (●, $\times 10^{-1}$), potassium (■) and calcium (◆) in the electron probe microanalyser. Counts per μA are plotted against acceleration potential (from Garland, 1973).

Table 7.3 Sodium, potassium, and calcium X-rays (in counts per second) emitted from various standard solutions (from Garland, Hopkins, Henderson, Haworth & Chester Jones, 1973).

Standard solution	*X-rays*		
	Sodium	*Potassium*	*Calcium*
(1) Na^+	*1037.9*	21.8	39.9
(2) K^+	36.2	*57.2*	36.3
(3) Ca^{2+}	36.9	19.2	*100.5*
(4) Na^+ and K^+	*902.5*	*48.8*	37.8
(5) Na^+ and Ca^{2+}	*946.7*	18.3	*101.7*
(6) Ca^{2+} and K^+	36.3	*50.2*	*95.7*
(7) Background	33.6	18.1	34.0

$K\beta_1$ second order wavelength of 6.18 nm. Ten samples of 5 mM calcium were counted on the $K\alpha_1$ wavelength of phosphorus. The mean value obtained (778±28 counts/70 s) was not significantly different from the background count for the same element (773±25 counts/70 s). Interference between elements is thus not significant, even in extreme cases where elements emit X-rays on adjacent wavelengths.

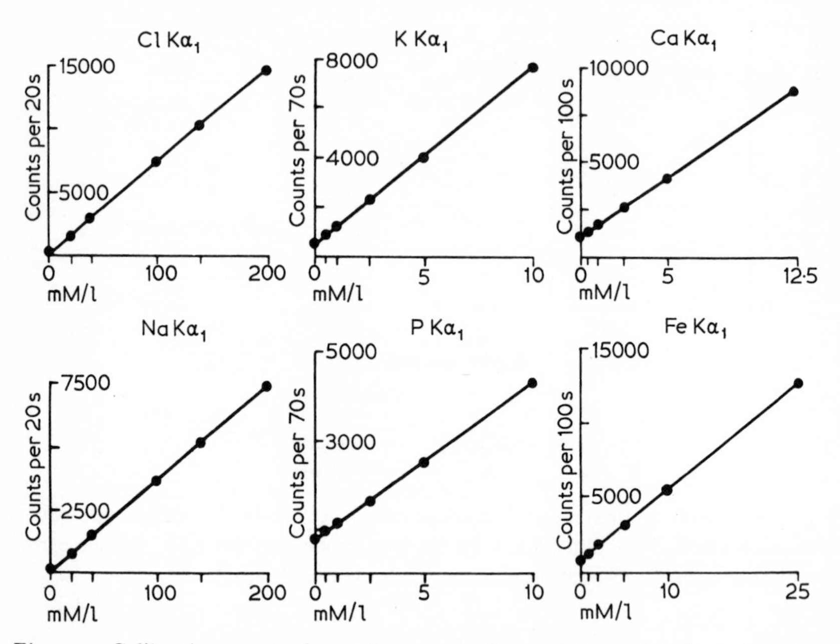

Fig. 7.7 Calibration curves for sodium, potassium, calcium, chlorine, phosphorus and iron using electron probe microanalysis. Mean count rates are plotted against concentration (from Morel, Roinel & Le Grimellec, 1969).

Linearity of standards

A number of investigations include 'typical' calibration data and curves for various elements (Ingram & Hogben, 1967; Cortney, 1969; Morel & Roinel, 1969; Morel, Roinel & Le Grimellec, 1969; Lechene, 1970; Garland, 1973; Garland, Hopkins, Henderson, Haworth & Chester Jones 1973; Le Grimellec, Roinel & Morel, 1973*a*; Roinel, 1975; Brown, 1976). Linearity has been demonstrated for all elements considered so far (Na, K, Ca, Cl, Mg, P, Fe, and Co) over the specific concentration ranges tested. Data from Morel, Roinel and Le Grimellec (1969) are given in Fig. 7.7.

Sample volatility

Traces for three of the elements tested for possible volatility during microprobe analysis (Garland, 1973) are shown in Fig. 7.8. Using a defocused electron beam, the sodium and potassium count rates from tubular fluid or plasma samples did not decline over a ten-minute period. It is interesting to note, however, that using a point-focused beam, (Fig. 7.8*b*), a considerable decrease in count rate occurs during the first minute. Count rates for magnesium, phosphorus, and calcium are also perfectly stable for more than 1000

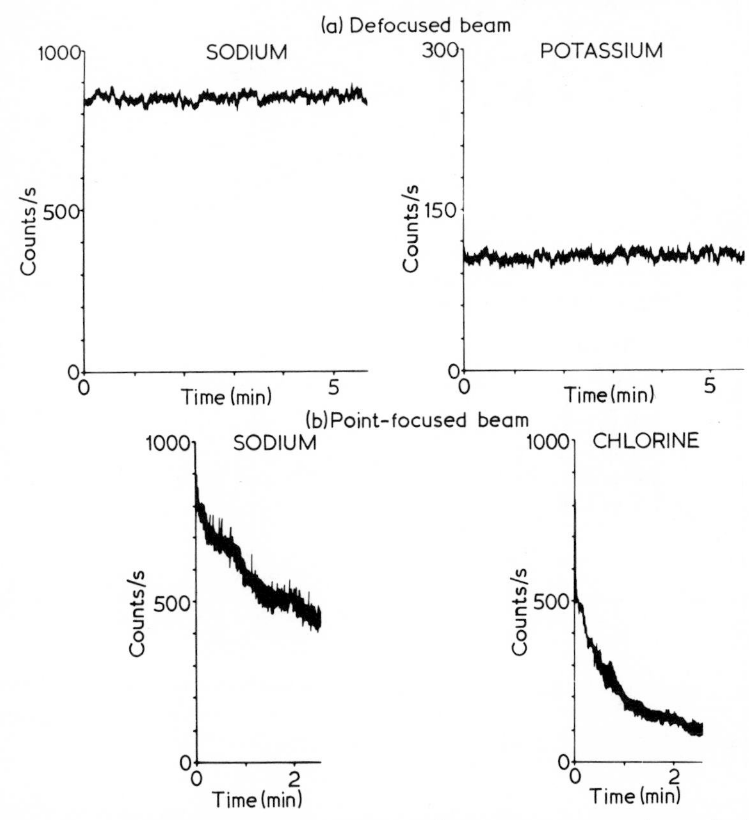

Fig. 7.8 Assessment of sample volatility during electron probe microanalysis; effects of defocused (sodium, potassium) and point-focused (sodium, chlorine) beams on count rates. Specimens were bombarded continuously for up to 5 minutes and X-rays were monitored by rate meter (from Garland, 1973).

seconds with a defocused electron beam (Roinel, 1975). Chlorine appears stable in urine, plasma, and tubular fluid samples, but may volatilize in pure mineral solutions. Roinel (1975) has investigated the problem further. The phenomenon would appear to depend upon sample quality, electron beam diameter, and probe current, but not kV. For some, as yet unexplained reason, the addition of urea to pure sodium chloride solutions helps prevent the decline in chlorine counts (Fig. 7.9). Urea (0.5 g/l) may be added to each standard mineral solution to stabilize chlorine counting rates.

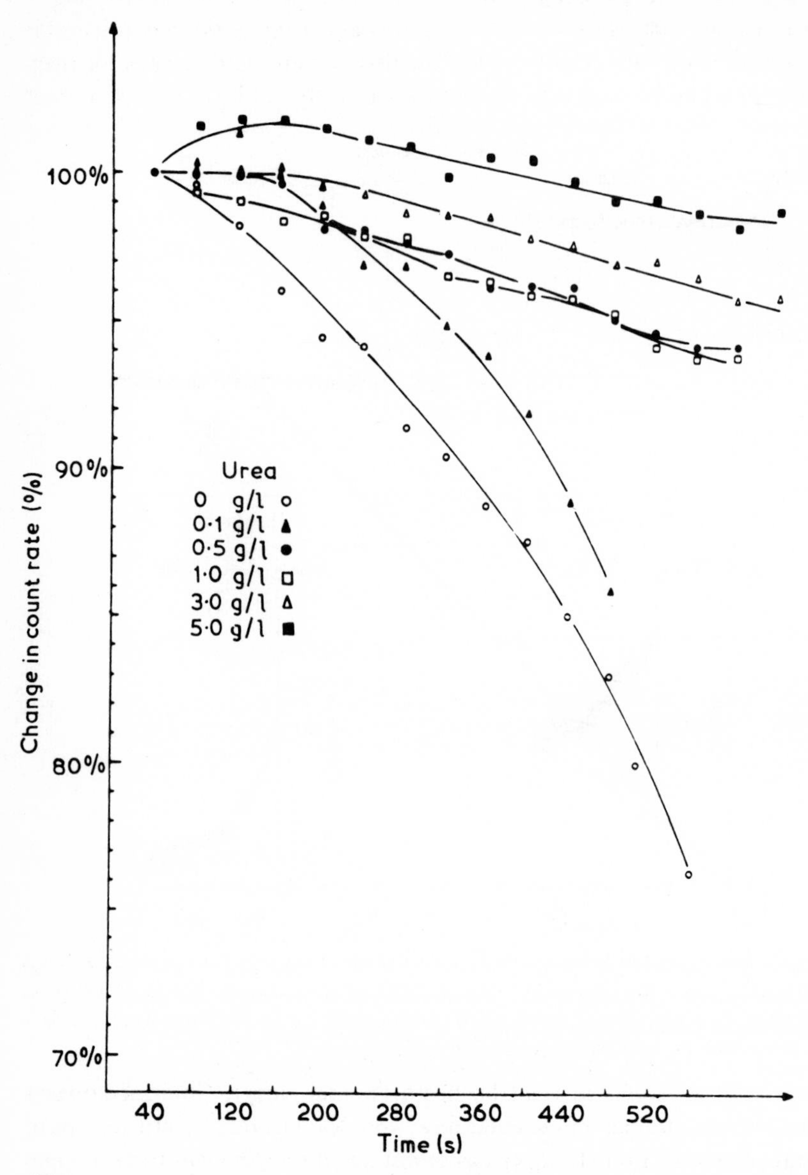

Fig. 7.9 The stabilization of chlorine count rates as a result of addition of urea (0.22 nl). Sodium chloride (200 mM) samples were bombarded for about 9 minutes in the presence of a range of urea concentrations (0–5 g/litre) (from Roinel, 1975).

7.5 Biological data

Understanding the delicate interactions between water and electrolyte transport across the renal tubule and the role of the kidney in body fluid homeostasis has been greatly furthered by electron probe microanalytical techniques. Most of the studies so far reported have concerned mammalian

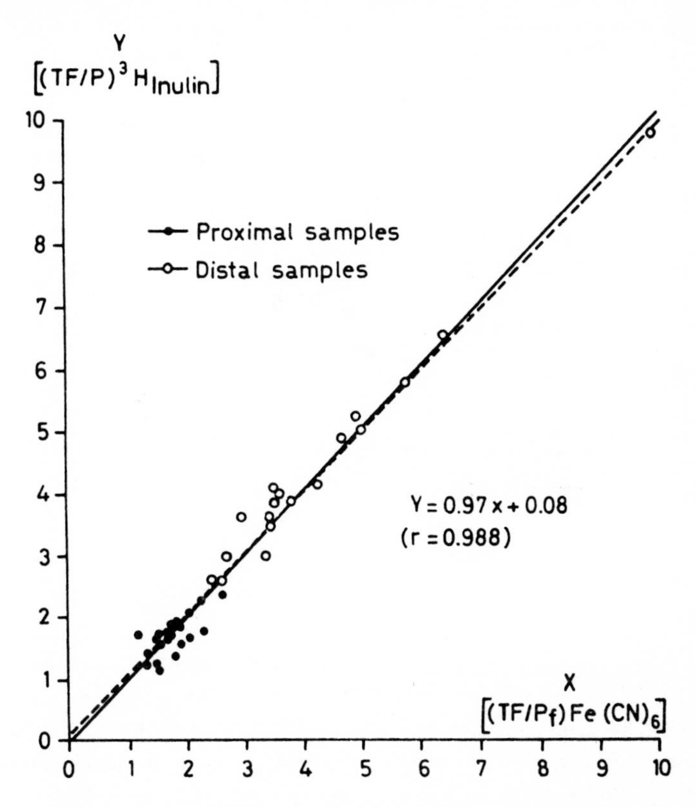

Fig. 7.10 The use of ferrocyanide as a monitor of glomerular activity and transtubular water movements. Tubular fluid: plasma ratios of ferrocyanide are plotted against those of inulin (from Morel, Roinel & Le Grimellec, 1969).

kidney function, but significant contributions have begun to emanate from studies on amphibians and fish. Inulin, the usual marker for glomerular filtration and transtubular water movements, has no suitable element for X-ray analysis. Alternative substances have therefore been sought and two solutions to this problem have been found. Ferrocyanide, containing iron, and cyanocobalamin (vitamin B_{12}), with a central cobalt atom, are substances which can be monitored by the electron probe and appear to be handled by the kidney in a similar manner to inulin (Figs. 7.10 and 7.11).

7.5.1 THE MAMMALIAN KIDNEY

The renal handling of water, phosphate, magnesium, calcium, potassium, sodium, chloride, and ferrocyanide by the rat kidney has been studied using microprobe techniques (Morel, Roinel & Le Grimellec, 1969; Le Grimellec,

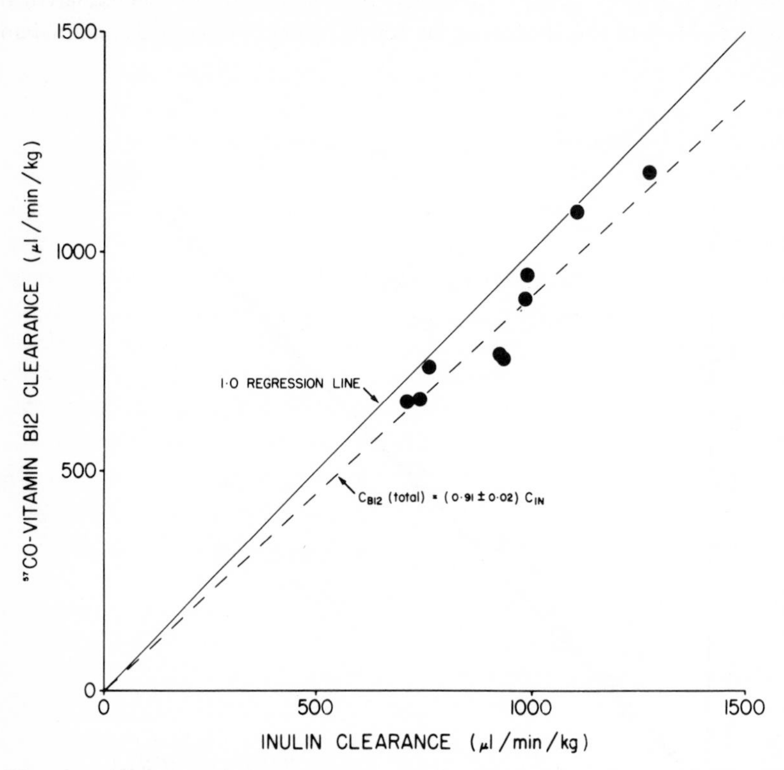

Fig. 7.11 Relationship between the simultaneously measured renal clearances of ^{14}C-inulin (abscissa) and ^{57}Co-vitamin B_{12} (ordinate) in *Necturus maculosus* (Garland, Henderson & Brown, 1975).

Roinel & Morel, 1973*a*, *b*, 1974*a*, *b;* Le Grimellec, 1975). In the normal kidney there is proximal reabsorption of phosphate, calcium, sodium, potassium, and chloride but limited reabsorption of magnesium. Comparison of late proximal and early distal tubular fluid composition indicates that in the loop of Henlé there is considerable reabsorption of magnesium, calcium, sodium, potassium, and chloride ions, but little phosphate reabsorption. Early distal tubular fluid and urine phosphate levels indicate phosphate reabsorption by the terminal nephron (distal segment and collecting duct) (Le Grimellec, Roinel & Morel, 1973*a*).

Recent analyses of tubular fluid specimens (TF) from convolutions close to the glomerulus (Le Grimellec, 1975) have shown an increase in TF: glomerular fluid (GF) chloride and potassium ratios, and a decrease in TF:GF phosphate ratios. Correlation between inulin and chloride TF:GF ratios suggests that water reabsorption accounts for the increased chloride ratios. Tubular fluid:glomerular fluid magnesium ratios indicate no reabsorption of this ion until TF:plasma (P) inulin ratios exceed 1.9.

The effects of acute increases in plasma concentrations of calcium, magnesium, and phosphate on electrolyte excretory rates have been investigated by whole kidney clearance studies (Heller, Hammarsten & Stutzman, 1953; Chesley & Tepper, 1958; Wallach & Carter, 1961; Duarte & Watson, 1967; Hulley, Goldsmith & Ingbar, 1969; Massry, Ahumada, Coburn & Kleeman, 1970). Such experiments reveal interactions between the renal handling of electrolytes such as calcium and magnesium and calcium and sodium (Chesley & Tepper, 1958; Walser, 1961; Coburn, Massry & Kleeman, 1970) but the sites of such interactions within the renal tubule cannot be positively evaluated. Microprobe analyses of micropuncture specimens gives insight into the intrarenal behaviour of electrolytes. Magnesium loading has little effect on the percentage magnesium reabsorption by the proximal segment (Le Grimellec, Roinel & Morel, 1973*a*), and analyses suggest that magnesium secretion may occur in the terminal part of the nephron. Hypermagnesaemia increases fractional calcium excretion and enhances phosphate reabsorption by the terminal segments (Le Grimellec, Roinel & Morel, 1973*a*).

Calcium loading (Le Grimellec, Roinel & Morel, 1974*a*) leads to a decrease in fractional calcium reabsorption by the proximal convoluted tubule, suggesting a saturable transport mechanism. Absolute and relative amounts of calcium delivered to the loop of Henlé are consequently increased. Distal collections indicate increased calcium reabsorption in the loop, but urinary output reflects its delivery to the distal segments, suggesting that calcium reabsorption in the terminal segments ceases. Fractional magnesium reabsorption is unchanged proximally, but inhibited in the loop, suggesting competition with calcium ions for transtubular transport. Increased magnesium delivery and suppression of its reabsorption in the terminal segments result in its increased excretion.

Phosphate loading of rats (Le Grimellec, Roinel & Morel, 1974*b*) reduces proximal reabsorption of water. Although both the filtered load and overall net reabsorption of phosphate increase during loading, relative reabsorption in the proximal segment decreases, suggesting that here phosphate is reabsorbed by a saturable mechanism. Net reabsorption in the terminal segments increases with load. Calcium TF:plasma filtrate (P_f) ratios fall below unity during hyperphosphataemia, but the change is comparable with the fall in TF:P inulin ratios, so that fractional reabsorption is unchanged.

Water and electrolyte movements along the loop of Henlé in the adult rat can only be estimated from late proximal and early distal tubular fluid collections, since the loop is inaccessible for micropuncture. In *Psammomys*, however, micropuncture of the loop is possible (de Rouffignac, Morel, Moss & Roinel, 1973). Microprobe analyses of loop fluid samples indicate addition

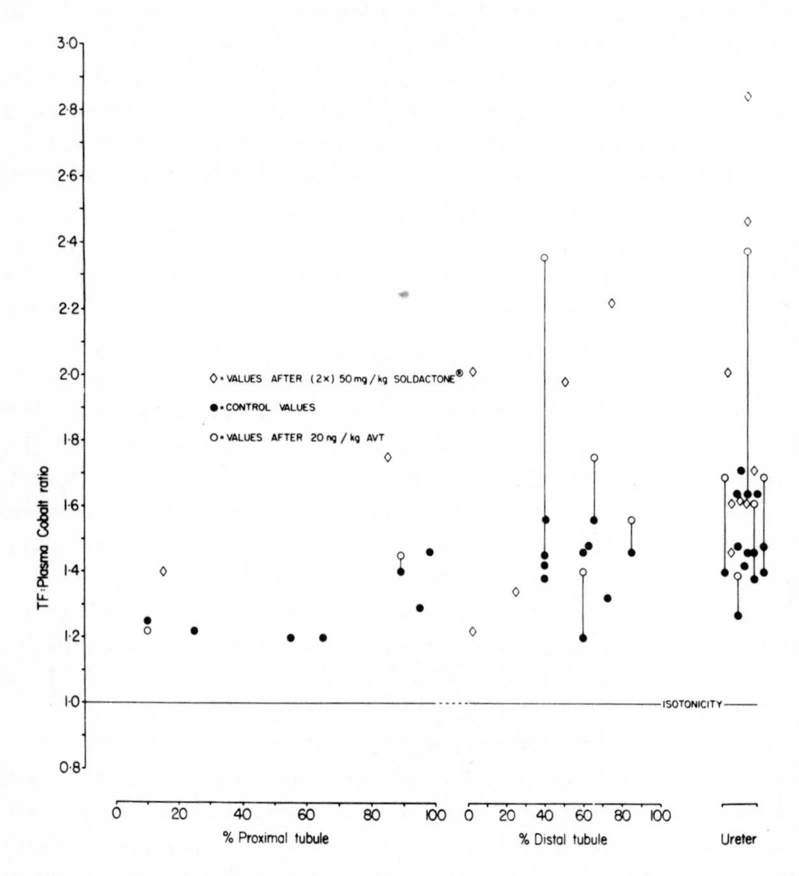

Fig. 7.12 The ratio of the concentration of cobalt (in vitamin B_{12}) in tubular fluid (TF) to that in the plasma at different sites along the nephron and in the ureter of *Necturus maculosus*. (●) indicate control values and (○) represent values after injection of arginine vasotocin; the latter are joined to their appropriate pre-injection control points by a vertical line. (◇) are values seen after SC 14266 (Soldactone) treatment. (see Garland, Henderson & Brown, 1975, for further details).

of sodium, potassium, chlorine, magnesium and to a lesser extent calcium along the descending limb. A constant load of these electrolytes is delivered to the distal segment indicating medullary recycling between the ascending and descending limbs. This suggests that the concentrating process along the loop of Henlé involves the addition of solute rather than withdrawal of water.

Cortney (1969) examined renal metabolism of adrenalectomized rats using X-ray microanalysis. Data suggest that the sodium-losing syndrome associated with adrenal insufficiency results from a defective reabsorptive mechanism in the distal tubule and collecting duct.

7.5.2 THE AMPHIBIAN KIDNEY

X-ray analysis has been used to investigate the renal handling of water and the electrolytes, sodium, potassium, and calcium in the urodele amphibian *Necturus maculosus* (Garland, 1973, 1974; Garland, Henderson & Chester Jones, 1973; Garland, Hopkins, Henderson, Haworth & Chester Jones, 1973;

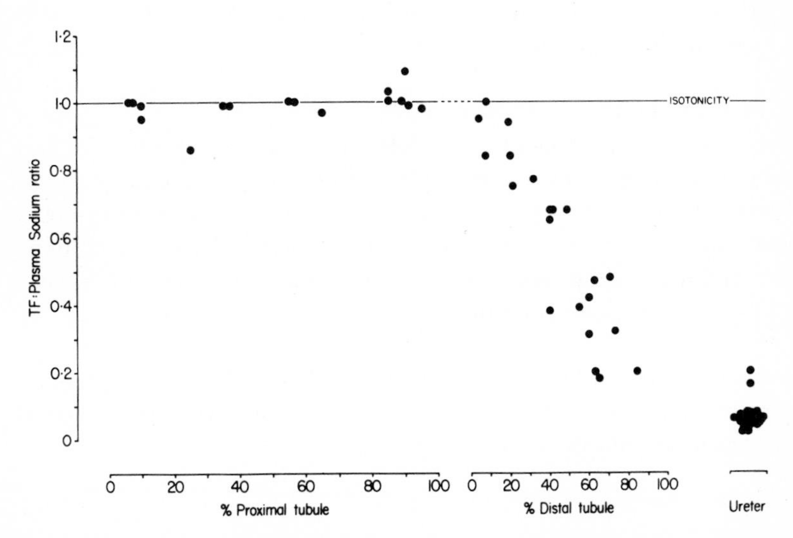

Fig. 7.13 Tubular fluid (TF) to plasma sodium concentration ratio along the nephron and in the ureter of *Necturus maculosus.* (see Fig. 7.11 and Garland, Henderson & Brown, 1975).

Brown, 1976). Cyanocobalamin was used as an indicator for glomerular filtration rates and transtubular water movement (Garland, Henderson & Brown, 1975). Tubular fluid: plasma cobalt ratios indicate that up to 30% of the filtered water is reabsorbed proximally (Fig. 7.12). Sodium concentrations in proximal tubular fluid samples are similar to those of plasma (Fig. 7.13). Distally a progressive decrease in tubular fluid sodium concentration occurs with the result that urine is extremely hypotonic to plasma for this ion.

Potassium is apparently actively secreted proximally since TF: P ratios are consistently greater than TF: P cobalt ratios. Distally there may be initial secretion followed by reabsorption (Giebisch, Boulpaep & Whittembury, 1971).

Micropuncture and microprobe techniques were used to analyse the effects of the neurohypophysial hormone, arginine vasotocin (AVT) and of an aldosterone antagonist, Soldactone (G. D. Searle, SC14266). AVT induces both a glomerular and a tubular antidiuresis. The latter effect derives from increased reabsorption of osmotically free water in the distal tubule (Fig. 7.12; Garland, Henderson & Brown, 1975).

Soldactone has no effect on glomerular filtration rate (GFR). Microprobe analyses of cobalt levels in TF and plasma samples indicated that the compound increases fluid reabsorption (Fig. 7.12; Garland, Henderson & Brown, 1975). Aldosterone blockade did not produce natriuresis, diuresis, and antikaliuresis as might have been expected; aldosterone may thus act as a 'glucocorticoid' in *Necturus*. In the rat, glucocorticoids maintain distal tubular water impermeability (Wiederholt & Wiederholt, 1968).

7.5.3 THE TELEOSTEAN FISH KIDNEY

The only micropuncture data for electrolyte levels in fish renal tubules have been obtained using X-ray microanalysis (Moss, 1976). Preliminary control data have indicated TF : P ratios approaching unity for sodium in presumptive proximal segments, with lower ratios in presumptive distal convolutions. Chlorine and magnesium ratios exceed 1.0 proximally, while calcium appears lower. All three ratios decrease distally.

The biological data which have been described are only a small part of the total information obtained by microprobe analysis. For further details readers should consult references listed in Section 7.1.1.

7.6 The results of X-ray analysis compared with other analytical methods

7.6.1 STUDIES IN WHICH SAMPLES HAVE BEEN ANALYSED BY MICROPROBE AND OTHER TECHNIQUES

Simultaneous determinations of electrolyte concentrations in urine, plasma, and tubular fluid specimens by X-ray analysis and conventional techniques have been carried out by several workers. Table 7.4 summarizes this information.

Electrolyte analyses by conventional methods have been carried out by photometry, fluorimetry, colorimetry, and by electrometric titration.

A high degree of correlation between microprobe results and those obtained by conventional techniques has been observed by Morel, Roinel and Le Grimellec (1969), and Kuntziger, Amiel, Roinel and Morel (1974) (Table 7.5;

Table 7.4 Summary of literature in which sodium, potassium, chlorine, calcium, and phosphorus in urine, tubular fluid and plasma have been simultaneously analysed by the electron probe and by more conventional techniques.

	Sample description		
Element analysed	*Urine*	*Tubular fluid*	*Plasma/plasma ultrafiltrate*
Sodium	Cortney (1969)	Morel *et al.* (1969)	Cortney (1969)
Potassium	Cortney (1969), Beck and Goldberg (1973*a*)	Morel *et al.* (1969)	Cortney (1969)
Chlorine	Cortney (1969)	–	Cortney (1969)
Calcium	–	–	Beck and Goldberg (1973*a*)
Phosphorus	–	Kuntziger *et al.* (1974)	Beck and Goldberg (1973*a*)

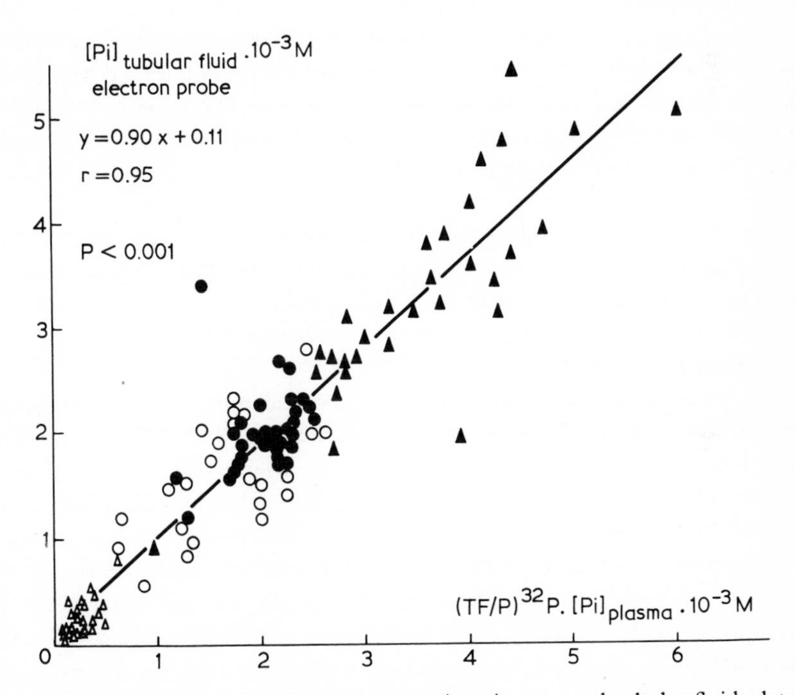

Fig. 7.14 The absolute phosphate concentrations in rat renal tubular fluids determined by electron probe microanalysis (ordinate) compared with concentrations extrapolated from ^{32}P and inoraganic concentrations. The various symbols represent different experimental conditions which are not relevant to the present discussion (from Kuntziger, Amiel, Roinel & Morel, 1974).

Table 7.5 Statistical comparison of electron probe (EPA) and flame photometric (flame) analysis of sodium and potassium in renal tubular fluid (TF) and plasma (P) (from Morel, Roinel & Le Grimellec, 1969).

x	y	*Regression line equation*	N	*Correlation coefficient*
(TF/P)Na (EPA)	(TF/P)Na (flame)	$y=0.99x+0.03$	39	0.957
(TF/P)K (EPA)	(TF/P)K (flame)	$y=1.00x+0.06$	30	0.965

Fig. 7.14). Similarly Cortney (1969) (Table 7.6), and Beck and Goldberg (1973*a*) found no significant difference between results obtained by X-ray analysis and by more usual techniques.

7.6.2 VALUES OBTAINED BY X-RAY ANALYSIS AND PREVIOUS DATA USING OTHER ANALYTICAL METHODS

Mammalian data

In the rat, values obtained by microprobe analyses for sodium, chlorine, magnesium, calcium, and phosphorus concentrations in plasma ultrafiltrates (Morel, Roinel & Le Grimellec, 1969) agree with those measured photometrically, colorimetrically, and using radioisotopes (Litchfield & Bott, 1962; Lassiter, Gottschalk & Mylle, 1963; Strickler, Thompson, Klose & Giebisch, 1964).

Data on the renal handling of chloride are similar to those obtained by conventional methods; for example, proximal tubular fluid chlorine concentration obtained by X-ray analysis is 141±2 mEq/l ($n=28$) (Morel, Roinel & Le Grimellec, 1969), while electrometric analysis gives 144.5±2.9 ($n=15$) (Litchfield & Bott, 1962).

The proximal tubular handling of calcium revealed by microprobe analyses (Morel, Roinel & Le Grimellec, 1969; Le Grimellec, Roinel & Morel, 1974*a*; Agus, Gardner, Beck & Goldberg, 1973; de Rouffignac, Morel, Moss & Roinel, 1973) agree with data obtained by other techniques (Lassiter, Gottschalk & Mylle, 1963; Duarte & Watson, 1967; Buerkert, Marcus & Jamison, 1972; Edwards, Sutton & Dirks, 1974). TF : P_f calcium ratios reported in the literature are presented in Table 7.7. All values whether determined by microprobe analyses or by other techniques were between 0.96 and 1.1. Increased ratios during calcium loading were noted both in rats using the electron probe (Le Grimellec, Roinel & Morel, 1974*a*, *b*) and in dogs using a helium glow photometer (Edwards, Sutton & Dirks, 1974). Increased calcium excretion during magnesium loading has been shown both by microprobe analyses (Le Grimellec, Roinel & Morel, 1973*b*) and conventional techniques (Chesley & Tepper, 1958; Samiy, Brown & Globus, 1960; Massry, Ahumada, Coburn & Kleeman, 1970).

Table 7.6 A comparison of micro (by electron probe microanalysis) and macro (flame photometry for sodium and potassium; chloridometer for chloride) analysis of urine and plasma (from Cortney, 1969).

Potassium, mEq/l			*Sodium, mEq/l*			*Chloride, mEq/l*		
Micro	*Macro*	*Macro less Micro*	*Micro*	*Macro*	*Macro less Micro*	*Micro*	*Macro*	*Macro less Micro*
				Urine				
33	33	0	75	74	−1	78	89	+11
33	34	+1	83	83	0	94	101	+7
18	19	+1	93	91	−2	87	94	+7
16	17	+1	117	112	−5	99	105	+6
19	19	0	97	95	−2	98	104	+6
17	17	0	60	64	+4	68	69	+1
23	23	0	58	60	+2	61	63	+2
21	20	−1	72	74	+2	77	78	+1
10	10	0	87	74	−13	84	74	−10
12	12	0	72	74	+2	70	70	0
16	16	0	88	94	+6	102	101	−1
16	16	0	95	93	−2	103	100	−3
24	23	−1	107	97	−10	99	102	+3
25	25	0	79	86	+7	99	98	−1
19	19	0	71	73	+2	72	84	+12
15	15	0	113	97	−16	100	103	+3
Mean 20	20	+0.06	85	84	−1.63	87	90	+2.75
±SD		±0.34			±4.20			±3.43
				Plasma				
	(×1.06)			(×1.06)			(×1.06)	
3.4	3.6	+0.2	145	150	+5	101	113	+12
3.2	3.5	+0.3	170	148	−22	102	110	+8
3.5	3.8	+0.3	148	146	−2	99	111	+12
3.6	3.8	+0.2	160	144	−16	98	108	+10
3.8	3.8	0	140	147	+7	120	113	−7
4.3	4.5	+0.2	152	147	−5	122	117	−5
3.7	4.1	+0.4	146	147	+1	119	115	−4
5.4	5.7	+0.3	142	143	+1	119	110	−9
4.3	4.5	+0.2	173	152	−21	113	108	−5
4.4	4.8	+0.4	153	153	0	113	110	−3
3.5	3.7	+0.2	168	153	−15	112	107	−5
3.6	3.6	0	181	155	−26	112	108	−4
3.6	3.8	+0.2	149	151	+2	109	108	−1
3.3	3.5	+0.2	153	152	−1	104	111	+7
3.7	3.8	+0.1	139	152	+13	121	112	−9
4.4	4.2	−0.2	143	150	+7	116	107	−9
Mean 3.6	4.0	+0.019	154	149	−4.5	111	111	−0.75
±SD		±0.106			±6.14			±3.70

Table 7.7 Tubular fluid calcium concentration/plasma ultrafiltrable calcium concentration (TF : P_f) in the proximal tubule of various species: a comparison of values obtained by electron probe microanalysis (EPMA) with those obtained by other methods.

Species	*Method*	*TF/P$_f$*	*Reference*
Rat	EPMA	1.03±0.02 (n = 25)	Le Grimellec *et al.* (1973*a*)
Dog	EPMA	1.1 ±0.02 (n = 46)	Agus *et al.* (1973)
Psammomys	EPMA	0.96±0.25 (n = 68)	de Rouffignac *et al.* (1973)
Rat	EPMA	1.0 ±0.16 (n = 29)	Morel *et al.* (1969)
Rat	Ca^{45}	1.0 (n = 31)	Lassiter *et al.* (1963)
Dog	Fluorometry	1.08±0.12 (n = 24)	Duarte and Watson (1967)
Rat	Helium glow photometry	1.06±0.02	Buerkert *et al.* (1972)

Tubular fluid phosphate levels obtained by X-ray analysis are similar to those determined by other means. A mean late proximal TF : P_f phosphate ratio of 0.71±0.03 (n = 21) (Le Grimellec, Roinel & Morel, 1974*b*) and an overall proximal phosphate TF : P_f ratio of 0.56 (Morel, Roinel & Le Grimellec, 1969), obtained by microprobe analyses, are similar to a mean value of 0.73±0.16 (n = 24) reported by Strickler, Thompson, Klose and Giebisch (1964). In addition, early distal TF : P_f phosphate ratios were greater than 1.0 after microcolorimeteric (Strickler, Thompson, Klose & Giebisch, 1964), and X-ray analyses (Le Grimellec, Roinel & Morel, 1974*b*). Phosphate reabsorption by the terminal segments (distal segment and collecting duct) observed in studies using microprobe analysis (Le Grimellec, Roinel & Morel, 1973*a*, *b*, 1974*a*, *b*; de Rouffignac, Morel, Moss & Roinel, 1973) was not found by Strickler, Thompson, Klose and Giebisch (1964) using other analytical methods.

Amphibian data

In *Necturus* the pattern of tubular concentration of cobalt obtained by microprobe analysis indicates reabsorption by the proximal segment of about 30% of the filtered volume. These data agree with those obtained previously using inulin (Bott, 1952, 1962; Giebisch, 1956; Oken & Solomon, 1963).

Microprobe data on sodium gradients along the *Necturus* nephron confirm those obtained from ultramicroflame photometry (Bott, 1962), and using cation-sensitive glass electrodes (Khuri, Goldstein, Maude, Edmonds & Solomon, 1963). The mean plasma sodium concentration obtained by X-ray analysis, 105.1±2.5 mM (n = 25) (Garland, Henderson & Brown, 1975), compares favourably with the mean values of 103.1±1.5 (n = 27) of Bott (1962), and 98.6±3.8 (n = 3) of Khuri, Goldstein, Maude, Edmonds and Solomon

(1963). Similarly a mean sodium concentration in the proximal tubule, 110.0±5.8 mM (Garland, Henderson & Brown, 1975), compares with values reported by Bott (1963) (101.2 mM) and Khuri, Goldstein, Maude, Edmonds and Solomon (1963) (96.7±4.8 mM). Proximal tubular potassium concentrations obtained by microprobe analyses (Garland, Henderson & Brown, 1975) agree with data of Oken and Solomon (1963), Khuri, Goldstein, Maude, Edmonds and Solomon (1963), and Watson, Clapp and Berliner (1964) obtained photometrically and using cation-sensitive glass electrodes.

7.7 Summary and conclusions

Electron probe microanalysis, a method now established in several laboratories for analysis of renal tubular fluid specimens, offers several advantages over alternative analytical techniques. Many elements can be simultaneously analysed in a single sample of less than 1 nl without dilution. Any element with an atomic number greater than 5 can be analysed; many biologically important elements not readily determined by other methods, can now be studied.

The accuracy of the method is largely dependent on sample deposition and subsequent production of a homogeneous distribution of small crystals. Of the various substrates used – quartz, carbon, beryllium and aluminium – aluminium is inexpensive, easily available, and non-toxic. Beryllium on the other hand, a more expensive and potentially toxic material, gives a much lower background count rate and is therefore the substrate of choice. Analyses of sample replicates indicate the standard error to be 2–3% of the mean. Sensitivity is estimated at 5 ± 10^{-13} M and varies with sample diameter and the accelerating potential. Interference between elements does not appear to be a problem for the elements at concentrations so far studied. Volatilization is not a source of error except for chlorine. Addition of urea to standard solutions obviates this problem.

Experimental investigations on renal function in a number of species (dog, rat, *Psammomys*, *Necturus* and trout) have shown electron probe microanalytical methods to give similar values to more conventional methods, and, because of the broader spectrum of analyses, more extensive information is gained.

Thus microprobe analysis is invaluable for elemental analysis of nanolitre volumes of biological fluids such as those obtained by renal tubular micropuncture.

7.8 References

AGUS, Z. S., GARDNER, L. B., BECK, L. H. & GOLDBERG, M. (1973) Effects of parathyroid hormone on renal tubular reabsorption of calcium, sodium, and phosphate. *American Journal of Physiology*, **224**, 1143–1148.

AMIEL, C., KUNTZIGER, H., ROINEL, N. & MOREL, F. (1972) Tubular handling of Ca and Mg in parathyroidectomized (PTX) and 3′,5′-cAMP infused PTX rats. *Vth Congreso Internacional de Nephrologia, Mexico*, p. 140.

BAUMANN, K., ROUFFIGNAC, C. DE, ROINEL, N., RUMRICH, G. & ULLRICH, K. J. (1975) Renal phosphate transport: inhomogeneity of local proximal transport rates and sodium dependence. *Pflügers Archiv. European Journal of Physiology*, **356**, 287–297.

BECK, L. H. & GOLDBERG, M. (1973*a*) Effects of acetazolamide and parathyroidectomy on renal transport of sodium, calcium, and phosphate. *American Journal of Physiology*, **224**, 1136–1142.

BECK, L. H. & GOLDBERG, M. (1973*b*) Mechanism of blunted phosphaturia in saline-loaded parathyroidectomized (PTX) dogs. *Clinical Research*, **21**, 676.

BECK, L. H., SENESKY, D. & GOLDBERG, M. (1973) Sodium-independent active potassium reabsorption in proximal tubule of the dog. *Journal of Clinical Investigation*, **52**, 2641–2645.

BOTT, P. A. (1952) Renal excretion of creatinine in *Necturus*. A reinvestigation by direct analysis of glomerular and tubular fluid for creatinine and inulin. *American Journal of Physiology*, **168**, 107–113.

BOTT, P. A. (1962) Micropuncture study of renal excretion of water, K, Na and Cl in *Necturus*. *American Journal of Physiology*, **203**, 662–666.

BROOKS, E. J., TOUSIMIS, A. J. & BIRKS, L. S. (1962) The distribution of calcium in the epiphyseal cartilage of rat tibia measured with the electron probe X-ray microanalyser. *Journal of Ultrastructural Research*, **7**, 56–60.

BROWN, J. A. (1976) Development and application of methods for the study of renal function in lower vertebrates. *Ph.D. Thesis*, University of Sheffield.

BUERKERT, J., MARCUS, D. & JAMISON, R. (1972) Renal tubule calcium reabsorption after parathyroidectomy. *Journal of Clinical Investigation*, **51**, 17a.

CHESLEY, L. C. & TEPPER, I. (1958) Some effects of magnesium loading upon renal excretion of magnesium and certain other electrolytes. *Journal of Clinical Investigation*, **37**, 1362–1372.

COBURN, J. W., MASSRY, S. G. & KLEEMAN, C. R. (1970) The effect of calcium infusion on renal handling of magnesium with normal and reduced glomerular filtration rate. *Nephron*, **7**, 131–143.

CORTNEY, M. A. (1969) Renal tubular transfer of water and electrolytes in adrenalectomized rats. *American Journal of Physiology*, **216**, 589–598.

DUARTE, C. G. & WATSON, J. F. (1967) Calcium reabsorption in proximal tubule of the dog nephron. *American Journal of Physiology*, **212**, 1355–1360.

EDWARDS, B. R., SUTTON, R. A. L. & DIRKS, J. H. (1974) Effect of calcium infusion on renal tubular reabsorption in the dog. *American Journal of Physiology*, **227**, 13–18.

FRONZO, R. A. DE, GOLDBERG, M. & AGUS, Z. S. (1976) The effects of glucose and inulin on renal electrolyte transport. *Journal of Clinical Investigation*, **58**, 83–90.

GARLAND, H. O. (1973) Renal micropuncture and general renal studies in amphibians and fish. *Ph.D. Thesis*, University of Sheffield.

GARLAND, H. O. (1974) The application of electron probe microanalysis to studies of kidney function in lower vertebrates. *Proceedings of the Royal Microscopical Society*, **9**, 45.

GARLAND, H. O., HENDERSON, I. W. & BROWN, J. A. (1975) Micropuncture study of the renal responses of the urodele amphibian *Necturus maculosus* to injections of arginine vasotocin and an antialdosterone compound. *Journal of Experimental Biology*, **63**, 249–264.

GARLAND, H. O., HENDERSON, I. W. & CHESTER JONES, I. (1973) Micropuncture study of the renal actions of arginine vasotocin in *Necturus maculosus. Journal of Endocrinology*, **58**, xxvi–xxvii.

GARLAND, H. O., HOPKINS, T. C., HENDERSON, I. W., HAWORTH, C. W. & CHESTER JONES, I. (1973) The application of quantitative electron probe microanalysis to renal micropuncture studies in amphibians. *Micron*, **4**, 164–176.

GIEBISCH, G. (1956) Measurements of pH, chloride and inulin concentrations in proximal tubule fluid of *Necturus. American Journal of Physiology*, **185**, 171–174.

GIEBISCH, G. (1972) (ed) Renal micropuncture techniques: A symposium. *Yale Journal of Biology and Medicine*, **45**, 187–456.

GIEBISCH, G., BOULPAEP, E. L. & WHITTEMBURY, G. (1971) Electrolyte transport in kidney tubule cells. *Philosophical Transactions of the Royal Society of London, Series B*, **262**, 175–196.

GOLDBERG, M., AGUS, Z. S. & BECK, L. H. (1974) Interrelationship of renal handling of sodium, calcium and phosphate. In *Recent Advances in Renal Physiology and Pharmacology*, ed. Wesson, L. G. & Fanelli, G. M. pp. 111–124. Lancaster: Medical and Technical Publishing Co. Ltd.

GOTTSCHALK, C. W. & LASSITER, W. E. (1973) Micropuncture Methodology. In *Renal Physiology – Handbook of Physiology Sect 8*, ed. Orloff, J. & Berliner, R. W. pp. 129–143, Washington, D.C.: American Physiological Society.

HELLER, B. I., HAMMARSTEN, J. F. & STUTZMAN, F. L. (1953) Concerning the effects of magnesium sulfate on renal function, electrolyte excretion and clearances of magnesium. *Journal of Clinical Investigation*, **32**, 858–861.

HULLEY, S. B., GOLDSMITH, R. S. & INGBAR, S. H. (1969) Effect of renal arterial and systemic infusion of phosphate on urinary Ca excretion. *American Journal of Physiology*, **217**, 1570–1575.

INGRAM, M. J. & HOGBEN, C. A. M. (1967) Electrolyte analysis of biological fluids with the electron microprobe. *Analytical Biochemistry*, **18**, 54–57.

KHURI, R. N., GOLDSTEIN, D. A., MAUDE, D. L., EDMONDS, C. & SOLOMON, A. K. (1963) Single proximal tubules of *Necturus* kidney. VIII. Na and K determinations by glass electrodes. *American Journal of Physiology*, **204**, 743–748.

KUNTZIGER, H., AMIEL, C., ROINEL, N. & MOREL, F. (1974) Effects of parathyroidectomy and cyclic AMP on renal transport of phosphate, calcium, and magnesium. *American Journal of Physiology*, **227**, 905–911.

LASSITER, W. E., GOTTSCHALK, C. W. & MYLLE, M. (1963) Micropuncture study of renal tubular reabsorption of calcium in normal rodents. *American Journal of Physiology*, **204**, 771–775.

LECHENE, C. (1970) The use of the electron microprobe to analyse very minute amounts of liquid samples. *Proceedings of the Vth National Conference on electron probe microanalysis*. 32A–32C.

LE GRIMELLEC, C. (1975) Micropuncture study along the proximal convoluted tubule. Electrolyte reabsorption in first convolutions. *Pflügers Archiv. European Journal of Physiology*, **354**, 133–150.

LE GRIMELLEC, C. & LECHENE, C. (1970) Renal effect of thyroid calcium infusion in parathyroidectomized rats – Electron probe analysis. In *IVth Annual meeting of American Society of Nephrology*, p. 45.

LE GRIMELLEC, C., POUJEOL, P. & ROUFFIGNAC, C. DE (1975) ^{3}H-inulin and electrolyte concentrations in Bowman's capsule in rat kidney. Comparison with artificial ultrafiltration. *Pflügers Archiv. European Journal of Physiology*, **354**, 117–131.

LE GRIMELLEC, C., ROINEL, N. & MOREL, F. (1973*a*) Simultaneous Mg, Ca, P, K, Na and Cl analysis in rat tubular fluid. I. During perfusion of either inulin or ferrocyanide. *Pflügers Archiv. European Journal of Physiology*, **340**, 181–196.

LE GRIMELLEC, C., ROINEL, N. & MOREL, F. (1973*b*) Simultaneous Mg, Ca, P, K, Na and Cl analysis in rat tubular fluid. II. During acute Mg plasma loading. *Pflügers Archiv. European Journal of Physiology*, **340**, 197–210.

LE GRIMELLEC, C., ROINEL, N. & MOREL, F. (1974*a*) Simultaneous Mg, Ca, P, K, Na and Cl analysis in rat tubular fluid. III. During acute Ca plasma loading. *Pflügers Archiv. European Journal of Physiology*, **346**, 171–188.

LE GRIMELLEC, C., ROINEL, N. & MOREL, F. (1974*b*) Simultaneous Mg, Ca P, K, Na and Cl analysis in rat tubular fluid. IV. During acute phosphate plasma loading. *Pflügers Archiv. European Journal of Physiology*, **346**, 189–204.

LITCHFIELD, J. B. & BOTT, P. A. (1962) Micropuncture study of renal excretion of water, K, Na and Cl in the rat. *American Journal of Physiology*, **203**, 667–670.

MASSRY, S. G., AHUMADA, J. J., COBURN, J. W. & KLEEMAN, C. R. (1970) Effect of $MgCl_2$ infusion on urinary Ca and Na during reduction in their filtered loads. *American Journal of Physiology*, **219**, 881–885.

MOREL, F. (1975) Application de la microsonde à la physiologie; microanalyse de la composition des fluides tubulaire intrarénaux. *Journal de Microscopie et de Biologie Cellulaire*, **22**, 479–482.

MOREL, F. & ROINEL, N. (1969) Application de la microsonde électronique à l'analyse élémentaire quantitative d'échantillons liquides d'un volume inférieur à 10^{-9} l. *Journal de Chimie Physique*, **66**, 1084–1091.

MOREL, F., ROINEL, N. & LE GRIMELLEC, C. (1969) Electron probe analysis of tubular fluid composition. *Nephron*, **6**, 350–364.

MOSS, N. G. (1976) Micropuncture studies on renal tubular function in selected vertebrates. *Ph.D. Thesis*, University College of Wales.

MOSS, N. G., MORIARTY, R. J. & RANKIN, J. C. (1974) The investigation of individual nephron function in the euryhaline teleost *Salmo gairdneri*. A new approach using micropuncture techniques. *INSERM Colloque Européen de Physiologie rénale*, **30**, Physiologie du Nephron: Mécanismes et régulation, p. 180, Paris.

MURAYAMA, Y., MOREL, F. & LE GRIMELLEC, C. (1972) Phosphate, calcium and magnesium transfers in proximal tubules and loops of Henlé, as measured by single nephron microperfusion experiments in the rat. *Pflügers Archiv. European Journal of Physiology*, **333**, 1–16.

OKEN, D. E. & SOLOMON, A. K. (1963) Single proximal tubules of *Necturus* kidney. VI. Nature of potassium transport. *American Journal of Physiology*, **204**, 377–380.

PRAGER, D. J., BOWMAN, R. L. & VUREK, G. G. (1965) Constant volume, self-filling nanolitre pipette; construction and calibration. *Science*, **147**, 606–608.

RICHARDS, A. N. (1938) Processes of urine formation. *Proceedings of the Royal Society, London*, **126**, 398–432.

RICHARDS, A. N. & WALKER, A. M. (1937) Methods of collecting fluid from known regions of the renal tubules of amphibia and of perfusing the lumen of a single tubule. *American Journal of Physiology*, **118**, 111–120.

ROBERTSON, A. J. (1968) The electron probe microanalyser and its applications in medicine. *Physics in Medicine and Biology*, **13**, 505–522.

ROINEL, N. (1975) Electron microprobe quanatitative analysis of lyophilised 10^{-10} l volume samples. *Journal de Microscopie et de Biologie Cellulaire*, **22**, 261–268.

RONEL, N., RICHARD, J. P., ROBIN, G. & MOREL, F. (1973) An automatic electron microprobe sample-feeder for the quantitative analysis of dried 10^{-10} l volume samples. *Journal de Microscopie et de Biologie Cellulaire*, **18**, 285–290.

ROUFFIGNAC, C. DE, MOREL, F., MOSS, N. G. & ROINEL, N. (1973) Micropuncture study of water and electrolyte movements along the loop of Henlé in *Psammomys* with special reference to magnesium, calcium and phosphorus. *Pflügers Archiv. European Journal of Physiology*, **344**, 309–326.

SAMIY, A. H. E., BROWN, J. L. & GLOBUS, D. L. (1960) Effects of magnesium and calcium loading on renal excretion of electrolytes in dogs. *American Journal of Physiology*, **198**, 595–598.

SHIRLEY, D. G., POUJEOL, P. & LE GRIMELLEC, C. (1976) Phosphate, calcium and magnesium fluxes into the lumen of the rat proximal convoluted tubule. *Pflügers Archiv. European Journal of Physiology*, **362**, 247–254.

STRICKLER, J. C., THOMPSON, D. D., KLOSE, R. M. & GIEBISCH, G. (1964) Micropuncture study of inorganic phosphate excretion in the rat. *Journal of Clinical Investigation*, **43**, 1596–1607.

WALKER, A. M. & OLIVER, J. (1941) Methods for the collection of fluid from single glomeruli and tubules of the mammalian kidney. *American Journal of Physiology*, **134**, 562–579.

WALLACH, S. & CARTER, A. C. (1961) Metabolic and renal effects of acute hypercalcaemia in dogs. *American Journal of Physiology*, **200**, 359–366.

WALSER, M. (1961) Calcium clearance as a function of sodium clearance in the dog. *American Journal of Physiology*, **200**, 1099–1104.

WATSON, J. F., CLAPP, J. R. & BERLINER, R. W. (1964) Micropuncture study of potassium concentration in proximal tubule of dog, rat and *Necturus*. *Journal of Clinical Investigation*, **43**, 595–605.

WIEDERHOLT, M. & WIEDERHOLT, B. (1968) Der Einfluss von dexamethason auf die Wasser- und Elektrolytausschiedung adrenalektomierter Ratten. *Pflügers Archiv. European Journal of Physiology*, **302**, 57–78.

WINDHAGER, E. E. (1968) *Micropuncture Techniques and Nephron Function*. London: Butterworths.

Index